FOUNDATIONS OF MODERN ANALYSIS

AVNER FRIEDMAN
Northwestern University

Dover Publications, Inc.
New York

To my father,
Moshe Simcha Friedman

Published in Canada by General Publishing Company, Ltd., 30 Lesmill Road, Don Mills, Toronto, Ontario.

Published in the United Kingdom by Constable and Company, Ltd., 10 Orange Street, London WC2H 7EG.

This Dover edition, first published in 1982, is an unabridged and corrected republication of the work originally published in 1970 by Holt, Rinehart and Winston, Inc., New York.

Manufactured in the United States of America
Dover Publications, Inc.
180 Varick Street
New York, N.Y. 10014

Library of Congress Cataloging in Publication Data

Friedman, Avner.
Foundations of modern analysis.

Reprint. Originally published: New York: Holt, Rinehart, and Winston, [1970]
Bibliography: p.
Includes index.
1. Mathematical analysis—Foundations. I. Title.
[QA299.8.F74 1982] 515 81-17312
ISBN 0-486-64062-0 (pbk.) AACR2

PREFACE

This book is aimed to serve as a first-year graduate course in modern analysis. The material includes measure and integration, metric spaces, the elements of functional analysis in Banach spaces, and spectral theory in Hilbert spaces. There are various books in the literature where all these topics are dealt with; however, we have not felt that we could use any of them systematically as a text book. They are either written at a level not consistent with that of a first-year graduate student or else they do not have the choice of topics we would like to see.

In Chapters 1 and 2 we present the theory of measure and integration. It has been our experience that there are too many definitions and too many theorems for the student to memorize, and that as a result of this, he often fails to grasp what is really essential in the theory. This motivated us to cut down on many details and to give a general, yet compact, presentation.

In Chapters 3 to 6 we present the most fundamental theorems of functional analysis. At the same time we give some applications to important problems in ordinary and partial differential equations, and in integral equations. This, we hope, will illustrate to the student the invaluable role played by functional analysis in solving concrete problems in analysis.

At the end of each section there are problems for the reader. The solution of some of these problems is used in the text itself.

The material of this book is somewhat more than can be covered in a one-year course. One might omit some topics that are rather independent of the rest of the material, such as Section 3.3, Theorem 3.6.3, Section 3.7, and Sections 4.2, 4.11, and 4.12 (with the exception of Theorem 4.12.3). The applications of functional analysis to differential and integral equations and to the Dirichlet problem are given in Sections 3.3, 3.8, 4.7, 4.9, 5.4, and 6.9. They are not used in the remaining sections of the book.

Evanston, Illinois
February 1970

Avner Friedman

CONTENTS

CHAPTER 1

MEASURE THEORY

1.1 RINGS AND ALGEBRAS

Let X be a set of elements. We call X also a *space* and its elements *points*. X will be fixed throughout the whole chapter. We shall refer to subsets of X briefly as sets (of X), and we shall refer to sets of subsets of X also as classes of sets. Given a set E, we denote by E^c the *complement* of E (in X)—that is,

$$E^c = \{x \in X;\, x \notin E\}.$$

The *union* of two sets E and F is the set $\{x;\, x \in E \text{ or } x \in F\}$, and it is denoted by $E \cup F$. The *difference* $E - F$ is defined as $\{x \in E;\, x \notin F\}$. Note that in this definition F is not required to be a subset of E. The *intersection* $E \cap F$ is defined as $\{x;\, x \in E \text{ and } x \in F\}$. We denote by $\varnothing$ the empty set. If $E \cap F = \varnothing$, then we say that the sets E and F are *disjoint*.

Let A be any set of elements and, for each $\alpha \in A$, let E_α be a set. The *union* $\bigcup_{\alpha \in A} E_\alpha$ is defined as the set consisting of those elements which belong to at least one of the sets E_α. The *intersection* $\bigcap_{\alpha \in A} E_\alpha$ is defined as the set consisting of those elements which belong to all the sets E_α. The following relations can easily be verified:

$$\left(\bigcup_{\alpha \in A} E_\alpha\right)^c = \bigcap_{\alpha \in A} E_\alpha^c, \tag{1.1.1}$$

$$\left(\bigcap_{\alpha \in A} E_\alpha\right)^c = \bigcup_{\alpha \in A} E_\alpha^c. \tag{1.1.2}$$

Let $\{E_n\}$ be a sequence of sets ($n = 1,2,...$). The *superior limit* of $\{E_n\}$ is the set consisting of those points which belong to infinitely many E_n. We denote it by

$$E^* = \limsup_{n\to\infty} E_n = \overline{\lim_{n\to\infty}}\, E_n = \overline{\lim_{n}}\, E_n.$$

Similarly, the *inferior limit*

$$E_* = \liminf_{n\to\infty} E_n = \underline{\lim_{n\to\infty}}\, E_n = \underline{\lim_{n}}\, E_n$$

is the set consisting of all those points which belong to all but a finite number of the E_n. If $E^* = E_*$, then we say that the sequence $\{E_n\}$ has a *limit*, and we denote the set E^* by $\lim_{n\to\infty} E_n$ or $\lim_n E_n$.

Note that if $\{E_n\}$ is *monotone increasing* (*decreasing*)—that is, if $E_n \subset E_{n+1}$ ($E_n \supset E_{n+1}$) for all $n \geq 1$, then $\lim_n E_n$ exists.

The following two definitions are fundamental in measure theory.

Definition 1.1.1 A class $\mathscr{R}$ of sets is called a *ring* if it has the following properties:

(a) $\varnothing \in \mathscr{R}$.
(b) If $A \in \mathscr{R}$ and $B \in \mathscr{R}$, then $A - B \in \mathscr{R}$.
(c) If $A \in \mathscr{R}$ and $B \in \mathscr{R}$, then $A \cup B \in \mathscr{R}$.

A ring $\mathscr{R}$ is called an *algebra* if it has the additional property:

(d) $X \in \mathscr{R}$.

Note that if $\mathfrak{R}$ is a ring and if $A, B \in \mathscr{R}$, then $A - B = A - (A - B)$ is in $\mathfrak{R}$. Note also that if $\mathfrak{R}$ is a ring, then any finite union and any finite intersection of sets in $\mathfrak{R}$ are in $\mathfrak{R}$.

Definition 1.1.2 A class $\mathscr{R}$ of sets is called a *σ-ring* if it has the properties (a), (b), and the additional property:

(e) If $A_n \in \mathscr{R}$ for $n = 1,2,...$, then $\bigcup_{n=1}^{\infty} A_n \in \mathscr{R}$.

A σ-ring is called a *σ-algebra* if it has the additional property (d).

If $\mathscr{R}$ is a σ-ring (σ-algebra), then we can take $A_n = \varnothing$ for $n \geq 3$ in (e), and thereby established the condition (c). Thus, a σ-ring (σ-algebra) is a ring (algebra).

If $\mathscr{R}$ is an algebra, then the following condition clearly holds:

(f) If $A \in \mathscr{R}$, then $A^c \in \mathscr{R}$.

Conversely, if the conditions (a), (c), and (f) hold, then $\mathscr{R}$ is an algebra—that is, (b) and (d) also hold. Indeed, (d) follows from (f) with $A = \emptyset$, and (b) follows from

$$A - B = A \cap B^c = (A^c \cup B)^c.$$

A *disjoint union* is a union of sets that are mutually disjoint. We introduce a condition that is weaker than (c), namely:

(c′) If $A \in \mathscr{R}$, $B \in \mathscr{R}$ and $A \cap B = \emptyset$, then $A \cup B \in \mathscr{R}$.

We then have: *A class* $\mathscr{R}$ *satisfying* (a), (b), (c′) *is a ring*. Indeed, (c) follows from $A \cup B = A \cup (B - A)$.

Next we introduce a condition weaker than (e):

(e′) Any countable disjoint union of sets A_n ($n = 1,2,\ldots$) which belong to $\mathscr{R}$ also belongs to $\mathscr{R}$.

Theorem 1.1.1 *If* $\mathscr{R}$ *satisfies* (a), (b), (e′) [*and* (d)], *then* $\mathscr{R}$ *is a σ-ring* (*σ-algebra*).

Proof. Taking, in (e′), $A_n = 0$ for $n \geq 3$, we see that (c′) holds. Hence (c) follows, and $\mathscr{R}$ is a ring. To prove (e), let $\{A_n\}$ be a sequence of sets in $\mathscr{R}$. Then the sets B_n, defined by

$$B_1 = A_1, \qquad B_{n+1} = A_{n+1} - \left[\bigcup_{j=1}^{n} A_j\right],$$

also belong to $\mathscr{R}$. Since $B_n \cap B_m = \emptyset$ if $n \neq m$ and $\bigcup_n B_n = \bigcup_n A_n$, (e′) implies that $\bigcup_n A_n \in \mathscr{R}$.

PROBLEMS

1.1.1. $\left(\underline{\lim}_n E_n\right)^c = \overline{\lim}_n E_n^c$, $\left(\overline{\lim}_n E_n\right)^c = \underline{\lim}_n E_n^c$.

1.1.2. $\overline{\lim}_n E_n = \bigcap_{k=1}^{\infty} \bigcup_{n=k}^{\infty} E_n$, $\underline{\lim}_n E_n = \bigcup_{k=1}^{\infty} \bigcap_{n=k}^{\infty} E_n$.

1.1.3. If $\mathscr{R}$ is a σ-ring and $E_n \in \mathscr{R}$, then

$$\bigcap_{n=1}^{\infty} E_n \in \mathscr{R}, \qquad \overline{\lim}_n E_n \in \mathscr{R}, \qquad \underline{\lim}_n E_n \in \mathscr{R}.$$

1.1.4. The intersection of any collection of rings (algebras, σ-rings, or σ-algebras) is also a ring (an algebra, σ-ring, or σ-algebra).

1.1.5. If $\mathscr{D}$ is any class of sets, then there exists a unique ring $\mathscr{R}_0$ such that (i) $\mathscr{R}_0 \supset \mathscr{D}$, and (ii) any ring $\mathscr{R}$ containing $\mathscr{D}$ contains also $\mathscr{R}_0$. $\mathscr{R}_0$ is called the *ring generated* by $\mathscr{D}$, and is denoted by $\mathscr{R}(\mathscr{D})$.

1.1.6. If $\mathscr{D}$ is any class of sets, then there exists a unique σ-ring $\mathscr{S}_0$ such that (i) $\mathscr{S}_0 \supset \mathscr{D}$, and (ii) any σ-ring containing $\mathscr{D}$ contains also $\mathscr{S}_0$. We call $\mathscr{S}_0$ the *σ-ring generated* by $\mathscr{D}$, and denote it by $\mathscr{S}(\mathscr{D})$. A similar result holds for σ-algebras, and we speak of the *σ-algebra generated* by $\mathscr{D}$.

1.1.7. If $\mathscr{D}$ is any class of sets, then every set in $\mathscr{R}(\mathscr{D})$ can be covered by (that is, is contained in) a finite union of sets of $\mathscr{D}$. [*Hint:* The class $\mathscr{K}$ of sets that can be covered by finite unions of sets of $\mathscr{D}$ forms a ring.]

1.1.8. If $\mathscr{D}$ is any class of sets, then every set in $\mathscr{S}(\mathscr{D})$ can be covered by countable union of sets of $\mathscr{D}$.

1.1.9. Let $\mathscr{D}$ consist of those sets which are either finite or have a finite complement. Then $\mathscr{D}$ is an algebra. If X is not finite, then $\mathscr{D}$ is not a σ-algebra.

1.1.10. Let $\mathfrak{A}$ consist of all sets of the form $\bigcup_{k=1}^{n} J_k$, where the J_k are mutually disjoint intervals on the real line, having the form $\{t; a < t \leq b\}$, where a,b are any real numbers. Then $\mathfrak{A}$ is a ring but not an algebra.

1.1.11. Find $\mathfrak{R}(\mathfrak{D})$ when $\mathfrak{D}$ consists of two distinct sets, each consisting of one point.

1.2 DEFINITION OF MEASURE

We introduce two symbols, "$+\infty$" (or, briefly, ∞) and "$-\infty$" (called "plus infinity" and "minus infinity"), and the following relations: $-\infty < x < \infty$ if x is real. Any entity that is either a real number or one of these two symbols will be called an *extended real number*. We also define arithmetic relations by: $x \pm \infty = \pm\infty$, $\pm\infty + (\pm\infty) = \pm\infty$, $x\cdot(\pm\infty) = \pm\infty$ if $x > 0$, $x\cdot(\pm\infty) = \mp\infty$ if $x < 0$, and so forth. The relations $+\infty + (-\infty)$, $0\cdot(\pm\infty)$ are not defined.

A *set function* μ is a function defined on a class $\mathscr{D}$ of sets. It is said to be extended real-valued, if its values are extended real numbers. $\mathscr{D}$ is called the *domain* (of definition) of μ.

Let μ be a set function defined on a ring $\mathscr{R}$. We say that μ is *completely additive* if

$$\mu\left(\bigcup_{n=1}^{\infty} E_n\right) = \sum_{n=1}^{\infty} \mu(E_n) \tag{1.2.1}$$

whenever the E_n are mutually disjoint sets of $\mathscr{R}$ such that $\bigcup_n E_n$ is also in $\mathscr{R}$.

Definition 1.2.1 A *measure* is an extended real-valued set function μ having the following properties:

(i) The domain $\mathfrak{A}$ of μ is a σ-algebra.
(ii) μ is nonnegative on $\mathfrak{A}$.
(iii) μ is completely additive on $\mathfrak{A}$.
(iv) $\mu(\varnothing) = 0$.

A real-valued set function μ is said to be *additive* if its domain $\mathscr{R}$ is a ring and if

$$\mu(E \cup F) = \mu(E) + \mu(F) \tag{1.2.2}$$

whenever $E \in \mathscr{R}$, $F \in \mathscr{R}$, $E \cap F = \varnothing$. μ is called *finitely additive* if, for any positive integer m,

$$\mu\left(\bigcup_{j=1}^{m} E_j\right) = \sum_{j=1}^{m} \mu(E_j) \tag{1.2.3}$$

whenever $E_j \in \mathscr{R}$ $(j = 1,\dots,m)$, $E_j \cap E_k = \varnothing$ if $j \neq k$. Taking in (1.2.1) $E_n = \varnothing$ for $n > m$ and using (iv), we see that a measure is finitely additive.

If $\mu(X) < \infty$, we say that μ is a *finite measure* (or a *totally finite measure*). If $X = \bigcup_{n=1}^{\infty} E_n$ and $\mu(E_n) < \infty$ for all n, then we say that μ is a *σ-finite measure.*

Theorem 1.2.1 *Let* μ *be a measure with domain* $\mathfrak{A}$. *Then*

(i) *If* $E \in \mathfrak{A}$, $F \in \mathfrak{A}$, $E \subset F$, *then* $\mu(E) \leq \mu(F)$.

(ii) *If* $E \in \mathfrak{A}$, $F \in \mathfrak{A}$, $E \subset F$, $\mu(F) < \infty$, *then*

$$\mu(F - E) = \mu(F) - \mu(E). \tag{1.2.4}$$

(iii) *If* $\{E_n\}$ *is a monotone-increasing sequence of sets of* $\mathfrak{A}$, *then*

$$\lim_{n\to\infty} \mu(E_n) = \mu\left(\lim_{n\to\infty} E_n\right). \tag{1.2.5}$$

(iv) *If* $\{E_n\}$ *is a monotone-decreasing sequence of sets of* $\mathfrak{A}$, *and if* $\mu(E_{n_0}) < \infty$ *for some* n_0, *then* (1.2.5) *holds.*

Proof. To prove (i), write F as a disjoint union $F = E \cup (F - E)$. Then

$$\mu(F) = \mu(E) + \mu(F - E). \tag{1.2.6}$$

Since $\mu(F - E) \geq 0$, (i) follows. If $\mu(F) < \infty$, then $\mu(E) < \infty$, and (1.2.4) follows from (1.2.6). To prove (iii) we first note, by Problem 1.1.3, that $\lim_n E_n$ belongs to $\mathfrak{A}$. Next, setting $E_0 = \varnothing$, we have

$$\mu(\lim_n E_n) = \mu\left[\bigcup_{n=1}^{\infty} (E_n - E_{n-1})\right] = \sum_{n=1}^{\infty} \mu(E_n - E_{n-1}) = \lim_m \sum_{n=1}^{m} \mu(E_n - E_{n-1})$$

$$= \lim_m \mu\left[\bigcup_{n=1}^{m} (E_n - E_{n-1})\right] = \lim_m \mu(E_m),$$

which proves (iii).

To prove (iv), we use (ii):

$$\mu\left[\lim_n (E_{n_0} - E_n)\right] = \mu\left[E_{n_0} - \lim_n E_n\right] = \mu(E_{n_0}) - \mu\left(\lim_n E_n\right).$$

Next, by (iii) and (ii),

$$\mu\left[\lim_n (E_{n_0} - E_n)\right] = \lim_n \mu(E_{n_0} - E_n) = \mu(E_{n_0}) - \lim_n \mu(E_n).$$

Comparing these two results, (iv) follows.

Theorem 1.2.2 *Let* μ *be a measure with domain* $\mathfrak{A}$ *and let* E_n (n = 1,2,...) *be sets of* $\mathfrak{A}$. *Then*

$$\mu\left(\bigcup_{n=1}^{\infty} E_n\right) \leq \sum_{n=1}^{\infty} \mu(E_n), \tag{1.2.7}$$

$$\mu\left(\varliminf_{n\to\infty} E_n\right) \leq \varliminf_{n\to\infty} \mu(E_n), \tag{1.2.8}$$

$$\mu\left(\varlimsup_{n\to\infty} E_n\right) \geq \varlimsup_{n\to\infty} \mu(E_n) \qquad \textit{provided } \mu\left(\bigcup_{n=1}^{\infty} E_n\right) < \infty. \tag{1.2.9}$$

Proof. Let

$$F_1 = E_1, \quad F_n = E_n - \left[\bigcup_{j=1}^{n-1} E_j\right] \qquad (n \geq 2). \tag{1.2.10}$$

Then $F_n \subset E_n$, $\bigcup_n F_n = \bigcup_n E_n$, and the F_n are mutually disjoint. Hence

$$\mu\left(\bigcup_{n=1}^{\infty} E_n\right) = \mu\left(\bigcup_{n=1}^{\infty} F_n\right) = \sum_{n=1}^{\infty} \mu(F_n) \leq \sum_{n=1}^{\infty} \mu(E_n).$$

To prove (1.2.8) we use Problem 1.1.2 and Theorem 1.2.1(iii). We then get

$$\mu\left(\varliminf_n E_n\right) = \lim_{k\to\infty} \mu\left(\bigcap_{n=k}^{\infty} E_n\right).$$

Since $\bigcap_{n=k}^{\infty} E_n \subset E_k$, Theorem 1.2.1(i) gives $\mu\left(\bigcap_{n=k}^{\infty} E_n\right) \leq \mu(E_k)$. Therefore

$$\mu\left(\varliminf_n E_n\right) \leq \varliminf_{k\to\infty} \mu(E_k).$$

Finally, (1.2.9) follows by applying (1.2.8) to the sets $E_n^* = \left[\bigcup_{k=1}^{\infty} E_k\right] - E_n$ and using Problem 1.1.1.

Corollary 1.2.3 *If* $\lim_n E_n$ *exists and if* $\mu\left(\bigcup_{n=1}^{\infty} E_n\right) < \infty$, *then*

$$\mu\left(\lim_{n\to\infty} E_n\right) = \lim_{n\to\infty} \mu(E_n). \tag{1.2.11}$$

PROBLEMS

1.2.1. If μ satisfies the properties (i)–(iii) in Definition 1.2.1, and if $\mu(E) < \infty$ for at least one set E, then (iv) is also satisfied.

1.2.2. Let X be an infinite space. Let $\mathfrak{A}$ be the class of all subsets of X. Define $\mu(E) = 0$ if E is finite and $\mu(E) = \infty$ if E is infinite. Then μ is finitely additive but not completely additive.

1.2.3. If μ is a measure on a σ-algebra $\mathfrak{A}$, and if E, F are sets of $\mathfrak{A}$, then

$$\mu(E) + \mu(F) = \mu(E \cup F) + \mu(E \cap F).$$

1.2.4. Let $\{\mu_n\}$ be a sequence (finite or infinite) of measures defined on the same σ-algebra $\mathfrak{A}$. Define $\sum_{n=1}^{\infty} \mu_n$ by $\left(\sum_{n=1}^{\infty} \mu_n\right)(E) = \sum_{n=1}^{\infty} \mu_n(E)$ for every $E \in \mathfrak{A}$. Prove that $\sum_{n=1}^{\infty} \mu_n$ is a measure.

1.2.5. Let X consist of a sequence $\{x_m\}$ and let $\{p_m\}$ be a sequence of nonnegative numbers. For any subset $A \subset X$, let

$$\mu(A) = \sum_{x_m \in A} p_m.$$

Then μ is a σ-finite measure.

1.2.6. Give an example of a measure μ and a monotone-decreasing sequence $\{E_n\}$ of $\mathfrak{A}$ such that $\mu(E_n) = \infty$ for all n, and $\mu\left(\lim_n E_n\right) = 0$.

1.3 OUTER MEASURE

An extended real-valued set function ν is said to be *subadditive* if

$$\nu(E \cup F) \le \nu(E) + \nu(F)$$

whenever E, F, and $E \cup F$ belong to the domain of ν. It is called *finitely subadditive* if, for any positive integer n,

$$\nu\left(\bigcup_{j=1}^{n} E_j\right) \le \sum_{j=1}^{n} \nu(E_j) \tag{1.3.1}$$

whenever the sets E_j and their union belong to the domain of ν. Finally, it is called *countably subadditive* if

$$\nu\left(\bigcup_{j=1}^{\infty} E_j\right) \le \sum_{j=1}^{\infty} \nu(E_j) \tag{1.3.2}$$

whenever the E_j and their union belong to the domain of ν.

An extended real-valued set function ν is said to be *monotone* if $\nu(E) \le \nu(F)$ whenever E, F are in the domain of ν and $E \subset F$.

Definition 1.3.1 An *outer measure* is an extended real-valued set function μ^* having the following properties:

(i) The domain of μ^* consists of all the subsets of X.
(ii) μ^* is nonnegative.
(iii) μ^* is countably subadditive.
(iv) μ^* is monotone.
(v) $\mu^*(\varnothing) = 0$.

Taking $E_j = \varnothing$ in (1.3.2), for $j \ge n$, and using (v), we see that an outer measure is finitely subadditive.

It is obvious that every measure whose domain consists of all the subsets of X is an outer measure.

The usefulness of outer measures is primarily in the role they play in the construction of measures. On one hand, it is fairly easy (see Section 1.4) to construct outer measures. On the other hand, one can always construct a measure from an outer measure. The latter assertion will be proved in this section (Theorem 1.3.1).

Definition 1.3.2 Given an outer measure μ^*, we say that a set E is *μ^*-measurable* (or, briefly, measurable) if

$$\mu^*(A) = \mu^*(A \cap E) + \mu^*(A - E) \tag{1.3.3}$$

for any subset A of X.

The (vague) motivation for this definition is that the sets we want to single out as measurable should be such that μ^* will become additive on them.

Since μ^* is subadditive, (1.3.3) is equivalent to the inequality:

$$\mu^*(A) \ge \mu^*(A \cap E) + \mu^*(A - E). \tag{1.3.4}$$

Theorem 1.3.1 *Let μ^* be an outer measure and denote by $\mathfrak{A}$ the class of all μ^*-measurable sets. Then $\mathfrak{A}$ is a σ-algebra, and the restriction of μ^* to $\mathfrak{A}$ is a measure.*

Proof. We shall give the proof in several steps.

(i) If $\mu^*(E) = 0$, then E is measurable. Indeed, for any set A we have

$$\mu^*(A \cap E) + \mu^*(A - E) \leq \mu^*(E) + \mu^*(A) = \mu^*(A)$$

—that is, (1.3.4) holds. Hence (1.3.3) is satisfied.

(ii) $\varnothing \in \mathfrak{A}$, since $\mu^*(\varnothing) = 0$.

(iii) If $E \in \mathfrak{A}$, then $E^c \in \mathfrak{A}$. Indeed,

$$\mu^*(A \cap E^c) + \mu^*(A - E^c) = \mu^*(A - E) + \mu^*(A \cap E) = \mu^*(A).$$

(iv) If E_1 and E_2 belong to $\mathfrak{A}$, then also $E_1 \cup E_2$ belongs to $\mathfrak{A}$. To prove it, we write (1.3.3) for $E = E_1$ and A and for $E = E_2$ and $A - E_1$:

$$\mu^*(A) = \mu^*(A \cap E_1) + \mu^*(A - E_1),$$
$$\mu^*(A - E_1) = \mu^*[(A - E_1) \cap E_2] + \mu^*[(A - E_1) - E_2].$$

Adding these qualities and noting that $(A - E_1) - E_2 = A - (E_1 \cup E_2)$,

$$[(A - E_1) \cap E_2] \cup [A \cap E_1] = A \cap (E_1 \cup E_2),$$

we get, after using the subadditivity of μ^*,

$$\mu^*(A) \geq \mu^*[A \cap (E_1 \cup E_2)] + \mu^*[A - (E_1 \cup E_2)].$$

Since A is an arbitrary set, the μ^*-measurability of $E_1 \cup E_2$ follows.

(v) If E_1, E_2 belong to $\mathfrak{A}$, then $E_1 - E_2$ belongs to $\mathfrak{A}$. Indeed, this follows from the relation

$$E_1 - E_2 = (E_1^c \cup E_2)^c,$$

using (iii) and (iv).

We have proved so far that $\mathfrak{A}$ is an algebra.

(vi) Let $\{E_n\}$ be a sequence of mutually disjoint sets of $\mathfrak{A}$ and let $S_n = \bigcup_{m=1}^{n} E_m$. Then, for any $n \geq 1$,

$$\mu^*(A \cap S_n) = \sum_{m=1}^{n} \mu^*(A \cap E_m) \tag{1.3.5}$$

for all sets A. The proof of (1.3.5) is by induction on n. The case $n = 1$ is obvious. The passage from n to $n + 1$ is given by

$$\begin{aligned}\mu^*(A \cap S_{n+1}) &= \mu^*(A \cap S_{n+1} \cap S_n) + \mu^*(A \cap S_{n+1} - S_n)\\ &= \mu^*(A \cap S_n) + \mu^*(A \cap E_{n+1})\\ &= \sum_{m=1}^{n} \mu^*(A \cap E_m) + \mu^*(A \cap E_{n+1}) = \sum_{m=1}^{n+1} \mu^*(A \cap E_m).\end{aligned}$$

Here we have used the measurability of S_n.

(vii) Let $\{E_n\}$ be a sequence of disjoint sets of $\mathcal{A}$, and let $S = \bigcup_{m=1}^{\infty} E_m$. Then, for any set A,

$$\mu^*(A \cap S) = \sum_{m=1}^{\infty} \mu^*(A \cap E_m). \tag{1.3.6}$$

Indeed, by the monotonicity of μ^* and (1.3.5),

$$\mu^*(A \cap S) \geq \mu^*(A \cap S_n) = \sum_{m=1}^{n} \mu^*(A \cap E_m).$$

Taking $n \to \infty$, we get

$$\mu^*(A \cap S) \geq \sum_{m=1}^{\infty} \mu^*(A \cap E_m).$$

Since the reverse inequality follows from the countable subadditivity of μ^*, (1.3.6) is completely proved.

(viii) A countable disjoint union of sets E_n $(n = 1,2,\ldots)$ of $\mathcal{A}$ is again in $\mathcal{A}$. Indeed, by (vi), for any set A,

$$\begin{aligned}\mu^*(A) &= \mu^*(A \cap S_n) + \mu^*(A - S_n)\\ &\geq \sum_{m=1}^{n} \mu^*(A \cap E_m) + \mu^*(A - S).\end{aligned}$$

Taking $n \to \infty$ and using (1.3.6), we get

$$\mu^*(A) \geq \mu^*(A \cap S) + \mu^*(A - S)$$

—that is, S is measurable.

We have proved that $\mathcal{A}$ is an algebra and that a countable, disjoint union of sets of $\mathcal{A}$ is again in $\mathcal{A}$. Hence, by Theorem 1.1.1, $\mathcal{A}$ is a σ-algebra. The restriction μ of μ^* to $\mathcal{A}$ clearly satisfies (i), (ii), and (iv) of Definition 1.2.1. It remains to prove that μ is completely additive. But this follows from (1.3.6) with $A = S$.

PROBLEMS

1.3.1. Define $\mu^*(E)$ as the number of points in E if E is finite and $\mu^*(E) = \infty$ if E is infinite. Show that μ^* is an outer measure. Determine the measurable sets.

1.3.2. Define $\mu^*(\emptyset) = 0$, $\mu^*(E) = 1$ if $E \neq \emptyset$. Show that μ^* is an outer measure, and determine the measurable sets.

1.3.3. Let X have a noncountable number of points. Set $\mu^*(E) = 0$ if E is countable, $\mu^*(E) = 1$ if E is noncountable. Show that μ^* is an outer measure, and determine the measurable sets.

1.3.4. If μ^* is an outer measure and B is a fixed set, then the set function ν^* given by $\nu^*(A) = \mu^*(A \cap B)$ is an outer measure. Find the relation between the measurable sets of μ^* and ν^*.

1.3.5. If $\{\mu_n^*\}$ is a sequence of outer measures, then $\sum_n \mu_n^*$, defined by $\left(\sum_n \mu_n^*\right)(A) = \sum_n \mu_n^*(A)$, is also an outer measure.

1.3.6. Prove that if an outer measure is finitely additive, then it is a measure.

1.4 CONSTRUCTION OF OUTER MEASURES

Let $\mathscr{K}$ be a class of sets (of X). We call $\mathscr{K}$ a *sequential covering class* (of X) if (i) $\varnothing \in \mathscr{K}$, and (ii) for every set A there is a sequence $\{E_n\}$ in $\mathscr{K}$ such that

$$A \subset \bigcup_{n=1}^{\infty} E_n.$$

For example, the bounded open intervals on the real line form a sequential covering class of the real line.

Let λ be an extended real-valued, nonnegative set function, with domain $\mathscr{K}$, such that $\lambda(\varnothing) = 0$. For each set A (of X), let

$$\mu^*(A) = \inf\left\{\sum_{n=1}^{\infty} \lambda(E_n);\ E_n \in \mathscr{K},\ \bigcup_{n=1}^{\infty} E_n \supset A\right\}. \tag{1.4.1}$$

Theorem 1.4.1 *For any sequential covering class $\mathscr{K}$ and for any nonnegative, extended real-valued set function λ with domain $\mathscr{K}$ and with $\lambda(\varnothing) = 0$, the set function μ^* defined by* (1.4.1) *is an outer measure.*

Proof. The conditions (i), (ii), and (v) in Definition 1.3.1 are trivially satisfied. The monotonicity condition (iv) is also fairly obvious. Thus it remains to verify (iii). Let $\{A_n\}$ be any sequence of sets. We have to prove that

$$\mu^*\left(\bigcup_{n=1}^{\infty} A_n\right) \le \sum_{n=1}^{\infty} \mu^*(A_n). \tag{1.4.2}$$

Let ε be any positive number. For each positive integer n there is a sequence $\{E_{nk}\}$ in $\mathscr{K}$ such that

$$A_n \subset \bigcup_{k=1}^{\infty} E_{nk},$$

$$\sum_{k=1}^{\infty} \lambda(E_{nk}) \le \mu^*(A_n) + \frac{\varepsilon}{2^n}. \tag{1.4.3}$$

The double sequence $\{E_{nk}\}$, with indices n and k, then satisfies

$$\bigcup_{n=1}^{\infty} A_n \subset \bigcup_{n,k=1}^{\infty} E_{nk}.$$

Hence, by the definition (1.4.1) and by (1.4.3),

$$\mu^*\left(\bigcup_{n=1}^{\infty} A_n\right) \leq \sum_{n,k=1}^{\infty} \lambda(E_{nk}) \leq \sum_{n=1}^{\infty} \left[\mu^*(A_n) + \frac{\varepsilon}{2^n}\right] \leq \sum_{n=1}^{\infty} \mu^*(A_n) + \varepsilon.$$

Since ε is arbitrary, (1.4.2) follows.

Theorems 1.4.1 and 1.3.1 enable us to construct a measure from any given pair $\mathscr{K}$, λ.

PROBLEMS

1.4.1. Let $\mathscr{K}$ consist of X, $\varnothing$, and all the one-point sets. Let $\lambda(X) = \infty$, $\lambda(\varnothing) = 0$, $\lambda(E) = 1$ if $E \neq X$, $E \neq \varnothing$. Describe the outer measure.

1.4.2. Let X be a noncountable space and $\mathscr{K}$ be as in Problem 1.4.1. Let $\lambda(X) = 1$, $\lambda(E) = 0$ if $E \neq X$. Describe the outer measure.

1.4.3. Show that, under the conditions of Theorem 1.4.1, $\mu^*(E) \leq \lambda(E)$ if $E \in \mathscr{K}$. Give an example where inequality holds.

1.4.4. If $\mathscr{K}$ is a σ-algebra and λ is a measure on $\mathscr{K}$, then $\mu^*(A) = \lambda(A)$ for any $A \in \mathscr{K}$. [*Hint:* $\mu^*(A) = \inf \{\lambda(E);\ E \in \mathscr{K}, E \supset A\}$.]

1.4.5. If $\mathscr{K}$ is a σ-algebra and λ is a measure on $\mathscr{K}$, then every set in $\mathscr{K}$ is μ^*-measurable.

The last two problems show that if we begin with a measure λ on a σ-algebra $\mathscr{K}$, then by applying Theorem 1.4.1 and then Theorem 1.3.1 we get another measure that is an extension of λ.

1.5 COMPLETION OF MEASURES

A measure μ with domain $\mathcal{A}$ is said to be *complete* if for any two sets N, E the following holds: If $N \subset E$, $E \in \mathcal{A}$ and $\mu(E) = 0$, then $N \in \mathcal{A}$. Note that *the measure constructed in Theorem* 1.3.1 *is complete.*

In the next theorem it is shown that any measure can be extended into a complete measure.

Theorem 1.5.1 *Let* μ *be a measure on a* σ*-algebra* $\mathcal{A}$ *and let* $\overline{\mathcal{A}}$ *denote the class of all sets of the form* $E \cup N$, *where* $E \in \mathcal{A}$ *and* N *is any subset of a set of* $\mathcal{A}$ *of measure zero. Then* $\overline{\mathcal{A}}$ *is a* σ*-algebra and the set function* $\bar{\mu}$ *defined by*

$$\bar{\mu}(E \cup N) = \mu(E) \tag{1.5.1}$$

is a complete measure on $\overline{\mathcal{A}}$.

We call $\bar{\mu}$ the *completion* of μ.

Proof. We first show that $\bar{\mathfrak{A}}$ is a σ-algebra. Let $E \in \mathfrak{A}$, $N \subset A \in \mathfrak{A}$, $\mu(A) = 0$. Then $E^c - A = E^c \cap A^c \subset E^c \cap N^c = (E \cup N)^c \subset E^c$. Hence, $(E \cup N)^c = (E^c - A) \cup N'$ where $N' \subset A$. Since $[E^c - A] \in \mathfrak{A}$, it follows that $(E \cup N)^c \in \bar{\mathfrak{A}}$. Since $E_1 - E_2 = (E_1^c \cup E_2)^c$, it remains to show that a countable union of sets of $\bar{\mathfrak{A}}$ is again in $\bar{\mathfrak{A}}$. Let these sets be $E_n \cup N_n$, where $N_n \subset A_n$, $\mu(A_n) = 0$. Then

$$\bigcup_{n=1}^{\infty} (E_n \cup N_n) = \left(\bigcup_{n=1}^{\infty} E_n \right) \cup N,$$

where $N = \bigcup_{n=1}^{\infty} N_n$. Since $N \subset \bigcup_{n=1}^{\infty} A_n = A$ and $\mu(A) = 0$, the assertion follows.

We shall how show that $\bar{\mu}$ is well defined on $\bar{\mathfrak{A}}$. Let

$$E_1 \cup N_1 = E_2 \cup N_2,$$

where $E_i \in \mathfrak{A}$, $N_i \subset A_i$, $\mu(A_i) = 0$. Then

$$E_1 \subset E_2 \cup A_2, \qquad E_2 \subset E_1 \cup A_1.$$

It follows that $\mu(E_1) = \mu(E_2)$. Hence the definition (1.5.1) is unambiguous. It is now easy to verify that $\bar{\mu}$ satisfies all the properties of a complete measure.

PROBLEMS

1.5.1. Let μ be a complete measure. A set for which $\mu(N) = 0$ is called a *null set*. Show that the class of null sets is a σ-ring. Is it also a σ-algebra?

1.5.2. Let the conditions of Theorem 1.5.1 hold and denote by $\mathfrak{A}^*$ the class of all sets of the form $E - N$ where $E \in \mathfrak{A}$ and N is any subset of $\mathfrak{A}$ having measure zero. Then $\mathfrak{A}^* = \bar{\mathfrak{A}}$.

1.6 THE LEBESGUE AND THE LEBESGUE-STIELTJES MEASURES

Denote by R^n the Euclidean space of n dimensions. The points of R^n are written in the form $x = (x_1, \ldots, x_n)$. By an *open interval* we shall mean a set of the form

$$I_{a,b} = \{x = (x_1, \ldots, x_n);\ a_i < x_i < b_i \text{ for } i = 1, \ldots, n\}, \tag{1.6.1}$$

where $a = (a_1, \ldots, a_n)$, $b = (b_1, \ldots, b_n)$ are points in R^n. The set $\mathscr{K}$ of all open intervals forms a sequential covering class of R^n. Let λ be given by $\lambda(\varnothing) = 0$ and

$$\lambda(I_{a,b}) = \prod_{i=1}^{n} (b_i - a_i) \qquad \text{if } a \neq b. \tag{1.6.2}$$

The outer measure determined by this pair $\mathscr{K},\lambda$ (in accordance with Theorem 1.4.1) is called the *Lebesgue outer measure*. The (complete) measure determined by this outer measure (in accordance with Theorem 1.3.1) is called the *Lebesgue measure*. The measurable sets are called *Lebesgue-measurable sets* or, simply, *Lebesgue sets*.

Let $f(x)$ be a real-valued, monotone-increasing function defined on the real line R^1. We assume that $f(x)$ is right continuous—that is,

$$\lim_{x \to y+} f(x) = f(y).$$

Let $\mathscr{K}$ consist of all the open intervals (a,b) of R^1 and define λ by $\lambda(\emptyset) = 0$ and

$$\lambda\{(a,b)\} = f(b) - f(a) \qquad \text{if } a < b. \tag{1.6.3}$$

The outer measure μ_f^* determined by this pair $\mathscr{K}$, λ is called the *Lebesgue-Stieltjes outer measure induced by f*. The measure μ_f corresponding to this outer measure is called the *Lebesgue-Stieltjes measure induced by f*.

It will be proved later on that closed sets and open sets are Lebesgue-measurable.

PROBLEMS

1.6.1. The Lebesgue measure of a point is zero.

1.6.2. The Lebesgue measure of a countable set of points is zero.

1.6.3. The outer Lebesgue measure of a closed bounded interval $[a,b]$ on the real line is equal to $b - a$. [*Hint:* Use the Heine-Borel theorem to replace a countable covering by a finite covering.]

1.6.4. The outer Lebesgue measure of each of the intervals (a,b), $[a,b)$,$(a,b]$ is equal to $b - a$.

1.6.5. Consider the transformation $Tx = \alpha x + \beta$ from the real line onto itself, where α,β are real numbers and $\alpha \neq 0$. It maps sets E onto sets $T(E)$. Denote by μ (μ^*) the Lebesgue measure (outer measure) on the real line. Prove:

(a) For any set E, $\mu^*(T(E)) = |\alpha| \, \mu^*(E)$.

(b) E is Lebesgue-measurable if and only if $T(E)$ is Lebesgue-measurable.

(c) If E is Lebesgue-measurable, then $\mu(T(E)) = |\alpha|\mu(E)$.

Additional information on the Lebesgue measure (in n dimensions) will be given in Problems 1.9.1–1.9.11. More information on the Lebesgue-Stieltjes measure will be given in Problems 1.9.13–1.9.15.

Theorem 1.6.1 *There exists a set on the real line that is not Lebesgue-measurable.*

Proof. Denote by ξ any irrational number and also the number 0. Denote by R a sequence $\{r_n\}$ of all the rational numbers. Finally, denote by E_ξ the set $\{\xi + r;\ r \in R\}$. Clearly $E_{\xi_1} \cap E_{\xi_2} = \varnothing$ if $E_{\xi_1} \neq E_{\xi_2}$. From each set E_ξ we now select one number η such that $0 \le \eta \le \frac{1}{2}$. (If two sets E_ξ, E_{ξ_0} are equal, then they have the same η.) The set of all these numbers η is denoted by F. We shall prove that F is not Lebesgue-measurable.

For each positive integer $k \ge 2$, denote by F_k the set $\{\eta + (1/k);\ \eta \in F\}$. Then $F_k \cap F_j = \varnothing$ if $k \neq j$. Indeed, otherwise we have $\eta_1 + 1/k = \eta_2 + 1/j$, where $\eta_1 \in E_{\xi_1}$, $\eta_2 \in E_{\xi_2}$ and $E_{\xi_1} \cap E_{\xi_2} = \varnothing$ (since $\eta_1 \neq \eta_2$). We get

$$\xi_1 + r_1 + \frac{1}{k} = \xi_2 + r_2 + \frac{1}{j} \qquad (r_1, r_2 \text{ rationals}),$$

which is impossible (since this would imply that $E_{\xi_1} \cap E_{\xi_2} \neq \varnothing$).

Suppose now that F is measurable. By Problem 1.6.5, each set F_k ($k \ge 2$) is then also measurable, and $\mu(F_k) = \mu(F)$. Since all the sets F_k are disjoint and are contained in $[0,1]$, it follows that

$$m\mu(F) = \sum_{k=2}^{m+1} \mu(F_k) \le 1 \qquad \text{for any integer } m.$$

Hence $\mu(F) = 0$.

Consider next the sets $G_r = \{r + \eta;\ \eta \in F\}$, where r is any rational number. By Problem 1.6.5, G_r is measurable and $\mu(G_r) = 0$. If r and s are distinct rational numbers, then $G_r \cap G_s = \varnothing$. Since, further

$$\bigcup_{n=1}^{\infty} G_{r_n} = (-\infty, \infty),$$

it follows that $\mu\{(-\infty,\infty)\} = 0$, which is impossible.

1.7 METRIC SPACES

Suppose that there exists a real-valued function ρ defined for every ordered pair (x,y) of points of X, having the following properties:

(i) $\rho(x,y) \ge 0$, and $\rho(x,y) = 0$ if and only if $x = y$.
(ii) $\rho(x,y) = \rho(y,x)$ (symmetry).
(iii) $\rho(x,z) \le \rho(x,y) + \rho(y,z)$ (the triangle inequality).

The function ρ is then called a *metric function* (or a *distance function*) on X, and the pair (X,ρ) is called a *metric space*. When there is no confusion about ρ, we denote the metric space simply by X.

Example. The function

$$\rho(x,y) \equiv |x - y| \equiv \left\{ \sum_{i=1}^{n} (x_i - y_i)^2 \right\}^{1/2} \tag{1.7.1}$$

is a metric function in the Euclidean space R^n. We call it the *Euclidean metric.* When we speak of R^n as a metric space, the metric we usually have in mind is the one given by (1.7.1).

We define the *distance between two sets A* and *B* by

$$\rho(A,B) = \inf_{\substack{x \in A \\ y \in B}} \rho(x,y).$$

If $A = \{x\}$, then we also write $\rho(x,B)$ for $\rho(A,B)$.

A set A in a metric space (X,ρ) is said to be *bounded* if its *diameter*,

$$d(A) = \sup_{\substack{x \in A \\ y \in A}} \rho(x,y),$$

is finite. For any $x \in X$, $\varepsilon > 0$, the set

$$B(y,\varepsilon) = \{y;\ \rho(x,y) < \varepsilon\}$$

is called the *open ball* with *center* y and *radius* ε, or an ε*-neighborhood of* y. The set

$$\bar{B}(y,\varepsilon) = \{y;\ \rho(x,y) \le \varepsilon\}$$

is called the *closed ball* with center y and radius ε.

A sequence $\{x_n\}$ is called a *convergent* sequence if there is a point y such that $\rho(x_n,y) \to 0$ as $n \to \infty$. We then say that $\{x_n\}$ *converges to* y and write

$$\lim_{n\to\infty} x_n = y \qquad \text{or} \qquad x_n \to y \quad \text{as } n \to \infty.$$

We call y the *limit* of the sequence $\{x_n\}$. It is easily seen that a sequence cannot have two distinct limits. It is also clear that a sequence $\{x_n\}$ is convergent to y if and only if every ε-neighborhood of y contains all but a finite number of the x_n.

A set E is called an *open set* if for any $y \in E$ there is a ball $B(y,\varepsilon)$ that is contained in E. Every open ball is an open set. It is easy to verify that a finite intersection of open sets is an open set, and a union of any class of open sets is an open set.

A point y is called a *point of accumulation* of a set D if any ball $B(y,\varepsilon)$ contains points of D other than y. It is easily seen that y is a point of accumulation of D if and only if there is a sequence $\{x_n\}$ in $D - \{y\}$ that converges to y.

Denote by D' the set of all the points of accumulation of D. The set $D \cup D'$ is called the *closure* of D, and it is denoted by $\bar{D}$.

A set is said to be *closed* if its complement is an open set. Note that the closed balls are closed sets.

The *interior* of a set E is the set of all points y in E such that there is an ε-neighborhood of y contained in E. We denote the interior of E by int E.

A sequence $\{x_n\}$ is called a *Cauchy sequence* if $\rho(x_n,x_m) \to 0$ as $n,m \to \infty$. Note that every convergent sequence is a Cauchy sequence. Also, a Cauchy sequence cannot have more than one point of accumulation. If every Cauchy sequence is convergent, then we say that the metric space is *complete*. The Euclidean space (with the Euclidean metric) is complete, but the subset consisting of all points with rational components forms an incomplete metric space (under the Euclidean metric).

In the next section we shall need the following definition:

Definition Denote by $\mathscr{B}$ the σ-ring generated by the class of all the open sets of X. The sets of $\mathscr{B}$ are called *Borel sets*.

Note that $\mathscr{B}$ coincides with the σ-algebra generated by the class of all the open sets of X.

PROBLEMS

1.7.1. A sequence $\{x_n\}$ is convergent to y if and only if every subsequence $\{x_{n_k}\}$ is convergent to y.

1.7.2. A set D is closed if and only if $\bar{D} = D$.

1.7.3. (a) $\overline{A \cup B} = \bar{A} \cup \bar{B}$; (b) $\bar{\bar{A}} = \bar{A}$.

1.7.4. A set E is open if and only if $E = \text{int } E$.

1.7.5. Let A be any set and x,y any two points. If $\rho(x,A) > \alpha$, $\rho(y,A) < \beta$ and $\alpha > \beta$, then $\rho(x,y) > \alpha - \beta$.

1.7.6. If A is a closed set and $x \notin A$, then $\rho(x,A) > 0$.

1.7.7. Denote by $\mathscr{D}$ the σ-ring generated by the class of all the closed sets of X. Show that $\mathscr{D} = \mathscr{B}$.

1.7.8. Denote by $\mathscr{P}$ the class of all half-open intervals $[a,b)$ on the real line. Show that the σ-ring generated by $\mathscr{P}$ coincides with the class $\mathscr{B}$ of the Borel sets on the real line.

1.8 METRIC OUTER MEASURE

Let X be a metric space with metric ρ.

Definition 1.8.1 An outer measure μ^* on X is called a *metric outer measure* if it satisfies the following property:

(vi) If $\rho(A,B) > 0$, then $\mu^*(A \cup B) = \mu^*(A) + \mu^*(B)$.

Lemma 8.1.1 *Let* μ^* *be a metric outer measure and let* A, B *be any sets such that* $A \subset B$, B *open*, $\mu^*(A) < \infty$. *For any positive integer* n, *let*

$$A_n = \left\{x;\, x \in A,\, \rho(x,B^c) \geq \frac{1}{n}\right\}.$$

Then

$$\lim_{n\to\infty} \mu^*(A_n) = \mu^*(A).$$

Proof. Since $A_n \subset A_{n+1}$ and μ^* is monotone, the sequence $\{\mu^*(A_n)\}$ is an increasing sequence. Also, $\mu^*(A_n) \leq \mu^*(A)$. It therefore suffices to show that

$$\lim_{n\to\infty} \mu^*(A_{2n}) \geq \mu^*(A). \tag{1.8.1}$$

Each point y of A lies in the open set B. Hence there is some ε-neighborhood of y that is contained in B. It follows that $\rho(y,B^c) > \varepsilon$. This shows that

$$A \subset \bigcup_{n=1}^{\infty} A_n.$$

Since $A_n \subset A$ for all n, we have

$$A = \bigcup_{n=1}^{\infty} A_n. \tag{1.8.2}$$

Let $G_n = A_{n+1} - A_n$ for $n \geq 1$. Then (1.8.2) implies that, for any $n \geq 1$,

$$A = A_{2n} \cup \left[\bigcup_{k=2n}^{\infty} G_k\right] = A_{2n} \cup \left[\bigcup_{k=n}^{\infty} G_{2k}\right] \cup \left[\bigcup_{k=n}^{\infty} G_{2k+1}\right].$$

Hence,

$$\mu^*(A) \leq \mu^*(A_{2n}) + \sum_{k=n}^{\infty} \mu^*(G_{2k}) + \sum_{k=n}^{\infty} \mu^*(G_{2k+1}). \tag{1.8.3}$$

From the definition of the sets A_n we have: if $x \in G_{2l}$, $y \in G_{2k+2}$, $l \leq k$, then

$$\rho(x,B^c) > \frac{1}{2l+1},\quad \rho(y,B^c) < \frac{1}{2k+2}.$$

Consequently (see Problem 1.7.5),

$$\rho(G_{2l}, G_{2k+2}) \geq \frac{1}{2k+1} - \frac{1}{2k+2} > 0.$$

Using the relation $A_{2n} \supset \bigcup_{k=1}^{n-1} G_{2k}$ and the property (vi) of metric outer measures, we get

$$\mu^*(A) \geq \mu^*(A_{2n}) \geq \mu^*\left(\bigcup_{k=1}^{n-1} G_{2k}\right) = \sum_{k=1}^{n-1} \mu^*(G_{2k}).$$

Since $\mu^*(A) < \infty$, this shows that the series

$$\sum_{k=1}^{\infty} \mu^*(G_{2k})$$

is convergent. Similarly one shows that the series

$$\sum_{k=1}^{\infty} \mu^*(G_{2k+1})$$

is convergent. Taking $n \to \infty$ in (1.8.3), we then obtain

$$\mu^*(A) \leq \lim_{n\to\infty} \mu^*(A_{2n}).$$

This proves (1.8.1).

Theorem 1.8.2 *If μ^* is a metric outer measure, then every closed set is measurable.*

Proof. Let F be a closed set. For any set A with $\mu^*(A) < \infty$, $A - F$ is contained in the open set F^c. Hence, by Lemma 1.8.1, there is a sequence E_n of subsets of $A - F$ such that

$$\rho(E_n, F) \geq \frac{1}{n} \qquad (n = 1,2,\ldots), \tag{1.8.4}$$

$$\lim_{n\to\infty} \mu^*(E_n) = \mu^*(A - F). \tag{1.8.5}$$

Using the property (vi) of a metric outer measure and (1.8.4), we get

$$\mu^*(A) \geq \mu^*[(A \cap F) \cup E_n] = \mu^*(A \cap F) + \mu^*(E_n).$$

Taking $n \to \infty$ and using (1.8.5), we obtain

$$\mu^*(A) \geq \mu^*(A \cap F) + \mu^*(A - F).$$

If $\mu^*(A) = \infty$, this inequality is trivial. Since A is an arbitrary set, F is measurable.

Corollary 1.8.3 *If μ^* is a metric outer measure, then every Borel set is measurable.*

Proof. Denote by $\mathfrak{A}$ the σ-algebra of all the μ^*-measurable sets. Every open set E has the form $E = X - F$, where $X \in \mathfrak{A}$ and $F \in \mathfrak{A}$ (since F is closed). Hence $E \in \mathfrak{A}$. It follows that $\mathfrak{A}$ contains the σ-ring generated by the class of all the open sets—that is, $\mathfrak{A} \supset \mathscr{B}$.

PROBLEMS

1.8.1. Prove the converse of Corollary 1.8.3—that is, if μ^* is an outer measure and if every open set is measurable, then μ^* is a metric outer measure.

1.8.2. Let μ be a measure with domain $\mathfrak{A}$. For any two sets E and F in $\mathfrak{A}$, let $\rho(E,F) = \mu[(E - F) \cup (F - E)]$. Prove that $\rho(E,F) = \rho(F,E)$ and $\rho(E,G) \le \rho(E,F) + \rho(F,G)$.

1.8.3. Let (X,ρ) be a metric space and let $\{x_n\}$ be a sequence of points in X. Define $\mu^*(E)$ to be the number of points x_n that belong to E. Prove that μ^* is a metric outer measure.

1.8.4. Let (X,ρ) be a metric space. Define $\mu^*(E) = 1$ if $E \neq \varnothing$ and $\mu^*(\varnothing) = 0$. Is μ^* a metric outer measure?

1.9 CONSTRUCTION OF METRIC OUTER MEASURES

Let $\mathscr{K}$ be a sequential covering class with $\varnothing \in \mathscr{K}$, and let

$$\mathscr{K}_n = \left\{A \in \mathscr{K}\,;\, d(A) \le \frac{1}{n}\right\} \qquad \text{[we define } d(\varnothing) = 0]$$

for any positive integer n. Then $\mathscr{K}_{n+1} \subset \mathscr{K}_n$. Assume that each $\mathscr{K}_n$ is a sequential covering class. Let λ be a nonnegative, extended real-valued set function defined on $\mathscr{K}$, with $\lambda(\varnothing) = 0$, and denote by μ_n^* the outer measure defined by each pair $\mathscr{K}_n,\lambda$ (according to Theorem 1.4.1). Clearly

$$\mu_n^*(A) \le \mu_{n+1}^*(A)$$

for any set A. Let

$$\mu_0^*(A) = \lim_{n\to\infty} \mu_n^*(A). \tag{1.9.1}$$

Theorem 1.9.1 *Let the foregoing assumptions on $\mathscr{K},\lambda$ hold. Then the set function μ_0^*, defined by* (1.9.1), *is a metric outer measure.*

Proof. The conditions (i), (ii), (iv), and (v) in Definition 1.3.1 hold for each μ_n^*. Taking $n \to \infty$, we see that they hold also for μ_0^*. To prove (iii), we note that $\mu_n^*(A) \le \mu_0^*(A)$. Hence,

$$\mu_n^*\left(\bigcup_{k=1}^{\infty} E_k\right) \le \sum_{k=1}^{\infty} \mu_n^*(E_k) \le \sum_{k=1}^{\infty} \mu_0^*(E_k).$$

Taking $n \to \infty$, we get

$$\mu_0^*\left(\bigcup_{k=1}^{\infty} E_k\right) = \lim_{n\to\infty} \mu_n^*\left(\bigcup_{k=1}^{\infty} E_k\right) \le \sum_{k=1}^{\infty} \mu_0^*(E_k).$$

It remains to prove the condition (vi) in Definition 1.8.1. If $\rho(A,B) > 0$, then $\rho(A,B) > 1/n$ for all $n \ge n_0$. For any $\varepsilon > 0$ and $n \ge n_0$, there is a sequence $\{E_{nk}\}$ in $\mathscr{K}_n$ such that

$$A \cup B \subset \bigcup_{k=1}^{\infty} E_{nk},$$

$$\sum_{k=1}^{\infty} \lambda(E_{nk}) \le \mu_n^*(A \cup B) + \varepsilon.$$

Since $d(E_{nk}) < 1/n$, no set E_{nk} can intersect both A and B. Hence the sequence $\{E_{nk}\}$ can be written as a disjoint union of two sequences $\{E_{nk'}\}$ and $\{E_{nk''}\}$, the first being a covering of A and the second being a covering B. It follows that

$$\mu_n^*(A) + \mu_n^*(B) \le \sum_{k=1}^{\infty} \lambda(E_{nk}) \le \mu_n^*(A \cup B) + \varepsilon.$$

Taking $n \to \infty$, and then $\varepsilon \to 0$, we get

$$\mu_0^*(A) + \mu_0^*(B) \le \mu_0^*(A \cup B).$$

Since the reverse inequality is obvious, the proof is complete.

Theorem 1.9.2 *Let the assumption on* $\mathscr{K}, \lambda$ *made in Theorem 1.9.1 hold. Assume further that for each set* $A \in \mathscr{K}$, *for any* $\varepsilon > 0$, *and for any positive integer* n, *there exists a sequence* $\{E_k\}$ *in* $\mathscr{K}_n$ *such that*

$$A \subset \bigcup_{k=1}^{\infty} E_k,$$

$$\sum_{k=1}^{\infty} \lambda(E_k) \le \lambda(A) + \varepsilon.$$

Then the outer measure μ^* *defined by* (1.4.1) *coincides with the metric outer measure constructed by* (1.9.1).

Proof. Since $\mathscr{K} \supset \mathscr{K}_n$ we have

$$\mu^*(A) \le \mu_n^*(A) \qquad \text{for any set } A \text{ and } n = 1,2,\ldots. \tag{1.9.2}$$

We shall now prove the reverse inequality. From the definition of μ^* it follows that for each set A and $\varepsilon > 0$, there exists a sequence $\{E_j\}$ in $\mathscr{K}$ such that

$$A \subset \bigcup_{j=1}^{\infty} E_j,$$

$$\sum_{j=1}^{\infty} \lambda(E_j) \le \mu^*(A) + \frac{\varepsilon}{2}.$$

By one of the hypotheses of the theorem, for each set E_j there is a sequence $\{B_{jk}\}$ of sets of $\mathscr{K}_n$ such that

$$E_j \subset \bigcup_{k=1}^{\infty} B_{jk},$$

$$\sum_{k=1}^{\infty} \lambda(B_{jk}) \le \lambda(E_j) + \frac{\varepsilon}{2^{j+1}}.$$

Since $\bigcup_{j,k} B_{jk}$ covers A, we have

$$\mu_n^*(A) \le \sum_{j,k=1}^{\infty} \lambda(B_{jk}) \le \sum_{j=1}^{\infty} \lambda(E_j) + \frac{\varepsilon}{2} \le \mu^*(A) + \varepsilon.$$

Taking $\varepsilon \to 0$, we get $\mu_n^*(A) \le \mu^*(A)$. Combining this with (1.9.2), we get $\mu_n^*(A) = \mu^*(A)$. Hence $\mu_0^*(A) = \mu^*(A)$.

PROBLEMS

1.9.1. Let $a_i < a_i + \delta_i < a_i + 2\delta_i < \cdots < a_i + r_i\delta_i = b_i$ for $i = 1,2,\ldots,n$. Denote by $I_{a,b,\gamma}$ the open interval

$$a_i + (\gamma_i - 1)\delta_i < x_i < a_i + \gamma_i\delta_i \qquad (1 \le i \le n)$$

when the γ_i vary from 1 to r_i. Let λ be defined by (1.6.2). Prove that

$$\lambda(I_{a,b}) = \sum_{\gamma} \lambda(I_{a,b,\gamma}).$$

[*Hint:* The number of terms in the sum is $r_1 r_2 \cdots r_n$ and $\lambda(I_{a,b,\gamma}) = \delta_1\delta_2\cdots\delta_n$.]

1.9.2. Denote by $I_{a-\varepsilon,b+\varepsilon}$ ($\varepsilon > 0$) the open interval determined by $(a_1 - \varepsilon, a_2 - \varepsilon, \ldots, a_n - \varepsilon)$, $(b_1 + \varepsilon, b_2 + \varepsilon, \ldots, b_n + \varepsilon)$. Then $\lambda(I_{a-\varepsilon,b+\varepsilon}) - \lambda(I_{a,b}) < c\varepsilon$ (c constant).

1.9.3. Show that every Borel set in R^n is a Lebesgue set. [*Hint:* We wish to apply Theorem 1.9.2 and Corollary 1.8.3. We thus have to verify the condition on $\mathscr{K},\lambda$ imposed in Theorem 1.9.2. Divide $I_{a,b}$ by hyperplanes $x_i = a_i + \gamma_i\delta_i$ with $\sum \delta_i < 1/m$, m any given positive integer. By Problem 1.9.1, $\sum \lambda(I_{a,b,\gamma}) = \lambda(I_{a,b})$. It remains to show that, for any $\varepsilon > 0$, each set $J = \{x \in I_{a,b} : x_i = a_i + \gamma_i\delta_i\}$ can be covered by a finite number of intervals

E_j with diameter $< 1/m$, such that $\sum \lambda(E_j) < \varepsilon$. J can be covered by a finite number of $(n-1)$-dimensional intervals E_j' lying on $x_i = a_i + \gamma_i\delta_i$ and having diameter $< 1/2m$. Let $C = \sum \lambda(E_j')$. Take $\eta < \varepsilon/2C$, $\eta < 1/4m$, and let $E_j = \{x;\ (x_1,\ldots,x_{i-1},\ a_i + \gamma_i\delta_i,\ x_{i+1},\ldots,x_n) \in E_j',\ a_i + \gamma_i\delta_i - \eta < x_i < a_i + \gamma_i\delta_i + \eta.$]

1.9.4. Let I_{c_k,e_k} $(k = 1,\ldots,m)$ form a covering of a closed interval $\bar{I}_{a,b}$. Let r be any positive integer and let $\delta_i = (b_i - a_i)/r$. Show that there is a covering of $\bar{I}_{a,b}$ by intervals I_{α_k,β_k} $(k = 1,\ldots,m)$, where $\alpha_k = (\alpha_{k1},\ldots,\alpha_{kn})$ and $\beta_k = (\beta_{k1},\ldots,\beta_{kn})$ have the form

$$\alpha_{ki} = a_i + \gamma_i'\delta_i \qquad (\gamma_i' \text{ integer}),$$

$$\beta_{ki} = a_i + \gamma_i''\delta_i \qquad (\gamma_i'' \text{ integer}),$$

and $I_{c_k,e_k} \subset I_{\alpha_k,\beta_k}$,

$$\lambda(I_{\alpha_k,\beta_k}) \le \lambda(I_{c_k,e_k}) + \frac{C}{r} \qquad (C \text{ independent of } r).$$

[*Hint:* Let $c_k = (c_{k1},\ldots,c_{kn})$, $e_k = (e_{k1},\ldots,e_{kn})$. Let α_{ki} be the largest number of the form $a_i + \gamma_i'\delta_i$ that is less than c_{ki}. Let β_{ki} be the smallest number of the form $a_{ki} + \gamma_i''\delta_i$ that is larger than e_{ki}. Then $I_{c_k,e_k} \subset I_{\alpha_k,\beta_k}$. Also

$$\beta_{ki} - \alpha_{ki} \le e_{ki} - c_{ki} + \frac{C'}{r} \qquad (C' \text{ constant}).$$

Now use Problem 1.9.2.]

1.9.5. Let $\bar{I}_{a,b}$ be the closure of a bounded interval $I_{a,b}$ in R^n. Show that

$$\mu^*(\bar{I}_{a,b}) = \mu(\bar{I}_{a,b}) = \lambda(I_{a,b}) = \prod_{i=1}^{n} (b_i - a_i).$$

[*Hint:* $\mu^*(\bar{I}_{a,b}) \le \mu^*(I_{a-\varepsilon,b+\varepsilon})$ for any $\varepsilon > 0$. Use Problem 1.9.2 to deduce

$$\mu^*(\bar{I}_{a,b}) \le \prod_{i=1}^{n} (b_i - a_i).$$

To prove the converse inequality, take any covering of $\bar{I}_{a,b}$ by intervals E_j. By the Heine-Borel theorem, a finite number, say $E_1,\ldots,E_m$, forms a covering of $\bar{I}_{a,b}$. It suffices to show that, for any $\varepsilon > 0$,

$$\prod_{i=1}^{n} (b_i - a_i) \le \sum_{k=1}^{m} \lambda(E_k) + \varepsilon.$$

Write $E_k = I_{c_k,e_k}$. Divide R^n by hyperplanes $x_i = a_i + \gamma_i\delta_i$, where $\delta_i = (b_i - a_i)/r$, γ_i integers. Let I_{α_k,β_k} be as in Problem 1.9.4, and choose r so large that $mC/r < \varepsilon$. Show that for any $k = 1,\ldots,m$, each interval $I_{a,b,\gamma}$ [$\gamma = (\gamma_1,\ldots\gamma_n)$,

γ_i integers] either lies entirely in I_{α_k,β_k} or does not intersect I_{α_k,β_k}. By Problem 1.9.1, $\lambda(I_{\alpha_k,\beta_k}) = \sum^{(k)} \lambda(I_{a,b,\gamma})$, $\lambda(I_{a,b}) = \sum \lambda(I_{a,b,\gamma})$. Since

$$\bigcup_k I_{\alpha_k,\beta_k} \supset I_{a,b},$$

it follows that every term $\lambda(I_{a,b,\gamma})$ that occurs in the last sum must also occur in at least one of the sums $\sum^{(k)}$. This gives

$$\prod_{i=1}^{n} (b_i - a_i) \leq \sum_{k=1}^{m} \lambda(I_{\alpha_k,\beta_k}) \leq \sum_{k=1}^{m} \lambda(I_{c_k,e_k}) + \varepsilon = \sum_{k=1}^{m} \lambda(E_k) + \varepsilon.]$$

1.9.6. Let $I_{a,b}$ be an open bounded interval in R^n. Show that

$$\mu(I_{a,b}) = \mu^*(I_{a,b}) = \mu(\bar{I}_{a,b}) = \lambda(I_{a,b}) = \prod_{i=1}^{n} (b_i - a_i).$$

1.9.7. If a set F in R^n is Lebesgue-measurable and if $\mu(F) < \infty$, then for any $\varepsilon > 0$ there exists an open set E such that $E \supset F$ and $\mu(E) < \mu(F) + \varepsilon$.

1.9.8. If a set F in R^n is Lebesgue-measurable, then there exists a Borel set E such that $E \supset F$ and $\mu(E - F) = 0$. [*Hint:* Consider first the case where $\mu(F) < \infty$.]

1.9.9. A straight line in R^n $(n \geq 2)$ has Lebesgue measure zero. More generally, a k-dimensional plane in R^n $(n \geq k + 1)$ has Lebesgue measure zero.

1.9.10. The boundary of a ball in R^n has Lebesgue measure zero.

1.9.11. Consider the transformation $Tx = Ax + k$ in R^n, where A is a nonsingular $n \times n$ matrix and x, k are column n-vectors. T maps sets E onto sets $T(E)$. Assume that $\lambda[T(I_{a,b})] = |\det A|\lambda(I_{a,b})$. (This will be proved in Problems 2.16.7 and 2.16.8.) Prove that T satisfies the properties (a)–(c) of Problem 1.6.5, with $|\alpha|$ replaced by $|\det A|$.

1.9.12. Extract from the interval (0,1) the middle third—that is, the interval $I_1 = (\frac{1}{3},\frac{2}{3})$. Next extract from the two remaining intervals, $(0,\frac{1}{3})$ and $(\frac{2}{3},1)$, the middle thirds I_2 and I_3 $[I_2 = (\frac{1}{9},\frac{2}{9}), I_3 = (\frac{7}{9},\frac{8}{9})]$, and so on. Let $I = I_1 \cup I_2 \cup I_3 \cup \cdots$. The set $C = [0,1] - I$ is called *Cantor's set*. Prove that the Lebesgue measure of C is zero. (It can be shown that C is not countable; in fact, it has the cardinal number of the continuum.)

1.9.13. The Lebesgue-Stieltjes outer measure μ_f, induced by a monotone function f (continuous on the right), satisfies

$$\mu_f(a,b] = f(b) - f(a).$$

1.9.14. The Lebesgue-Stieltjes outer measure is a metric outer measure.

1.9.15. Let $f(x) = 0$ if $x < 0$, $f(x) = 1$ if $x \geq 0$. Prove that

$$\mu_f\{(-1,0)\} < f(0) - f(-1).$$

As we have seen, every Borel set in R^n is a Lebesgue (measurable) set. From Problem 1.9.8 it also follows that every Lebesgue set has the form $E - N$, where E is a Borel set and N is a null set [that is, $\mu(N) = 0$]. Thus (compare Problem 1.5.2) *the Lebesgue measure is the completion of its restriction to the Borel sets.*

We conclude this section with a few results that we shall not prove. (They will not be used in the sequel.)

It is known that not every Lebesgue set is a Borel set. In fact, the two classes of sets have different cardinal numbers.

Denote by μ^* the Lebesgue outer measure on R^n. Then for any bounded interval $I_{a,b}$ the following is true: a set $E \subset I_{a,b}$ is measurable if and only if

$$\mu(I_{a,b}) = \mu^*(E) + \mu^*(I_{a,b} - E).$$

The *inner measure* μ_* is defined by

$$\mu_*(E) = \sup \{\mu(F);\ F \subset E,\ F \text{ Lebesgue-measurable}\}.$$

It is known that

$$\mu_*(E) = \mu(I_{a,b}) - \mu^*(I_{a,b} - E)$$

for any subset E of $I_{a,b}$. Consequently, *E is measurable if and only if $\mu^*(E) = \mu_*(E)$.*

1.10 SIGNED MEASURES

Definition An extended real-valued set function μ is called a *signed measure* if it assumes at most one of the values $+\infty$ and $-\infty$, and if it satisfies the conditions (i), (iii), (iv) of Definition 1.2.1.

The motivation for this notion comes by looking at the difference $\mu = \mu_1 - \mu_2$ of two measures μ_1, μ_2, one of which is finite, defined by

$$\mu(E) = \mu_1(E) - \mu_2(E) \qquad (E \in \mathfrak{A}), \tag{1.10.1}$$

where μ_1, μ_2 have the same domain $\mathfrak{A}$. It is clear that μ is a signed measure. More generally, any linear combination (with real coefficients) of measures having the same domain, defined in the obvious way (analogously to 1.10.1) is a signed measure, provided all but one of the measures are finite.

In this section we shall prove that every signed measure is necessarily of the form (1.10.1).

Let μ be a signed measure with domain $\mathfrak{A}$. A measurable set E is called *positive* (*negative*) with respect to μ if $\mu(E \cap A) \geq 0$ (≤ 0) for any measurable set A.

Theorem 1.10.1 *If μ is a signed measure, then there exist two measurable sets A and B such that A is positive and B is negative with respect to μ, and $X = A \cup B$, $A \cap B = \varnothing$.*

We say that A and B form a *Hahn decomposition* of X with respect to μ. The decomposition is not unique, in general.

We shall need the following lemma.

Lemma 1.10.2 *Let* μ *be a signed measure. If* E *and* F *are measurable sets such that* $E \subset F$, $|\mu(F)| < \infty$, *then* $|\mu(E)| < \infty$.

The proof follows easily from the relation

$$\mu(F) = \mu(F - E) + \mu(E).$$

To prove Theorem 1.10.1 we shall assume that

$$-\infty < \mu(E) \le \infty \qquad \text{for all } E \in \mathfrak{A}. \tag{1.10.1}$$

The case where $-\infty \le \mu(E) < \infty$ for all $E \in \mathfrak{A}$ can be treated by looking at $-\mu$.

Observe that the difference of two negative sets is again a negative set and that the union of two negative sets is a negative set. Since also the countable disjoint union of negative sets is a negative set, we conclude (by Theorem 1.1.1) that any countable union of negative sets is a negative set.

Define

$$\beta = \inf \mu(B_0), \qquad B_0 \text{ varies over all the negative sets.}$$

Then there is a sequence $\{B_j\}$ of negative sets such that $\mu(B_j) \to \beta$. The set $B = \bigcup_{j=1}^{\infty} B_j$ is then a negative set and, therefore, $\mu(B) \le \mu(B_j)$. It follows that

$$\beta = \mu(B). \tag{1.10.2}$$

Note, by (1.10.1), that $\beta > -\infty$.

We shall show that the set $A = X - B$ is positive; this will complete the proof of the theorem.

If A is not positive, then there is a measurable set $E_0 \subset A$ such that $\mu(E_0) < 0$. E_0 cannot be a negative set, for otherwise $B \cup E_0$ is a negative set and thus

$$\beta \le \mu(B \cup E_0) = \mu(B) + \mu(E_0) = \beta + \mu(E_0) < \beta.$$

Therefore E_0 contains a set E_1 with $\mu(E_1) > 0$. Let m_1 be the smallest positive integer for which such a set E_1 exists with $\mu(E_1) \ge 1/m_1$. By Lemma 1.10.2, $\mu(E_1) < \infty$. Since

$$\mu(E_0 - E_1) = \mu(E_0) - \mu(E_1) \le \mu(E_0) - \frac{1}{m_1} < 0,$$

we can apply to $E_0 - E_1$ the argument that has just been applied to E_0. Thus there is a smallest positive integer m_2 with the property that $E_0 - E_1$

contains a measurable subset E_2 with $\mu(E_2) \geq 1/m_2$. Continuing this construction step by step indefinitely, we obtain at the kth step an integer m_k with the property that m_k is the smallest positive integer for which there exists a subset E_k of $E_0 - \left[\bigcup_{i=1}^{k-1} E_i\right]$ such that

$$\mu(E_k) \geq \frac{1}{m_k}. \tag{1.10.3}$$

We fix E_k as a subset of $E_0 - \left[\bigcup_{i=1}^{k-1} E_i\right]$ for which (1.10.3) holds.

By Lemma 1.10.2, $\mu\left(\bigcup_{k=1}^{\infty} E_k\right) < \infty$. Since the sets E_k are mutually disjoint, it follows that $\mu(E_k) \to 0$ as $k \to \infty$. Hence

$$\lim_{k\to\infty} \frac{1}{m_k} = 0.$$

It follows that every measurable subset F of $F_0 = E_0 - \bigcup_{k=1}^{\infty} E_k$ satisfies

$$\mu(F) \leq \frac{1}{m_k - 1} \to 0 \qquad \text{if } k \to \infty$$

—that is, $\mu(F) = 0$. Thus F_0 is a negative set. Since

$$\mu(F_0) = \mu(E_0) - \sum_{k=1}^{\infty} \mu(E_k) < \mu(E_0) < 0,$$

we get $\mu(B \cup F_0) = \mu(B) + \mu(F_0) < \mu(B)$. Since $\mu(B \cup F_0) \geq \beta$ (for $B \cup F_0$ is a negative set), we get a contradiction to (1.10.2).

Set

$$\mu^+(E) = \mu(E \cap A), \qquad \mu^-(E) = -\mu(E \cap B). \tag{1.10.4}$$

Then μ^+ and μ^- are measures, and at least one of them is finite. They are independent of the particular Hahn decomposition (see Problem 1.10.4). Also, $\mu(E) = \mu^+(E) - \mu^-(E)$. We can now state the following result, known as the *Jordan decomposition* of a signed measure.

Theorem 1.10.3 *Let μ be a signed measure. Then at least one of the measures μ^+, μ^-, given by* (1.10.4), *is finite, and*

$$\mu = \mu^+ - \mu^-. \tag{1.10.5}$$

Note that the decomposition of a signed measure into a difference of two measures is not unique. The Jordan decomposition is a particular decomposition.

The measures μ^+, μ^- given by (1.10.4) are called the *upper* and *lower variations* of μ (respectively). The measure $|\mu|$ given by $|\mu| = \mu^+ + \mu^-$ is called the *total variation* of μ. Note that $|\mu|(E) = \mu^+(E) + \mu^-(E)$ is generally different from $|\mu(E)| = |\mu^+(E) - \mu^-(E)|$.

A signed measure is said to be *finite* (*σ-finite*) if $|\mu|$ is finite (σ-finite).

PROBLEMS

1.10.1. If $X = \bigcup_{n=1}^{\infty} A_n$ and μ is a signed measure with $|\mu(A_n)| < \infty$ for all n, then μ^+ and μ^- are σ-finite. (Hence, by definition, μ is σ-finite.)

1.10.2. Let μ be a signed measure and let $\{E_n\}$ be a monotone sequence of measurable sets [with $|\mu(E_1)| < \infty$ if $\{E_n\}$ is decreasing]. Then

$$\mu\left(\lim_n E_n\right) = \lim_n \mu(E_n).$$

1.10.3. Give an example of a signed measure for which the Hahn decomposition is not unique.

1.10.4. If $X = A_1 \cup B_1$, $X = A_2 \cup B_2$ are two Hahn decompositions of a signed measure μ, then for any measurable set E,

$$\mu(E \cap A_1) = \mu(E \cap A_2), \qquad \mu(E \cap B_1) = \mu(E \cap B_2).$$

1.10.5. A *complex measure* is, by definition, a finite complex-valued set function satisfying the properties (i), (iii), (iv) of Definition 1.2.1. Prove that μ is a complex measure with domain $\mathfrak{A}$ if and only if their exist finite measures $\mu_1, \mu_2, \mu_3, \mu_4$ with the same domain $\mathfrak{A}$ such that

$$\mu = \mu_1 - \mu_2 + i\mu_3 - i\mu_4.$$

CHAPTER 2

INTEGRATION

2.1 DEFINITION OF MEASURABLE FUNCTIONS

By a *measure space* we mean a triple $(X,\mathfrak{A},\mu)$, where X is a space, $\mathfrak{A}$ is a σ-algebra of subsets of X, and μ is a measure with domain $\mathfrak{A}$. When there is no confusion about $\mathfrak{A}$ and μ, we refer to $(X,\mathfrak{A})$, or just to X, as the measure space. The sets of $\mathfrak{A}$ are called *μ-measurable* (or just *measurable*) sets. The *Lebesgue measure space* (*Borel measure space*) R^n is the triple $(X,\mathfrak{A},\mu)$ where $X = R^n$, $\mathfrak{A}$ is the class of the Lebesgue (Borel) sets, and μ is the Lebesgue measure.

The measure space is said to be *finite* (or *totally finite*) if $\mu(X) < \infty$, and *σ-finite* if μ is σ-finite. The measure space is *complete* if μ is a complete measure.

Let f be a real-valued function defined on a measurable set X_0 of a measure space. We say that f is a *measurable function* (on X_0, or in X_0) if the inverse image of any open set in R^1 is measurable—that is, for any open set M in R^1, the set $f^{-1}(M) = \{y \in X_0; f(y) \in M\}$ is a measurable set. When X is the Lebesgue space (Borel space), then we say that f is *Lebesgue-measurable* (*Borel-measurable*) on X_0.

We shall need the concept of measurability also for extended real-valued functions. In this case we add the requirement that the sets $f^{-1}(+\infty)$, $f^{-1}(-\infty)$ be measurable.

Theorem 2.1.1 *Let* f *be an extended real-valued function defined on a measurable set* $X_0 \subset X$, X *a measure space.* f *is measurable if and only if* (i) *for every real number* c, *the set* $f^{-1}\{(-\infty,c)\}$ *is measurable, and* (ii) *the sets* $f^{-1}(+\infty)$, $f^{-1}(-\infty)$ *are measurable.*

Proof. Since $(-\infty,c)$ is an open set in R^1, if f is measurable, then $f^{-1}\{(-\infty,c)\}$ is a measurable set. The condition (ii) also holds for a measurable function f. Suppose conversely that f is a function satisfying (i) and (ii). We shall prove that f is measurable. Since

$$f^{-1}\{(-\infty,c]\} = \bigcap_{n=1}^{\infty} f^{-1}\left\{\left(-\infty, c+\frac{1}{n}\right)\right\},$$

the set $f^{-1}\{(-\infty,c]\}$ is measurable for any real number c. Writing

$$f^{-1}\{(a,b)\} = f^{-1}\{(-\infty,b)\} - f^{-1}\{(-\infty,a]\},$$

we see that $f^{-1}\{(a,b)\}$ is measurable. Since every open set M is a countable union of open intervals, $f^{-1}(M)$ is also a measurable set. This, together with (ii), shows that f is measurable.

PROBLEMS

2.1.1. Let $(X,\mathfrak{A},\mu)$ be a measure space and let Y be a measurable set. Denote by $\mathfrak{A}_Y$ the class of all the measurable sets that are subsets of Y, and denote by μ_Y the restriction of μ to $\mathfrak{A}_Y$. Prove that $(Y,\mathfrak{A}_Y,\mu_Y)$ is a measure space. It is called a *measure subspace* of $(X,\mathfrak{A},\mu)$.

2.1.2. Let Z be a subset of a space X, and let $(Z,\mathcal{B},\nu)$ be a measure space. Denote by $\mathfrak{A}$ the class of all sets E in X such that $E \cap Z$ is in $\mathcal{B}$, and define a set function μ by $\mu(E) = \nu(E \cap Z)$. Then $(X,\mathfrak{A},\mu)$ is a measure space and $(Z,\mathcal{B},\nu)$ coincides with $(Z,\mathfrak{A}_Z,\mu_Z)$.

2.1.3. A function f defined on a measurable set Y of a measure space $(X,\mathfrak{A},\mu)$ is measurable if and only if it is measurable as a function on Y considered as a set in the measure space $(Y,\mathfrak{A}_Y,\mu_Y)$.

Let Ω be a Lebesgue-measurable set in R^n. The measure subspace of the Lebesgue space R^n corresponding to Ω is called the *Lebesgue measure space* of Ω. By Problem 2.1.3, f is measurable on the measure space Ω if and only if f is Lebesgue-measurable on Ω (as a subset of the Lebesgue space R^n).

2.1.4. Prove that a function f is measurable if and only if: (i) the sets $f^{-1}\{(-\infty,c]\}$ are measurable for all real numbers c, and (ii) the sets $f^{-1}(+\infty)$, $f^{-1}(-\infty)$ are measurable.

2.1.5. Prove that a function f is measurable if and only if (i) the sets $f^{-1}\{(c,\infty)\}$ (or the sets $f^{-1}\{[c,\infty)\}$) are measurable for any real number c, and (ii) the sets $f^{-1}(+\infty), f^{-1}(-\infty)$ are measurable.

A real-valued function f defined on an open set X_0 of a metric space X is said to be *continuous* if the inverse image of any open set in R^1 is an open set in X. Corollary 1.8.3 implies:

Theorem 2.1.2 *Let X be a metric space and let μ be the measure corresponding (by Theorem 1.3.1) to a metric outer measure μ*. If f is a continuous function defined on an open subset X_0, then f is a measurable function on X_0.*

Theorem 2.1.3 *Let X be a measure space. If f is a measurable function on X, then, for any Borel set B of R^1, $f^{-1}(B)$ is a measurable set.*

Proof. Consider the class $\mathscr{D}$ of all the sets E on the real line for which $f^{-1}(E)$ is measurable. It is easy to verify that $\mathscr{D}$ is a σ-ring. Since $\mathscr{D}$ contains all the open sets, it also contains all the Borel sets.

PROBLEMS

2.1.6. The *characteristic function* of a set E is the function χ_E defined by

$$\chi_E(x) = \begin{cases} 1, & \text{if } x \in E, \\ 0, & \text{if } x \notin E. \end{cases}$$

Prove that the set E is measurable if and only if the function χ_E is measurable.

2.1.7. Let $\{E_n\}$ be a sequence of sets and let $E^* = \overline{\lim_n} E_n$. Prove that

$$\chi_{E^*}(x) = \overline{\lim_{n\to\infty}} \chi_{E_n}(x).$$

2.1.8. The *positive part* f^+ of a function f is defined by

$$f^+(x) = \begin{cases} f(x), & \text{if } f(x) > 0, \\ 0, & \text{if } f(x) \le 0. \end{cases}$$

Similarly, the *negative part* f^- of f is defined by

$$f^-(x) = \begin{cases} 0, & \text{if } f(x) > 0, \\ -f(x), & \text{if } f(x) \le 0. \end{cases}$$

Prove that f is measurable if and only if f^+ and f^- are measurable.

2.1.9. If f is measurable, then $|f|$ and $|f|^2$ are measurable.

2.1.10. A monotone function defined on the real line is Lebesgue-measurable.

2.1.11. A function $f(X)$, defined in a metric space X, is said to be *upper-semicontinuous* (*lower-semicontinuous*) if for any point $x \in X$,

$$\overline{\lim_{y\to x}} f(y) \le f(x) \qquad \left[\underline{\lim_{y\to x}} f(y) \ge f(x)\right].$$

Prove that upper (or lower) semicontinuous functions in a metric measure space as in Corollary 2.1.2 are measurable.

2.1.12. A complex-valued function f is said to be *measurable* if, for any open set M in the complex plane, $f^{-1}(M)$ is a measurable set. Prove that f is measurable if and only if both its real and imaginary parts are measurable (real-valued) functions.

2.2 OPERATIONS ON MEASURABLE FUNCTIONS

In this section and in the following ones, X is a fixed measure space. Furthermore, unless the contrary is explicitly stated, all measurable functions are defined on the whole space and are extended real-valued functions.

***Lemma* 2.2.1** *If* f *and* g *are measurable functions, then the set* $\mathrm{E} = \{\mathrm{x};\ \mathrm{f}(\mathrm{x}) < \mathrm{g}(\mathrm{x})\}$ *is a measurable set.*

Proof. Let $\{r_n\}$ be a sequence of all the rational numbers. Then

$$E = \{x;\ f(x) < g(x)\} = \bigcup_{r_n} [\{x;\ f(x) < r_n\} \cap \{x;\ g(x) > r_n\}].$$

Since each set on the right is measurable, E is also measurable.

When we consider the sum $f + g$ of two measurable functions f and g, we always tacitly assume that at no point x, $f(x) = \infty$, $g(x) = -\infty$ or $f(x) = -\infty$, $g(x) = \infty$. Similarly, when we consider the product fg, we always assume that at no point x, $f(x) = 0$, $g(x) = \pm\infty$ or $f(x) = \pm\infty$, $g(x) = 0$.

***Theorem* 2.2.2** *If* f *and* g *are measurable functions, then* $\mathrm{f} \pm \mathrm{g}$ *and* fg *are measurable functions.*

Proof. Using Problem 2.1.5, we easily verify that, for any real number c, $c - g(x)$ is a measurable function. We next have

$$E \equiv \{x; f(x) + g(x) < c\} = \{x; f(x) < c - g(x)\}.$$

Since the set on the right is measurable (by Lemma 2.2.1), also E is measurable. Next, the sets $(f+g)^{-1}(\infty) = f^{-1}(\infty) \cup g^{-1}(\infty)$ and $(f+g)^{-1}(-\infty) = f^{-1}(-\infty) \cup g^{-1}(-\infty)$ are also measurable. Now apply Theorem 2.1.1 to conclude that $f + g$ is measurable. Similarly one proves that $f - g$ is measurable.

The measurability of fg follows from the measurability of $|f + g|^2$ and $|f - g|^2$ (compare Problem 2.1.9), and the relation

$$fg = \tfrac{1}{4}\{(f+g)^2 - (f-g)^2\}.$$

For any sequence $\{f_n\}$ of functions, we define $\sup_n f_n$, $\overline{\lim_n} f_n$ by

$$\left(\sup_n f_n\right)(x) = \sup_n f_n(x), \qquad \left(\overline{\lim_n} f_n\right)(x) = \overline{\lim_{n\to\infty}} f_n(x).$$

Similarly, we define $\inf_n f_n$, $\underline{\lim_n} f_n$.

Theorem 2.2.3 *If* $\{f_n\}$ *is a sequence of measurable functions, then* $\sup_n f_n$, $\inf_n f_n$, $\overline{\lim_n} f_n$ *and* $\underline{\lim_n} f_n$ *are measurable functions.*

Proof. The assertion for $\sup_n f_n$ follows from Problem 2.1.4 and the relations

$$\left\{x;\, \sup_n f_n(x) \le c\right\} = \bigcap_{n=1}^{\infty} \{x;\, f_n(x) \le c\},$$

$$\left(\sup_n f_n\right)^{-1}(\infty) = \bigcap_{m=1}^{\infty} \bigcup_{n=1}^{\infty} f_n^{-1}(m, \infty], \qquad \left(\sup_n f_n\right)^{-1}(-\infty) = \bigcap_{n=1}^{\infty} f_n^{-1}(-\infty).$$

The assertion for $\inf_n f_n$ follows from $\inf_n f_n = -\sup_n(-f_n)$. Now use the relations

$$\overline{\lim_n} f_n = \inf_k \sup_{n\ge k} f_n,$$

$$\underline{\lim_n} f_n = \sup_k \inf_{n\ge k} f_n.$$

Definition A property P concerning the points x of the measure space X is said to be true *almost everywhere* (briefly, *a.e.*) if the set E of points for which P is not true has measure zero. If $E = \varnothing$, then we say that P is true *everywhere*.

Note that if f and g are two extended real-valued functions such that f is measurable and $g = f$ a.e., and if X is complete, then g is also measurable.

Definition A sequence $\{f_n\}$ of functions is said to be (*pointwise*) *convergent almost everywhere* (or *everywhere*) if there exists a function g such that

$$\lim_{n\to\infty} f_n(x) = g(x) \qquad \text{a.e. (or everywhere).}$$

We then say that $\{f_n\}$ *converges* (*pointwise*) *almost everywhere* (or *everywhere*) to g, and write $\lim_n f_n = g$ a.e.

Corollary 2.2.4 *If a sequence* $\{f_n\}$ *of measurable functions converges everywhere to a function* g*, then* g *is measurable. If* X *is complete and* $\{f_n\}$ *converges to* g *a.e., then* g *is measurable.*

Indeed, by Theorem 2.2.3, $g_0 = \overline{\lim_n} f_n$ is a measurable function. From this the assertions of the corollary immediately follow.

A function f is called a *simple function* if there is a finite number of mutually disjoint measurable sets $E_1,\dots,E_m$ and real numbers $\alpha_1,\dots,\alpha_m$ such that

$$f(x) = \begin{cases} \alpha_i, & \text{if } x \in E_i \quad (i = 1,2,\dots,m), \\ 0, & \text{if } x \notin \bigcup_{i=1}^{m} E_i. \end{cases}$$

We can write

$$f(x) = \sum_{i=1}^{m} \alpha_i \chi_{E_i}(x).$$

Theorem 2.2.5 *Let* f *be a nonnegative measurable function. Then there exists a monotone-increasing sequence* $\{f_n\}$ *of simple nonnegative functions, such that* $\{f_n\}$ *converges to* f *everywhere.*

Proof. For each integer $n \geq 1$ and for each $x \in X$, let

$$f_n(x) = \begin{cases} \dfrac{i-1}{2^n}, & \text{if } \dfrac{i-1}{2^n} \leq f(x) < \dfrac{i}{2^n}, \quad i = 1,2,\dots,n2^n, \\ n, & \text{if } f(x) \geq n. \end{cases}$$

Then the f_n are simple nonnegative functions, and $f_{n+1}(x) \geq f_n(x)$. At a point x where $f(x) < \infty$,

$$0 \leq f(x) - f_n(x) \leq \frac{1}{2^n}, \qquad \text{if } n > f(x).$$

At a point x where $f(x) = \infty$, $f_n(x) = n$. This shows that, for any point x, $f_n(x) \to f(x)$ as $n \to \infty$.

PROBLEMS

2.2.1. If g is an extended real-valued Borel-measurable function on the real line and if f is a real-valued measurable function on a measure space X, then the composite function $h(x) = g[f(x)]$ is a measurable function on X.

2.2.2. Let $g(u_1,\dots,u_k)$ be a continuous function in R^k, and let $\varphi_1,\dots,\varphi_k$ be measurable functions. Prove that the composite function $h(x) = g[\varphi_1(x),\dots,\varphi_k(x)]$ is a measurable function. Note that as a special case we may conclude that

$$\max(\varphi_1,\dots,\varphi_n) \quad \text{and} \quad \min(\varphi_1,\dots,\varphi_n)$$

are measurable functions.

2.2.3. Let $f(x)$ be a measurable function and define

$$g(x) = \begin{cases} \dfrac{1}{f(x)}, & \text{if } f(x) \neq 0, \\ 0, & \text{if } f(x) = 0. \end{cases}$$

Prove that g is measurable.

2.2.4. The nth *Baire class* $\mathscr{B}_n$ is a class of functions f defined as follows: $f \in \mathscr{B}_n$ if and only if f is the pointwise limit everywhere of a sequence of functions belonging to $\mathscr{B}_{n-1}$, and $\mathscr{B}_0$ is the class of all continuous functions. Prove that all the functions of $\mathscr{B}_n$ are measurable.

2.2.5. Show that the characteristic function of the set of rational numbers in R^1 belongs to $\mathscr{B}_2$.

2.2.6. Let f be a measurable function, and define

$$g(x) = \begin{cases} 0, & \text{if } f(x) \text{ is rational}, \\ 1, & \text{if } f(x) \text{ is irrational}. \end{cases}$$

Prove that g is measurable.

2.2.7. A measurable function is the pointwise limit everywhere of a sequence of simple functions.

2.2.8. If f is a bounded measurable function, then f is the uniform limit of a sequence of simple functions.

2.3 EGOROFF'S THEOREM

A measurable function f is said to be *a.e. real-valued* if the set $\{x; |f(x)| = \infty\}$ has measure zero.

Definition A sequence $\{f_n\}$ of a.e. real-valued, measurable functions, in a measure space $(X,\mathfrak{A},\mu)$, is said to *converge almost uniformly* to a measurable function f if for any $\varepsilon > 0$ there exists a measurable set E such that $\mu(E) < \varepsilon$ and $\{f_n\}$ converges to f uniformly on $X - E$. Note that f is necessarily a.e. real-valued.

Theorem 2.3.1 *If a sequence* $\{f_n\}$ *of a.e. real-valued, measurable functions converges almost uniformly to a measurable function* f, *then* $\{f_n\}$ *converges to* f *a.e.*

Proof. For every positive integer m there is a measurable set E_m such that $\mu(E_m) < 1/m$ and $\{f_n\}$ converges uniformly to f on $E_m^c = X - E_m$. The sequence $\{f_n\}$ is then convergent to f on the set

$$F = \bigcup_{m=1}^{\infty} E_m^c = \left(\bigcap_{m=1}^{\infty} E_m\right)^c.$$

Since $\mu(F^c) = \mu\left(\bigcap_{n=1}^{\infty} E_n\right) \le \mu(E_m) < 1/m$ for any positive integer m, $\mu(F^c) = 0$. Hence $\{f_n\}$ converges to f a.e.

If X is a finite measure space, then the following converse of Theorem 2.3.1 is also true, and is known as *Egoroff's theorem.*

Theorem 2.3.2 *Let* X *be a finite measure space. If a sequence* $\{f_n\}$ *of a.e. real-valued, measurable functions converges a.e. to an a.e. real-valued, measurable function* f, *then* $\{f_n\}$ *converges to* f *almost uniformly.*

Proof. We may assume that f and the f_n are real-valued everywhere. For any two positive integers k and n, let

$$E_n^k = \bigcap_{m=n}^{\infty} \left\{x;\, |f_m(x) - f(x)| < \frac{1}{k}\right\}.$$

Then $E_n^k \subset E_{n+1}^k$. Since $\{f_n\}$ converges a.e. to f, $\lim_n E_n^k = E$, where E is a set with $\mu(X - E) = 0$. By Theorem 1.2.1(iv),

$$\lim_{n\to\infty} \mu(X - E_n^k) = \mu(X - E) = 0.$$

Hence, for any $\varepsilon > 0$ there is an integer n_k such that

$$\mu(X - E_n^k) < \frac{\varepsilon}{2^k} \qquad \text{if } n \ge n_k.$$

Let $F = \bigcap_{k=1}^{\infty} E_{n_k}^k$. This is a measurable set, and

$$\mu(X - F) = \mu\left[\bigcup_{k=1}^{\infty} (X - E_{n_k}^k)\right] \le \sum_{k=1}^{\infty} \mu(X - E_{n_k}^k) < \varepsilon.$$

The sequence $\{f_n\}$ is uniformly convergent, on F, to f. In fact, this follows from the inequalities

$$|f_n(x) - f(x)| < \frac{1}{k} \qquad \text{if } x \in F \subset E_{n_k}^k,\quad n \ge n_k.$$

The proof is thereby completed.

PROBLEMS

2.3.1. Let X be the Lebesgue measure space on the real line and let f_n be the characteristic function of (n,∞). Show that $\{f_n\}$ is convergent to zero a.e., but not almost uniformly.

2.3.2. Let $\{f_n\}$ be a sequence of measurable functions in a finite measure space X. Suppose that for almost every x, $\{f_n(x)\}$ is a bounded set. Then for any $\varepsilon > 0$ there exist a positive number c and a measurable set E with $\mu(X - E) < \varepsilon$, such that $|f_n(x)| \leq c$ for all $x \in E$, $n = 1,2,\ldots$.

2.4 CONVERGENCE IN MEASURE

Definition A sequence $\{f_n\}$ of a.e. real-valued, measurable functions is said to be *convergent in measure* if there is a measurable function f such that, for any $\varepsilon > 0$,

$$\lim_{n\to\infty} \mu[\{x; |f_n(x) - f(x)| \geq \varepsilon\}] = 0. \tag{2.4.1}$$

We then say that $\{f_n\}$ *converges in measure to* f.

It is easily seen that if $\{f_n\}$ converges in measure to two functions, f and g, then $f = g$ a.e. Also, if $\{f_n\}$ converges to f in measure, then f is a.e. real-valued.

Theorem 2.4.1 *If a sequence* $\{f_n\}$ *of a.e. real-valued, measurable functions converges almost uniformly to a measurable function* f, *then* $\{f_n\}$ *converges in measure to* f.

Proof. For any $\varepsilon > 0$, $\delta > 0$, there is a set E with $\mu(E) < \delta$ such that $|f_n(x) - f(x)| < \varepsilon$ for all $x \in X - E$ and n sufficiently large (independently of x). But this immediately implies convergence in measure.

Corollary 2.4.2 *If* $\mu(X) < \infty$, *then any sequence* $\{f_n\}$ *of a.e. real-valued, measurable functions that converges a.e. to an a.e. real-valued, measureable function* f *is also convergent to* f *in measure.*

This follows from Theorem 2.4.1 and Egoroff's theorem.

Definition A sequence $\{f_n\}$ of a.e. real-valued, measurable functions is called a *Cauchy sequence in measure* if, for any $\varepsilon > 0$,

$$\mu[\{x; |f_n(x) - f_m(x)| \geq \varepsilon\}] \to 0 \qquad \text{as } m,n \to \infty.$$

It is easily seen that if $\{f_n\}$ converges to f in measure, then $\{f_n\}$ is Cauchy in measure. The converse will be proved later on in this section.

We have seen so far that almost uniform convergence implies both convergence a.e. and convergence in measure. Furthermore, if $\mu(X) < \infty$, then almost uniform convergence is equivalent to convergence a.e. We shall now prove that convergence in measure of a sequence implies almost uniform convergence of some subsequence. In fact, the following theorem gives a somewhat stronger result.

Theorem 2.4.3 *If a sequence* $\{f_n\}$ *of a.e. real-valued, measurable functions is Cauchy in measure, then there is a measurable function* f *and a subsequence* $\{f_{n_k}\}$ *such that* $\{f_{n_k}\}$ *converges to* f *almost uniformly.*

Proof. For any positive integer k there is an integer n_k such that

$$\mu\left[\left\{x; |f_n(x) - f_m(x)| \geq \frac{1}{2^k}\right\}\right] < \frac{1}{2^k} \qquad \text{if } n,m \geq n_k.$$

We may assume that $n_1 < n_2 < \cdots$. Then $\{f_{n_k}\}$ is an infinite subsequence of $\{f_n\}$. Let

$$E_k = \left\{x; |f_{n_k}(x) - f_{n_{k+1}}(x)| \geq \frac{1}{2^k}\right\},$$

$$F_m = \bigcup_{k=m}^{\infty} E_k.$$

Then

$$|f_{n_h}(x) - f_{n_j}(x)| < \frac{1}{2^{j-1}} \qquad \text{if } x \in X - F_m, \quad j \geq h \geq m.$$

Thus, on each set $X - F_m$, $\{f_{n_j}\}$ satisfies the Cauchy criterion for uniform convergence. It follows that there exists a function f defined on

$$A = \bigcup_{m=1}^{\infty} (X - F_m)$$

such that $\{f_{n_j}\}$ converges to f uniformly on every set $X - F_m$. By the first part of Corollary 2.2.4, f is measurable on A. Define $f(x) = 0$ for $x \in X - A$. Then, as easily seen, f is a measurable function on X. Since

$$\mu(F_m) \leq \sum_{k=m}^{\infty} \mu(E_k) \leq \sum_{k=m}^{\infty} \frac{1}{2^k} = \frac{1}{2^{m-1}},$$

and since $\{f_{n_j}\}$ converges to f uniformly on any set $X - F_m$, it follows that $\{f_{n_j}\}$ converges to f almost uniformly.

Corollary 2.4.4 *If* $\{f_n\}$ *is a sequence of a.e. real-valued, measurable functions that is Cauchy in measure, then there is an a.e. real-valued, measurable function* f *such that* $\{f_n\}$ *converges to* f *in measure.*

This follows from the relation

$$\{x; |f_n(x) - f(x)| \geq \varepsilon\} \subset \left\{x; |f_n(x) - f_{n_k}(x)| \geq \frac{\varepsilon}{2}\right\} \cup \left\{x; |f_{n_k}(x) - f(x)| \geq \frac{\varepsilon}{2}\right\},$$

and from the fact that $\{f_{n_k}\}$ converges to f almost uniformly (and thus in measure).

PROBLEMS

2.4.1. Let X be a finite measure space. Let $\{f_n\}$ be a sequence of a.e. real-valued, measurable functions that is convergent a.e. to f, and let $f \neq 0$ a.e., $f_n \neq 0$ a.e. for all n. For any $\varepsilon > 0$ there is a positive number b and a measurable set E_n such that $|f_n(x)| \geq b$ on E_n, and $\mu(E_n^c) < \varepsilon$. [*Hint:* Prove the assertion first for f.]

2.4.2. Let X be a finite measure space. Let $\{f_n\}$ and $\{g_n\}$ be sequences of a.e. real, measurable functions that converge in measure to f and g, respectively. Let α and β be any real numbers. Then:

(a) $\{\alpha f_n + \beta g_n\}$ converges in measure to $\alpha f + \beta g$.
(b) $\{|f_n|\}$ converges in measure to $|f|$.
(c) $\{f_n g\}$ converges in measure to fg.
(d) $\{f_n g_n\}$ converges in measure to fg. [*Hint:* Consider first the case $f = 0$, $g = 0$.]
(e) If $f_n \neq 0$ a.e. and if $f \neq 0$ a.e., then $\{1/f_n\}$ converges in measure to $1/f$. [*Hint:* Use Problem 2.4.1.]

2.4.3. Prove the following result (which immediately yields another proof of Corollary 2.4.2): Let f_n $(n = 1,2,\ldots)$ and f be a.e. real-valued measurable functions in a finite measure space. For any $\varepsilon > 0$, $n \geq 1$, let

$$E_n(\varepsilon) = \{x; |f_n(x) - f(x)| \geq \varepsilon\}.$$

Then $\{f_n\}$ converges a.e. to f if and only if

$$\lim_{n\to\infty} \mu\left[\bigcup_{m=n}^{\infty} E_m(\varepsilon)\right] = 0 \qquad \text{for any } \varepsilon > 0. \tag{2.4.2}$$

[*Hint:* Let $F = \{x; \{f_n(x)\}$ is not convergent to $f(x)\}$. Then $F = \bigcup_{k=1}^{\infty} \overline{\lim_n}\, E_n(1/k)$. Show that $\mu(F) = 0$ if and only if (2.4.2) holds.]

2.4.4. Let X be the set of all positive integers, $\mathfrak{A}$ the class of all subsets of X, and $\mu(E)$ (for any $E \in \mathfrak{A}$) the number of points in E. Prove that in this measure space, convergence in measure is equivalent to uniform convergence.

2.4.5. Define functions $f_m^n(x)$ on $[0,1)$ by $f_m^n(x) = 1$ if $x \in [(m-1)/n, m/n)$ and $f_m^n(x) = 0$ otherwise. Enumerate the f_m^n in any way, and call the sequence thus obtained $\{\varphi_j\}$. For instance, $\varphi_1 = f_1^1$, $\varphi_2 = f_2^1$, $\varphi_3 = f_1^2$, $\varphi_4 = f_3^1, \ldots$. Prove that $\{\varphi_j\}$ converges in measure to zero, but $\lim \varphi_j(x)$ does not exist for any $x \in [0,1)$.

2.4.6. If $\{f_n\}$ is a sequence of a.e. real-valued measurable functions in X, and if $\mu(X) < \infty$, then there exist positive constants λ_n such that $\{f_n/\lambda_n\}$ converges a.e. to zero. [*Hint:* $|f_n(x)| \leq$ const. on a set E_n with $\mu(X - E_n) < 2^{-n}$.]

2.5 INTEGRALS OF SIMPLE FUNCTIONS

A simple function $f = \sum_{i=1}^{n} \alpha_i \chi_{E_i}$ on a measure space $(X, \mathcal{A}, \mu)$ is said to be *integrable* if $\mu(E_i) < \infty$ for all the indices i for which $\alpha_i \neq 0$. The *integral* of f is the number $\sum_{i=1}^{n} \alpha_i \mu(E_i)$, where we agree to take $\alpha_i \mu(E_i) = 0$ for all the indices i for which $\alpha_i = 0$, $\mu(E_i) = \infty$. We denote this sum also by $\int f(x)\, d\mu(x)$ or $\int f\, d\mu$. Thus,

$$\int f(x)\, d\mu(x) = \int f\, d\mu = \sum_{i=1}^{n} \alpha_i \mu(E_i). \tag{2.5.1}$$

Suppose f is written in form: $\sum_{j=1}^{m} \beta_j \chi_{F_j}$. Then clearly $\alpha_i = \beta_j$ if $E_i \cap F_j \neq \varnothing$. Also,

$$F_j \subset \bigcup_{i=1}^{n}{}^{*} E_i \quad \text{if } \beta_j \neq 0, \qquad E_i \subset \bigcup_{j=1}^{m}{}^{*} F_j \quad \text{if } \alpha_i \neq 0.$$

where the asterisk by the first (second) union symbol indicates that the union is taken only over the sets $E_i (F_j)$ for which $\alpha_i \neq 0 \, (\beta_j \neq 0)$. We conclude that $\mu(F_j) < \infty$ if $\beta_j \neq 0$. Hence f is integrable according to the representation $\sum \beta_j \chi_{F_j}$. Write $\gamma_{ij} = \alpha_i = \beta_j$ if $E_i \cap F_j \neq \varnothing$. It follows that

$$\sum_{j=1}^{m} \beta_j \mu(F_j) = \sum_{j=1}^{m} \beta_j \sum_{i=1}^{n} \mu(F_j \cap E_i) = \sum_{i=1}^{n} \sum_{j=1}^{m} \gamma_{ij} \mu(E_i \cap F_j)$$

$$= \sum_{i=1}^{n} \alpha_i \sum_{j=1}^{m} \mu(E_i \cap F_j) = \sum_{i=1}^{n} \alpha_i \mu(E_i).$$

Consequently, the definition (2.5.1) of the integral is independent of the representation of f, and is thus unambiguously defined.

If E is any measurable set and f is an integrable simple function, then

$\chi_E f$ is also an integrable simple function. We define the *integral of f over E* by

$$\int_E f\,d\mu = \int \chi_E f\,d\mu. \tag{2.5.2}$$

In particular we have

$$\int \chi_E\,d\mu = \int_E d\mu = \mu(E)$$

if E is a measurable set with finite measure.

The following theorem contains some basic results on integrals. The proofs are all rather obvious and are left for the reader.

Theorem 2.5.1 *Let* f *and* g *be integrable simple functions, and let* α, β *be real numbers. Then*:

(a) $\alpha f + \beta g$ *is an integrable simple function, and*

$$\int (\alpha f + \beta g)\,d\mu = \alpha \int f\,d\mu + \beta \int g\,d\mu.$$

(b) *If* $f \geq 0$ *a.e., then* $\int f\,d\mu \geq 0$.

(c) *If* $f \geq g$ *a.e., then* $\int f\,d\mu \geq \int g\,d\mu$.

(d) $|f|$ *is an integrable simple function, and*

$$\left|\int f\,d\mu\right| \leq \int |f|\,d\mu.$$

(e) $\int |f + g|\,d\mu \leq \int |f|\,d\mu + \int |g|\,d\mu$.

(f) *If* $m \leq f \leq M$ *a.e. on a measurable set* E *with* $\mu(E) < \infty$, *then*

$$m\,\mu(E) \leq \int_E f\,d\mu \leq M\,\mu(E).$$

(g) *If* $f \geq 0$ *a.e., and if* E *and* F *are measurable sets with* $E \subset F$, *then*

$$\int_E f\,d\mu \leq \int_F f\,d\mu.$$

(h) *If* E *is a disjoint union* $\bigcup_{m=1}^{\infty} E_m$ *of measurable sets, then*

$$\int_E f\,d\mu = \sum_{m=1}^{\infty} \int_{E_m} f\,d\mu.$$

Definition A sequence $\{f_n\}$ of integrable simple functions is said to be a *Cauchy sequence in the mean* if

$$\int |f_n - f_m|\,d\mu \to 0 \qquad \text{as } n,m \to \infty. \tag{2.5.3}$$

Lemma 2.5.2 *If* $\{f_n\}$ *is a sequence of integrable simple functions that is Cauchy in the mean, then there is an a.e. real-valued, measurable function* f *such that* $\{f_n\}$ *converges in measure to* f.

Proof. For any $\varepsilon > 0$, let

$$E_{nm} = \{x; |f_n(x) - f_m(x)| \geq \varepsilon\}.$$

Since $f_n - f_m$ is integrable, E_{nm} has finite measure. From Theorem 2.5.1 we then get

$$\int |f_n - f_m|\, d\mu \geq \varepsilon\mu(E_{nm}).$$

It follows that $\mu(E_{nm}) \to 0$ if $n,m \to \infty$. Thus $\{f_n\}$ is a Cauchy sequence in measure. The assertion of the lemma now follows from Corollary 2.4.4.

PROBLEMS

2.5.1. If f and g are integrable simple functions, then fg is also an integrable simple function.

2.5.2. An integrable simple function f is equal a.e. to zero if and only if $\int_E f\, d\mu = 0$ for any measurable set E.

2.5.3. If f is a nonnegative integrable simple function and if $\int f\, d\mu = 0$, then $f = 0$ a.e.

2.6 DEFINITION OF THE INTEGRAL

Definition 2.6.1 An extended real-valued, measurable function f, on a measure space $(X,\mathfrak{A},\mu)$ is said to be *integrable* if there exists a sequence $\{f_n\}$ of integrable simple functions having the following properties:

(a) $\{f_n\}$ is a Cauchy sequence in the mean.

(b) $\lim_{n\to\infty} f_n(x) = f(x)$ a.e.

We shall compare this definition with another one whereby (b) is replaced by

(b′) $\{f_n\}$ converges in measure to f.

Theorem 2.6.1 *A function* f *is integrable according to the first definition* [*with* (a), (b)] *if and only if it is integrable according to the second definition* [*with* (a), (b′)].

Proof. Suppose f is integrable according to the first definition. By Lemma 2.5.2, $\{f_n\}$ converges in measure to an a.e. real-valued, measurable function g. Theorems 2.4.3 and 2.3.1 imply that there is a subsequence $\{f_{n_k}\}$ of $\{f_n\}$ that converges a.e. to g. Since (b) holds, $g = f$ a.e. Hence $\{f_n\}$ converges in measure to f—that is, (a) and (b′) hold.

Suppose conversely that f is integrable according to the definition (a), (b′). That means that there is a sequence $\{g_n\}$ of integrable simple functions such that $\{g_n\}$ is Cauchy in the mean and $\{g_n\}$ converges in measure to f. By Theorems 2.4.3, 2.3.1 there is a subsequence $\{g_{n_k}\}$ of $\{g_n\}$ that is convergent to f a.e. Denote this subsequence by $\{f_k\}$. Then $\{f_k\}$ satisfies (a) and (b). Thus f is integrable according to the first definition [with (a), (b)].

From (b′) we conclude: *If f is integrable, then f is a.e. real-valued.*

Definition 2.6.2 Let f be an integrable function and let (a), (b) hold. The *integral* of f is defined to be the number $\lim_{n\to\infty} \int f_n \, d\mu$, and is denoted by $\int f(x) \, d\mu(x)$ or $\int f \, d\mu$. Thus

$$\int f(x) \, d\mu(x) = \int f \, d\mu = \lim_{n\to\infty} \int f_n \, d\mu. \tag{2.6.1}$$

Note that

$$\left| \int f_n \, d\mu - \int f_m \, d\mu \right| \le \int |f_n - f_m| \, d\mu \to 0 \quad \text{if } n,m \to \infty,$$

so that the limit on the right-hand side of (2.6.1) indeed exists.

Theorem 2.6.2 *The definition of the integral, as given by* (2.6.1), *is independent of the choice of the sequence* $\{f_n\}$ [*satisfying* (a), (b)].

The proof depends upon two lemmas.

Lemma 2.6.3 *Let* $\{f_n\}$ *be a sequence of integrable simple functions satisfying* (a), (b), *and let*

$$\lambda(E) = \lim_{n\to\infty} \int_E f_n \, d\mu$$

for any measurable set E. *Then* $\lambda(E)$ *is a completely additive set function.*

Proof. The limit defining $\lambda(E)$ exists uniformly with respect to E. Indeed, this follows from

$$\left| \int_E f_n \, d\mu - \int_E f_m \, d\mu \right| \le \int_E |f_n - f_m| \, d\mu \le \int |f_n - f_m| \, d\mu \tag{2.6.2}$$

and (a). Note also that $\lambda(E)$ is finitely additive. Now let E be a disjoint union $\bigcup_{i=1}^{\infty} E_i$ of measurable sets E_i. Then

$$\left|\lambda(E) - \sum_{i=1}^{m} \lambda(E_i)\right| \le \left|\lambda(E) - \int_E f_n\, d\mu\right| + \left|\int_E f_n\, d\mu - \sum_{i=1}^{m} \int_{E_i} f_n\, d\mu\right| + \left|\int_{\bigcup_{i=1}^{m} E_i} f_n\, d\mu - \lambda\left(\bigcup_{i=1}^{m} E_i\right)\right|.$$

For any $\varepsilon > 0$, the first and third terms on the right are less than $\varepsilon/2$ for all $n \ge n_0$, where n_0 is sufficiently large, depending on ε, but not on m. Hence,

$$\left|\lambda(E) - \sum_{i=1}^{m} \lambda(E_i)\right| \le \varepsilon + \left|\int_E f_{n_0}\, d\mu - \sum_{i=1}^{m} \int_{E_i} f_{n_0}\, d\mu\right|.$$

Taking $m \to \infty$ and using Theorem 2.5.1(h), we get

$$\overline{\lim_{m\to\infty}} \left|\lambda(E) - \sum_{i=1}^{m} \lambda(E_i)\right| \le \varepsilon.$$

Since ε is arbitrary, the assertion follows.

Lemma 2.6.4 *Let $\{f_n\}$ and $\{g_n\}$ be two Cauchy sequences in the mean of integrable simple functions, such that*

$$\lim_{n\to\infty} f_n = \lim_{n\to\infty} g_n = f \qquad a.e.$$

For any measurable set E, let

$$\lambda(E) = \lim_{n\to\infty} \int_E f_n\, d\mu, \qquad \nu(E) = \lim_{n\to\infty} \int_E g_n\, d\mu.$$

If E is σ-finite, then $\lambda(E) = \nu(E)$.

Proof. Consider first the case where $\mu(E) < \infty$. By the proof of Theorem 2.6.1 we have that $\{f_n\}$ and $\{g_n\}$ converge in measure to f. For any $\varepsilon > 0$,

$$E_n = \{x;\, |f_n(x) - g_n(x)| \ge \varepsilon\} \subset \left\{x;\, |f_n(x) - f(x)| \ge \frac{\varepsilon}{2}\right\} \cup \left\{x;\, |g_n(x) - f(x)| \ge \frac{\varepsilon}{2}\right\}.$$

Hence $\mu(E_n) \to 0$ if $n \to \infty$. Next,

$$\int_E |f_n - g_n|\, d\mu \le \int_{E-E_n} |f_n - g_n|\, d\mu + \int_{E\cap E_n} |f_n|\, d\mu + \int_{E\cap E_n} |g_n|\, d\mu. \quad (2.6.3)$$

The first integral on the right is less than $\varepsilon\mu(E)$. The second integral is estimated by

$$\int_{E\cap E_n} |f_n|\,d\mu \le \int_{E\cap E_n} |f_n - f_{n_0}|\,d\mu + \int_{E\cap E_n} |f_{n_0}|\,d\mu$$

$$\le \varepsilon + \int_{E\cap E_n} |f_{n_0}|\,d\mu$$

if n_0 is sufficiently large and $n \ge n_0$. With n_0 fixed, we have $|f_{n_0}| \le c$ a.e., where c is a finite number. Since $\mu(E_n) \to 0$ as $n \to \infty$, we can choose n_1 sufficiently large so that

$$\int_{E\cap E_n} |f_{n_0}|\,d\mu \le c\mu(E \cap E_n) < \varepsilon \qquad \text{if } n \ge n_1.$$

We conclude that

$$\int_{E\cap E_n} |f_n|\,d\mu \le 2\varepsilon \qquad \text{if } n \ge n_2 \quad [n_2 = \max(n_0, n_1)].$$

The third integral on the right-hand side of (2.6.3) is estimated in the same way.

Combining the estimates for the integrals on the right-hand side of (2.6.3), we get

$$\int_E |f_n - g_n|\,d\mu \le \varepsilon\mu(E) + 4\varepsilon.$$

Hence

$$|\lambda(E) - \nu(E)| \le \overline{\lim_{n\to\infty}} \int_E |f_n - g_n|\,d\mu \le \varepsilon\mu(E) + 4\varepsilon.$$

Since ε is arbitrary, $\lambda(E) = \nu(E)$ provided $\mu(E) < \infty$.

Consider now the case where E is σ-finite. Then we can write E as a countable union $\bigcup_{j=1}^{\infty} E_j$ of sets with finite measure. Taking $F_1 = E_1$, $F_j = E_j - \left[\bigcup_{i=1}^{j-1} E_i\right]$, we see that E is a countable union of the sets F_j, which are mutually disjoint and have a finite measure. By what we have already proved, $\lambda(F_j) = \nu(F_j)$ for all j. Since λ and ν are completely additive (by Lemma 2.6.3), it follows that $\lambda(E) = \nu(E)$.

We shall now complete the proof of Theorem 2.6.2. We introduce the sets

$$N(f_n) = \{x;\, f_n(x) \ne 0\}, \qquad N(g_n) = \{x;\, g_n(x) \ne 0\},$$

$$N = \left[\bigcup_{n=1}^{\infty} N(f_n)\right] \cup \left[\bigcup_{n=1}^{\infty} N(g_n)\right].$$

Since each of the sets $N(f_n)$, $N(g_n)$ has a finite measure, N is σ-finite. Lemma 2.6.4 then gives

$$\lim_{n\to\infty} \int_N f_n \, d\mu = \lim_{n\to\infty} \int_N g_n \, d\mu. \tag{2.6.4}$$

But, clearly, $f_n = \chi_N f_n$, $g_n = \chi_N g_n$ everywhere on X. Hence,

$$\int_N f_n \, d\mu = \int f_n \, d\mu, \qquad \int_N g_n \, d\mu = \int g_n \, d\mu.$$

Combining this with (2.6.4), we get

$$\lim_{n\to\infty} \int f_n \, d\mu = \lim_{n\to\infty} \int g_n \, d\mu.$$

This completes the proof of Theorem 2.6.2.

Definition 2.6.3 Let f be an integrable function and let E be any measurable set. The *integral of f over E* is defined by

$$\int_E f \, d\mu = \int \chi_E f \, d\mu. \tag{2.6.5}$$

Note that $\chi_E f$ is integrable, so that the expression on the right is well defined.

Definition 2.6.4 Let E be a measurable set and let f be a measurable function on E. If $\chi_E f$ is *integrable*, then we say that f is *integrable on E*. We define the *integral of f over E* by (2.6.5).

Definition 2.6.5 If $(X, \mathfrak{A}, \mu)$ is the Lebesgue measure space, then we say that f is *Lebesgue-integrable* (*on E*) if and only if f is integrable (on E). The integral $\int_E f \, d\mu$ is then called the *Lebesgue integral of f over E*. It is often denoted by $\int_E f(x) \, dx$.

Similarly we define the *Lebesgue-Stieltjes integral* over E, when μ is the Lebesgue-Stieltjes measure μ_g induced by a monotone-increasing function g. It is denoted by

$$\int_E f \, dg \qquad \text{or} \qquad \int_E f(x) \, dg(x).$$

We finally give the following useful definition.

Definition 2.6.6 If f is a nonnegative measurable function (on a measurable set E) and if f is not integrable (on E), then we say that the integral of f (over E) *is equal to infinity*, and write

$$\int f \, d\mu = \infty \qquad \left(\int_E f \, d\mu = \infty \right).$$

If f and g are integrable functions, then they are a.e. real-valued. Hence $f+g$ and $f \cdot g$ are defined a.e. In what follows we shall think of $f+g$ and $f \cdot g$ as everywhere defined measurable functions. All we have to do is define $f(x)+g(x)$ [or $f(x) \cdot g(x)$] on the set

$$E = [f^{-1}(\infty) \cap g^{-1}(-\infty)] \cup [f^{-1}(-\infty) \cap g^{-1}(\infty)]$$

(or $E = [f^{-1}(0) \cap g^{-1}\{\infty, -\infty\}] \cap [f^{-1}\{\infty, -\infty\} \cap g^{-1}(0)]$) such that the resulting function $f+g$ (or $f \cdot g$) is measurable. This can be done, for instance, by setting $f+g=0$ (or $f \cdot g = 0$) on E. Theorems concerning convergence a.e., almost uniformly, and in measure are obviously unaffected by the particular definition of $f+g$ (or $f \cdot g$) on E. Also, by Problem 2.6.1, the integrals $\int (f+g)\, d\mu$, $\int (f \cdot g)\, d\mu$ are independent of the definition of $f+g$, $f \cdot g$ on E. In fact, none of the results given in this book will depend on the particular definition used for $f+g$, $f \cdot g$ on the sets E.

PROBLEMS

2.6.1. If f is integrable and if g is measurable and $g = f$ a.e., then g is integrable and $\int f\, d\mu = \int g\, d\mu$.

2.6.2. If f is integrable, then the set $N(f) = \{x; f(x) \neq 0\}$ is σ-finite.

2.6.3. Let f be a measurable function. Prove that f is integrable if and only if f^+ and f^- are integrable, or if and only if $|f|$ is integrable.

2.6.4. Let X be the measure space described in Problem 2.4.4. Then f is integrable if and only if the series $\sum_{n=1}^{\infty} |f(n)|$ is convergent. If f is integrable, then

$$\int f\, d\mu = \sum_{n=1}^{\infty} f(n).$$

2.6.5. A function φ on a real interval (a,b) is called a *step function* if there exist points x_i $(i = 1,\ldots,n)$ and real numbers c_i $(i = 1,\ldots,n)$ such that $a = x_0 < x_1 < \cdots < x_{n-1} < x_n = b$ and $\varphi(x) = c_i$ if $x_{i-1} < x < x_i$ $(i = 1,\ldots,n)$. Prove that if f is Lebesgue-integrable over a bounded interval (a,b), then there exists a sequence $\{\varphi_m\}$ of step functions such that $\int |f - \varphi_m|\, d\mu \to 0$ as $m \to \infty$, and $\lim_m \varphi_m = f$ a.e. [*Hint:* Prove it first when f is a simple function using Problem 1.9.7.]

2.6.6. Let f be a Lebesgue-integrable function over a bounded interval (a,b). Prove that for any $\varepsilon > 0$, $\delta > 0$, there exists a continuous function g in $[a,b]$ such that $|g(x) - f(x)| < \varepsilon$ for all x in a subset E of (a,b), and $\mu[(a,b) - E] < \delta$.

2.6.7. A complex-valued function f is said to be *simple* if it has the form $\sum_{j=1}^{n} c_j \chi_{E_j}$, where E_i are measurable sets and c_j are complex numbers.

f is then said to be *integrable* if $\mu(E_i) < \infty$ for any i for which $c_i \neq 0$. The *integral* of f is defined by $\sum_{i=1}^{n} c_i \mu(E_i)$. Now let f be any measurable complex-valued function. f is said to be *integrable* if there exists a sequence $\{f_m\}$ of complex-valued, integrable simple functions, such that (a) and (b) in Definition 2.6.1 hold. Prove that $f = f_1 + if_2$ (f_1, f_2 are the real and imaginary parts of f) is integrable if and only if f_1 and f_2 are integrable, and $\int f\,d\mu = \int f_1\,d\mu + i\int f_2\,d\mu$.

2.7 ELEMENTARY PROPERTIES OF INTEGRALS

We begin by extending Theorem 2.5.1 to integrable functions.

Theorem 2.7.1 *Let* f *and* g *be integrable functions and let* α, β *be real numbers. Then*

(a) $\alpha f + \beta g$ *is integrable and*

$$\int (\alpha f + \beta g)\,d\mu = \alpha \int f\,d\mu + \beta \int g\,d\mu.$$

(b) *If* $f \geq 0$ *a.e., then* $\int f\,d\mu \geq 0$.
(c) *If* $f \geq g$ *a.e., then* $\int f\,d\mu \geq \int g\,d\mu$.
(d) $|f|$ *is integrable and*

$$\left|\int f\,d\mu\right| \leq \int |f|\,d\mu.$$

(e) $\int |f + g|\,d\mu \leq \int |f|\,d\mu + \int |g|\,d\mu$.
(f) *If* $m \leq f \leq M$ *a.e. on a measurable set* E, *and if* $\mu(E) < \infty$, *then*

$$m\mu(E) \leq \int_E f\,d\mu \leq M\,\mu(E).$$

(g) *If* $f \geq 0$ *a.e. and if* E *and* F *are two measurable sets with* $E \subset F$, *then*

$$\int_E f\,d\mu \leq \int_F f\,d\mu.$$

(h) *If* $f \geq m > 0$ *on a measurable set* E, *then* $\mu(E) < \infty$.

Proof. (a) follows from Theorem 2.5.1 by taking limits. To prove (b), let $\{f_n\}$ be a sequence of integrable simple functions that is Cauchy in the mean and that satisfies: $\lim f_n = f$ a.e. Then $\{|f_n|\}$ is also a Cauchy sequence in the mean, and $\lim_n |f_n| = f$ a.e. Hence

$$\int f\, d\mu = \lim_{n\to\infty} \int |f_n|\, d\mu \geq 0.$$

(c) follows from (b). The integrability of $|f|$ follows from Problem 2.6.3. The inequality in (d) follows from (c). Next, (e)–(g) follow easily from (c). Finally, to prove (h), suppose $\mu(E) = \infty$. By Problem 2.6.2, E is σ-finite. Hence there exists a monotone-increasing sequence of sets E_j having finite measure, such that $\lim E_j = E$. By (g), (f),

$$\int_E f\, d\mu \geq \int_{E_j} f\, d\mu \geq m\mu(E_j) \to \infty$$

as $j \to \infty$—a contradiction.

Definition 2.7.1 A sequence $\{f_n\}$ of integrable functions $\{f_n\}$ is said to be a *Cauchy sequence in the mean* if

$$\int |f_n - f_m|\, d\mu \to 0 \qquad \text{as } n,m \to \infty.$$

If there is an integrable function f such that

$$\int |f_n - f|\, d\mu \to 0 \qquad \text{as } n \to \infty,$$

then we say that $\{f_n\}$ *converges in the mean to* f.

It is clear that if a sequence $\{f_n\}$ is convergent in the mean (to some integrable function f), then it is also Cauchy in the mean. The converse is also true and will be proved in Section 2.8.

Theorem 2.7.2 *If* $\{f_n\}$ *is a sequence of integrable functions that converges in the mean to an integrable function* f, *then* $\{f_n\}$ *converges in measure to* f.

Proof. For any $\varepsilon > 0$, let

$$E_n = \{x;\, |f_n(x) - f(x)| \geq \varepsilon\}.$$

By Theorem 2.7.1(h), $\mu(E_n) < \infty$. Applying Theorem 2.7.1(g) and (f), we get

$$\int |f_n - f|\, d\mu \geq \int_{E_n} |f_n - f|\, d\mu \geq \varepsilon\mu(E_n).$$

Hence $\mu(E_n) \to 0$ as $n \to \infty$. This proves the assertion.

Theorem 2.7.3 *If* f *is an a.e. nonnegative, integrable function, then* $\int f\,d\mu = 0$ *if and only if* f = 0 *a.e.*

Proof. If $f = 0$ a.e., then, by Problem 2.6.1, $\int f\,d\mu = 0$. Suppose conversely that $\int f\,d\mu = 0$. There exists a sequence $\{f_n\}$ that is Cauchy in the mean and that is convergent in measure to f. Since $f \geq 0$, the same is true of the sequence $\{|f_n|\}$. Hence

$$\lim_{n\to\infty} \int |f_n|\,d\mu = \int f\,d\mu = 0.$$

It follows that $\{f_n\}$ converges in the mean to zero. Theorem 2.7.2 then implies that $\{f_n\}$ converges in measure to zero. Since this sequence is also convergent to f in measure, we get $f = 0$ a.e.

Theorem 2.7.4 *Let* f *be a measurable function and let* E *be a set of measure zero. Then* f *is integrable on* E *and*

$$\int_E f\,d\mu = 0.$$

Proof. $\chi_E f = 0$ a.e. By Problem 2.6.1, $\chi_E f$ is integrable and $\int \chi_E f\,d\mu = 0$. Now use Definition 2.6.4.

Theorem 2.7.5 *Let* f *be an integrable function that is positive everywhere on a measurable set* E. *If* $\int_E f\,d\mu = 0$, *then* $\mu(E) = 0$.

Proof. Let $E_n = \{x \in E; f(x) \geq 1/n\}$. Then $\{E_n\}$ is an increasing sequence of sets and $E - \bigcup_{n=1}^{\infty} E_n$ has measure zero. Hence $\mu(E) = \lim \mu(E_n)$. Since $\mu(E_n) < \infty$,

$$0 = \int f\,d\mu \geq \int_{E_n} f\,d\mu \geq \frac{1}{n}\,\mu(E_n) \geq 0,$$

so that $\mu(E_n) = 0$ for all n. Hence $\mu(E) = 0$.

Theorem 2.7.6 *Let* f *be an integrable function. If* $\int_E f\,d\mu = 0$ *for every measurable set* E, *then* f = 0 *a.e.*

Proof. By Theorem 2.7.5, the set E where $f(x) > 0$ has measure zero. Similarly, the set E where $f(x) < 0$ has measure zero. Hence $f = 0$ a.e.

PROBLEMS

2.7.1. Let $(X,\mathfrak{A},\mu_1)$ and $(X,\mathfrak{A},\mu_2)$ be two measure spaces, and let $\mu = \mu_1 + \mu_2$. Then:

(a) $(X,\mathfrak{A},\mu)$ is a measure space.

(b) A simple function f is μ_i-integrable (that is, integrable with respect to μ_i) for $i = 1$ and $i = 2$ if and only if f is μ-integrable, and we then have

$$\int f\,d\mu = \int f\,d\mu_1 + \int f\,d\mu_2. \tag{2.7.1}$$

(c) A measurable function f is μ_i-integrable for $i = 1$ and $i = 2$ if and only if f is μ-integrable, and then (2.7.1) holds.

2.7.2. If f is an integrable function, g is a simple function, and $|f(x)| \geq |g(x)|$, then g is integrable.

2.7.3. Let f be an integrable function. Prove: (a) if $\int_E f\,d\mu \geq 0$ for all measurable sets E, then $f \geq 0$ a.e.; (b) if $\mu(X) < \infty$ and if $\int_E f\,d\mu \leq \mu(E)$ for all measurable sets E, then $f \leq 1$ a.e.

2.7.4. Show that the function $f(x) = \sin x + \cos x$ is not Lebesgue-integrable on the real line.

2.7.5. Show that the function $f(x) = (\sin x)/x$ is not Lebesgue-integrable over the interval $(1,\infty)$.

2.7.6. Show that the function $f(x) = 1/x$ is not Lebesgue-integrable in the interval $(0,1)$.

2.7.7. Prove that Theorem 2.7.1(f) is valid for any measurable set E, provided $m > 0$.

2.8 SEQUENCES OF INTEGRABLE FUNCTIONS

We begin with a lemma that is concerned with Definition 2.6.1.

Lemma 2.8.1 *If* f *is an integrable function and if* $\{f_n\}$ *is a sequence of integrable simple functions satisfying* (a),(b) [*or* (a),(b′)] *then*

$$\lim_{n\to\infty} \int |f - f_n|\,d\mu = 0.$$

Proof. For each fixed positive integer n, the sequence $\{|f_n - f_m|\}$ is Cauchy in the mean, since $\int ||f_n - f_m| - |f_n - f_j||\,d\mu \leq \int |f_m - f_j|\,d\mu \to 0$ if $m, j \to \infty$. It also converges a.e. [or, in measure, if (b′) holds] to $|f_n - f|$. Hence, by Definition 2.6.2,

$$\int |f_n - f|\,d\mu = \lim_{m\to\infty} \int |f_n - f_m|\,d\mu.$$

From this the assertion follows.

The following theorem shows that if we try to extend the definition of the integral by starting with integrable functions (instead of integrable simple functions), then we do not get anything new.

Theorem 2.8.2 *If* $\{f_n\}$ *is a sequence of integrable functions satisfying the conditions* (a),(b) [*or* (a),(b′)] *in Definition* 2.6.1, *then* f *is integrable and* $\int f\, d\mu = \lim_n \int f_n\, d\mu$.

Proof. Suppose that (a) and (b′) hold. By Lemma 2.8.1, for each n there is an integrable simple function $\tilde{f}_n$ such that

$$\int |\tilde{f}_n - f_n|\, d\mu < \frac{1}{n^2}.$$

It follows (cf. the proof of Theorem 2.7.2) that

$$\mu\left[\left\{x;\, |\tilde{f}_n(x) - f_n(x)| \geq \frac{1}{n}\right\}\right] < \frac{1}{n}.$$

It is clear that $\{\tilde{f}_n\}$ satisfies the conditions (a),(b′) (with the same f). Hence f is integrable, and

$$\int f\, d\mu = \lim_{n\to\infty} \int \tilde{f}_n\, d\mu = \lim_{n\to\infty} \int f_n\, d\mu.$$

If (a),(b) hold, then we take a subsequence $\{f_{n_k}\}$ satisfying (a) and (b′). We conclude that f is integrable, and $\int f\, d\mu = \lim \int f_{n_k}\, d\mu = \lim \int f_n\, d\mu$.

We now state an important theorem.

Theorem 2.8.3 *If* $\{f_n\}$ *is a sequence of integrable functions that is Cauchy in the mean, then there is an integrable function* f *such that* $\{f_n\}$ *converges in the mean to* f.

Proof. The proof of Lemma 2.5.2 extends word by word to sequences of any integrable functions. We thus conclude that $\{f_n\}$ is convergent in measure to a measurable function f. By Theorem 2.8.2, f is integrable. The sequence $\{|f_n - f|\}$ is then a sequence of integrable functions that is Cauchy in the mean, and that converges in measure to 0. By Theorem 2.8.1,

$$\lim_{n\to\infty} \int |f_n - f|\, d\mu = 0.$$

Definition 2.8.1 A real-valued set function λ is said to be *absolutely continuous* if for any $\varepsilon > 0$ there exists a number $\delta > 0$ such that, for any measurable set E with $\mu(E) < \delta$, $|\lambda(E)| < \varepsilon$.

Theorem 2.8.4 *Let* f *be an integrable function and let* λ *be the set function defined by*

$$\lambda(E) = \int_E f \, d\mu \tag{2.8.1}$$

for all the measurable sets E. *Then* λ *is completely additive and absolutely continuous.*

The set function λ is called the *indefinite integral* of f.

Proof. The complete additivity of λ follows from Lemma 2.6.3. To prove absolute continuity, let $\{f_n\}$ be a sequence of integrable simple functions satisfying (a),(b) in Definition 2.6.1. Then

$$\left| \int_E f \, d\mu \right| \le \left| \int_E f_n \, d\mu \right| + \left| \int_E f_n \, d\mu - \int_E f \, d\mu \right|.$$

By Lemma 2.8.1,

$$\left| \int_E f_n \, d\mu - \int_E f \, d\mu \right| \le \int_E |f_n - f| \, d\mu \le \int |f_n - f| \, d\mu < \frac{\varepsilon}{2}$$

if $n \ge n_0$, where n_0 is a number depending on ε, but not on E. The simple function f_{n_0} satisfies $|f_{n_0}(x)| \le c$ a.e., where c is a finite number. Taking $\delta \le \varepsilon/2c$, we thus find that

$$\left| \int_E f \, d\mu \right| \le \varepsilon \qquad \text{if } \mu(E) < \delta.$$

This completes the proof.

Theorem 2.8.4 shows that if f is an integrable function, then $\lambda(E)$ is a finite signed measure. A Hahn decomposition for λ is given by $A = \{x; f(x) \ge 0\}$, $B = \{x; f(x) < 0\}$. Applying Corollary 1.2.3 to the measures λ^+, λ^-, we conclude:

Corollary 2.8.5 *If* f *is integrable and* $\{E_n\}$ *is a sequence of measurable sets with* $\lim_n E_n = E$, $\mu(E) = 0$, *then*

$$\lim_{n\to\infty} \int_{E_n} f \, d\mu = 0.$$

We conclude this section with a few definitions that will be needed in Section 2.14.

A function $g(x)$ defined on a real interval (a,b) is said to be *absolutely continuous* if for any $\varepsilon > 0$ there is a $\delta > 0$ such that for any sequence of mutually disjoint intervals (a_i,b_i) in (a,b), with $\sum (b_i - a_i) < \delta$,

$$\sum |g(b_i) - g(a_i)| < \varepsilon.$$

A function $f(x)$ on a real interval $[a,b]$ is said to be of *bounded variation* if there is a constant K such that for any partition $a = x_0 < x_1 < \cdots < x_{n-1} < x_n = b$ of (a,b),

$$\sum |f(x_i) - f(x_{i-1})| \le K.$$

The g.l.b. of all the constants K is called the *total variation* of f in (a,b).

PROBLEMS

2.8.1. A measurable function f is called a *null function* if $f = 0$ a.e. We shall say that f is *equivalent* to g (and write $f \sim g$) if $f - g$ is a null function. Denote by $\bar{f}$ the class of all measurable functions that are equivalent to f. We denote by $L^1(X,\mathfrak{A},\mu)$, or, more briefly, by $L^1(X,\mu)$, the set of all classes $\bar{f}$ for which f is integrable, and define on it the function

$$\rho(\bar{f},\bar{g}) = \rho(f,g) = \int |f - g| \, d\mu.$$

[Note that if $f_0 \in \bar{f}$, $g_0 \in \bar{g}$, then $\rho(f_0,g_0) = \rho(f,g)$.] Prove that $L^1(X,\mu)$ is a complete metric space with the metric ρ.

2.8.2. For any Lebesgue-integrable function $g(y)$ on a bounded interval (a,b), the function

$$f(x) = \int_{(a,x)} g(y) \, dy$$

is absolutely continuous.

2.8.3. A function $f(x)$ is of bounded variation in a bounded interval (a,b) if and only if it is the difference $g_1(x) - g_2(x)$ of monotone-increasing functions g_1,g_2. If $f(x)$ is also continuous, then $g_1(x)$ and $g_2(x)$ can be taken to be continuous monotone-increasing functions. [*Hint:* Take $g_1(x)$ to be the total variation of f in the interval (a,x).]

2.8.4. If $f(x)$ is absolutely continuous, then it is of bounded variation.

2.8.5. Prove that the function $f(x) = x \sin(1/x)$, for $0 < x < 1$, is uniformly continuous but not of bounded variation.

2.8.6. If f is not integrable, then the absolute continuity of $\lambda(E)$ (in Theorem 2.8.4) can still be proved, provided f is integrable on sets E of finite measure (see Corollary 2.12.3). However, the same extension does not hold for the property of complete additivity. Give an example.

2.9 LEBESGUE'S BOUNDED CONVERGENCE THEOREM

One of the most important theorems in the theory of integration is the following one, known as *Lebesgue's bounded convergence theorem.*

Theorem 2.9.1 *Let* $\{f_n\}$ *be a sequence of integrable functions that* *converges either in measure or a.e. to a measurable function* f. *If* $|f_n(x)| \le g(x)$ *a.e. for all* n, *where* g *is an integrable function, then* f *is integrable and*

$$\lim_{n\to\infty} \int |f_n - f| \, d\mu = 0. \tag{2.9.1}$$

Proof. Consider first the case where $\{f_n\}$ converges to f in measure. We shall prove that $\{f_n\}$ is a Cauchy sequence in the mean. Let

$$E = \bigcup_{n=1}^{\infty} \{x;\, f_n(x) \neq 0\}.$$

Then, by Problem 2.6.2, E is σ-finite. We can therefore write $E = \bigcup_{k=1}^{\infty} E_k$, where $E_k \subset E_{k+1}$ and $\mu(E_k) < \infty$, for all k. Let $F_k = E - E_k$. Then

$$\int_{F_k} |f_m - f_n| \, d\mu \le \int_{F_k} (|f_m| + |f_n|) \, d\mu \le 2 \int_{F_k} g \, d\mu.$$

Noting that $F_k \supset F_{k+1}$, $\lim F_k = \varnothing$, we can apply Corollary 2.8.5 and conclude that, for any $\eta > 0$,

$$\int_{F_k} |f_m - f_n| \, d\mu < \eta \qquad \text{if } k \ge k_0, \tag{2.9.2}$$

where k_0 is a large positive integer that depends on η but not on m,n.

For any $\varepsilon > 0$, introduce the sets

$$G_{mn} = \{x;\, |f_m(x) - f_n(x)| \ge \varepsilon\}.$$

We have

$$\begin{aligned} \int_{E_k} |f_m - f_n| \, d\mu &= \int_{E_k - G_{mn}} |f_m - f_n| \, d\mu + \int_{E_k \cap G_{mn}} |f_m - f_n| \, d\mu \\ &\le \varepsilon\mu(E_k) + 2\int_{E_k \cap G_{mn}} g \, d\mu. \end{aligned} \tag{2.9.3}$$

Since $\{f_m\}$ converges to f in measure, $\mu(G_{mn}) \to 0$ if $m,n \to \infty$. Theorem 2.8.4 then gives

$$2\int_{E_k \cap G_{mn}} g \, d\mu \le 2\int_{G_{mn}} g \, d\mu < \varepsilon \qquad \text{if } m \ge n \ge n_0, \tag{2.9.4}$$

where n_0 is a large positive integer depending only on ε and g. Combining (2.9.2)–(2.9.4), we get

$$\int |f_m - f_n| \, d\mu \le \eta + \varepsilon + \varepsilon\mu(E_k) \qquad \text{if } m \ge n \ge n_0,\ k \ge k_0.$$

Taking $m,n \to \infty$, we get

$$\overline{\lim_{m,n\to\infty}} \int |f_m - f_n|\, d\mu \le \eta + \varepsilon + \varepsilon\mu(E_k).$$

Since ε is arbitrary, we get

$$\overline{\lim_{m,n\to\infty}} \int |f_m - f_n|\, d\mu \le \eta.$$

Since η is arbitrary, we find that $\{f_m\}$ is indeed a Cauchy sequence in the mean.

We now use Theorem 2.8.3. We conclude that there exists an integrable function h such that

$$\lim_{n\to\infty} \int |f_n - h|\, d\mu = 0. \tag{2.9.5}$$

Theorem 2.7.2 shows that $\{f_n\}$ converges in measure to h. Since f is also the limit in measure of $\{f_n\}$, we get $f = h$ a.e. Thus, f is integrable. Finally, (2.9.1) follows from (2.9.5).

Suppose next that $\{f_n\}$ converges a.e. to f (instead of in measure). If we prove that $\{f_n\}$ converges to f in measure, then the proof of the theorem is completed, by the previous part. Denote by N the set of points for which either $|f(x)| > g(x)$ or $|f_n(x)| > g(x)$ for some n. Then, for any $\varepsilon > 0$,

$$E_n = \bigcup_{j=n}^{\infty} \{x;\, |f_j(x) - f(x)| \ge \varepsilon\} \subset \left\{x;\, g(x) \ge \frac{\varepsilon}{2}\right\} \cup N, \qquad \mu(N) = 0.$$

Since g is integrable, $\mu(E_n) < \infty$ (by Problem 2.7.2). The hypothesis that $\lim_n f_n = f$ a.e. implies that $\mu\left(\bigcap_{n=1}^{\infty} E_n\right) = 0$. Hence, by Theorem 1.2.1(iv), $\lim_n \mu(E_n) = 0$. Since

$$\{x;\, |f_n(x) - f(x)| \ge \varepsilon\} \subset E_n,$$

it follows that $\{f_n\}$ converges to f in measure.

PROBLEMS

2.9.1. Let $f_n(x) = n$ if $0 \le x < 1/n$, $f_n(x) = 0$ if $1 \ge x \ge 1/n$. Then $\lim_n f_n(x) = 0$ a.e. on $[0,1]$, but $\lim_n \int f_n\, d\mu = 1$ (μ the Lebesgue integral). This example shows that the boundedness condition $|f_n| \le g$ in Theorem 2.9.1 is essential.

2.9.2. Let $(X, \mathfrak{A}, \mu)$ be the measure space introduced in Problem 2.4.4.

Let

$$f_n(k) = \begin{cases} \dfrac{1}{n}, & \text{if } 1 \le k \le n, \\ 0, & \text{if } k > n. \end{cases}$$

Then $\{f_n\}$ is a sequence of integrable functions that converges uniformly to $f \equiv 0$, but $\lim \int f_n \, d\mu \neq \int f \, d\mu$. This example again shows that the boundedness condition $|f_n| \le g$ in Theorem 2.9.1 is essential.

2.9.3. Let $(X,\mathfrak{A},\mu)$ be the measure space introduced in Problem 2.4.4. Let

$$f_n(k) = \begin{cases} \dfrac{1}{k}, & \text{if } 1 \le k \le n, \\ 0, & \text{if } k > n, \end{cases}$$

and $f(k) = 1/k$ for all $k \ge 1$. Then $\{f_n\}$ is a sequence of integrable functions that converges uniformly to f, but f is not integrable.

2.9.4. If $\mu(X) < \infty$ and if $\{f_n\}$ is a sequence of measurable functions that converges uniformly to a function f, then f is integrable and $\lim \int f_n \, d\mu = \int f \, d\mu$.

2.10 APPLICATIONS OF LEBESGUE'S BOUNDED CONVERGENCE THEOREM

Theorem 2.10.1 *Let* f *and* g *be measurable functions. If* $|f| \le g$ *a.e. and* g *is integrable, then* f *is integrable.*

Proof. It suffices to prove that $|f|$ is integrable (see Problem 2.6.3). By Theorem 2.2.5 there is an increasing sequence $\{h_n\}$ of simple nonnegative functions such that $\lim h_n = |f|$ a.e. The relations $h_n \le g$ show that the h_n are integrable simple functions (by the result of Problem 2.7.2). Now use the Lebesgue bounded convergence theorem to conclude that $|f|$ is integrable.

Definition A measurable function $f(x)$ is said to be *essentially bounded* if there is a (finite) constant c such that $|f(x)| \le c$ a.e. The g.l.b. of these constants c is called the *essential supremum* of f and is denoted by

$$\operatorname*{ess\,sup}_{x \in X} |f(x)| \qquad \text{or} \qquad \operatorname*{ess\,sup}_{X} |f|.$$

Corollary 2.10.2 *If* f(x) *is integrable and* g *is a measurable, essentially bounded function, then* fg *is integrable.*

This follows from Theorem 2.10.1, noting that $|fg| \le c|f|$ a.e. and $c|f|$ is integrable.

Corollary 2.10.3 *If a measurable function* f *is essentially bounded on a measurable set* E *with finite measure, then* f *is integrable over* E.

Indeed, the function $|\chi_E f|$ is bounded a.e. by $c\chi_E$, where c is a (finite) real number, and $c\chi_E$ is an integrable function. Now apply Theorem 2.10.1.

The next theorem is often referred to as the *Lebesgue monotone convergence theorem.*

Theorem 2.10.4 *Let* $\{f_n\}$ *be a monotone increasing sequence of nonnegative integrable functions, and let* $f(x) = \lim_n f_n(x)$. *Then*

$$\lim_{n\to\infty} \int f_n \, d\mu = \int f \, d\mu. \tag{2.10.1}$$

Recall that if f is not integrable then the integral on the right of (2.10.1) is taken in the sense of Definition 2.6.6.

Proof. If f is integrable, then the inequality $f_n \le f$ implies that

$$\int f_n \, d\mu \le \int f \, d\mu. \tag{2.10.2}$$

If f is not integrable, then (2.10.2) is also true, since the integral on the right is equal to ∞ (by Definition 2.6.6). Thus, in either case (2.10.2) holds. Taking $n \to \infty$, we get

$$\lim_{n\to\infty} \int f_n \, d\mu \le \int f \, d\mu. \tag{2.10.3}$$

It remains to show that if $\lim_n \int f_n \, d\mu < \infty$, then f is integrable and the equality (2.10.1) holds. If $m \ge n$, then $f_m \ge f_n$. Consequently

$$\int \left| f_m - f_n \right| d\mu = \int f_m \, d\mu - \int f_n \, d\mu \to 0 \qquad \text{if } m \ge n \to \infty.$$

Thus $\{f_m\}$ is a Cauchy sequence in the mean and $f_m \to f$ a.e. By Theorem 2.8.2 it follows that f is integrable and (2.10.1) holds.

The next result, known as *Fatou's lemma,* is very useful.

Theorem 2.10.5 *Let* $\{f_n\}$ *be a sequence of nonnegative integrable functions, and let* $f(x) = \underline{\lim}_n f_n(x)$. *Then*

$$\int f\,d\mu \le \underline{\lim}_{n\to\infty} \int f_n\,d\mu. \tag{2.10.4}$$

If, in particular, $\underline{\lim}_{n} \int f_n\,d\mu < \infty$, *then* f *is integrable.*

Proof. If $\underline{\lim} \int f_n\,d\mu = \infty$, then (2.10.4) is obviously true. It thus remains to consider the case where $\underline{\lim} \int f_n\,d\mu < \infty$. Let $g_n(x) = \inf_{j\ge n} f_j(x)$. Then $\{g_n\}$ is a monotone-increasing sequence of nonnegative integrable functions, and $g_n \le f_n$. It follows that

$$\lim_{n\to\infty} \int g_n\,d\mu \le \underline{\lim}_{n\to\infty} \int f_n\,d\mu < \infty. \tag{2.10.5}$$

Since $\lim g_n = \underline{\lim} f_n = f$, the Lebesgue monotone convergence theorem implies that f is integrable and

$$\int f\,d\mu = \lim_{n\to\infty} \int g_n\,d\mu.$$

Combining the last equality with (2.10.5), we obtain the inequality (2.10.4).

PROBLEMS

2.10.1. Let f and g be measurable functions. Prove: (a) f and g are integrable if $(f^2 + g^2)^{1/2}$ is integrable. (b) If f^2 and g^2 are integrable, then fg is integrable.

2.10.2. Derive the Lebesgue monotone convergence theorem from Fatou's lemma.

2.10.3. In a finite measure space, a sequence $\{f_n\}$ of a.e. real-valued measurable functions is convergent in measure to zero if and only if

$$\int \frac{|f_n|}{1+|f_n|}\,d\mu \to 0 \qquad \text{as } n\to\infty.$$

2.10.4. Denote by Z the space of all classes $\bar{f}$ of a.e. real-valued measurable functions f on a finite measure space $(X,\mathfrak{A},\mu)$, with $\bar{f} = \bar{g}$ if and only if $f = g$ a.e. (cf. Problem 2.8.1). Define, on Z,

$$\rho(\bar{f},\bar{g}) = \rho(f,g) = \int \frac{|f-g|}{1+|f-g|}\,d\mu.$$

Prove that Z is a complete metric space with the metric ρ.

2.10.5. Let $\mu(X) < \infty$ and let $f(x,t)$ be a function of $x \in X$, $t \in (a,b)$. Assume that for each fixed t, $f(x,t)$ is integrable, and that the partial derivative

$\partial f(x,t)/\partial t$ exists and is uniformly bounded for $x \in X$, $t \in (a,b)$. Then, for each $t \in (a,b)$, $\partial f(x,t)/\partial t$ is integrable, $\int f(x,t)\,d\mu(x)$ is differentiable, and

$$\frac{d}{dt}\int f(x,t)\,d\mu(x) = \int \frac{\partial}{\partial t} f(x,t)\,d\mu(x).$$

2.10.6. Let $f(x)$ be Lebesgue-measurable on the real line, and let $f(x) \geq 0$, $\int_{(0,\infty)} f(x)\,dx < \infty$. Prove that the function

$$g(t) = \int_{(0,\infty)} e^{-tx} f(x)\,dx \qquad (0 < t < \infty)$$

is differentiable, and $g'(t) = -\int_{(0,\infty)} xe^{-tx} f(x)\,dx$.

2.10.7. Let $\{f_n\}$ be a sequence of integrable functions. Prove that if $\sum_{n=1}^{\infty} \int |f_n|\,d\mu < \infty$, then the series $\sum_{n=1}^{\infty} f_n(x)$ is convergent to an integrable function $f(x)$, and

$$\int f\,d\mu = \sum_{n=1}^{\infty} \int f_n\,d\mu.$$

2.10.8. Let $\{f_n\}$ be a sequence of nonnegative integrable functions. Prove that if the series $f(x) = \sum f_n(x)$ is an integrable function, then $\sum_{n=1}^{\infty} \int f_n\,d\mu < \infty$.

2.10.9. Let f and f_n $(n = 1,2,\ldots)$ be integrable functions such that $0 \leq f_n(x) \leq f(x)$ a.e. Then

$$\int \left(\overline{\lim_n}\, f_n\right) d\mu \geq \overline{\lim_n} \int f_n\,d\mu \geq \underline{\lim_n} \int f_n\,d\mu \geq \int \left(\underline{\lim_n}\, f_n\right) d\mu.$$

2.10.10. Extend the result of Problem 2.10.9 to the case where $|f_n(x)| \leq f(x)$ a.e. [instead of $0 \leq f_n(x) \leq f(x)$ a.e.].

2.10.11. Let $X = \bigcup_{n=1}^{\infty} E_n$, $E_n \subset E_{n+1}$ for all n. Let f be a nonnegative measurable function. Prove that

$$\int f\,d\mu = \lim_{n\to\infty} \int_{E_n} f\,d\mu.$$

2.10.12. Let f be a nonnegative measurable function and let

$$f_n(x) = \begin{cases} f(x), & \text{if } f(x) \leq n, \\ n, & \text{if } f(x) > n. \end{cases}$$

Prove that $\lim_n \int f_n\,d\mu = \int f\,d\mu$.

2.10.13. If $f_1,\ldots,f_k$ are integrable functions, then $\max(f_1,\ldots,f_k)$ is an integrable function.

2.10.14. Give an example where Fatou's lemma holds with strict inequality.

2.10.15. Let f be a real-valued measurable function in a finite measure space. Let

$$s_n = \sum_{k=-\infty}^{\infty} \frac{k}{2^n} \mu\left[\left\{x; \frac{k}{2^n} < f(x) \le \frac{k+1}{2^n}\right\}\right] \qquad (n = 1,2,\ldots).$$

If f is integrable, then each series is absolutely convergent and

$$\int f \, d\mu = \lim_{n\to\infty} s_n. \tag{2.10.6}$$

Conversely, if one of the series s_n is absolutely convergent, then f is integrable and (2.10.6) holds. [*Hint:* If f is nonnegative and integrable, then apply Theorem 2.10.4 with

$$f_n(x) = \frac{k}{2^n} \qquad \text{if } \frac{k}{2^n} \le f(x) < \frac{k+1}{2^n} \qquad (k = 0,1,2,\ldots).$$

If $f \ge 0$ and one series, say that of s_n, is convergent, then use the inequality $f(x) \le f_n(x) + 1/2^n$.]

2.10.16. With the notation of Problem 2.8.1, we denote by $L^\infty(X,\mathcal{A},\mu)$ or, more briefly, $L^\infty(X,\mu)$, the space of all classes $\bar{f}$ of measurable and essentially bounded functions f. Prove that $L^\infty(X,\mu)$ is a complete metric space with the metric

$$\rho(\bar{f},\bar{g}) = \rho(f,g) = \operatorname*{ess\,sup}_{x\in X} |f(x) - g(x)|.$$

2.11 THE RIEMANN INTEGRAL

Let $[a,b]$ be a bounded closed interval on the real line and let $f(x)$ be a bounded function on $[a,b]$. For any partition Π: $a = x_0 < x_1 < \cdots < x_{n-1} < x_n = b$ of $[a,b]$, form the sums

$$S_\Pi = \sum_{i=1}^{n} M_i(x_i - x_{i-1}),$$

$$s_\Pi = \sum_{i=1}^{n} m_i(x_i - x_{i-1}),$$

where

$$M_i = \sup_{x_{i-1}\le x\le x_i} f(x), \qquad m_i = \inf_{x_{i-1}\le x\le x_i} f(x).$$

The number $|\Pi| = \max_{1 \le i \le n} (x_i - x_{i-1})$ is called the *mesh* of the partition, and S_π, s_π are called the *upper* and *lower Darboux sums* associated with the partition Π. Note that $S_\Pi \ge s_\Pi$.

Definition f is said to be *Darboux-integrable* if $\lim S_\Pi$ and $\lim s_\Pi$ exist as $|\Pi| \to 0$, and if

$$\lim_{|\Pi| \to 0} S_\Pi = \lim_{|\Pi| \to 0} s_\Pi. \tag{2.11.1}$$

The common limit is then denoted by

$$\int_a^b f(x)\, dx$$

and is called the *integral of* f *from* a *to* b, *in the sense of Darboux.*

Definition Consider the *Riemann sums*

$$T = \sum_{i=1}^{n} f(y_i)(x_i - x_{i-1}) \qquad \text{for any} \quad y_i \in [x_{i-1}, x_i].$$

f is said to be *Riemann-integrable* on (a, b) if there exists a number K such that $\lim_{|\Pi| \to 0} T = K$. The number K is then called the *integral of* f *from* a *to* b *in the sense of Riemann.*

It is well known that f is Riemann-integrable if and only if it is Darboux-integrable, and then the two integrals coincide. From now on we shall refer to the common integral as the Riemann integral.

The Lebesgue integral of f over $[a,b]$ will be denoted in this section by

$$\int_{[a,b]} f(x)\, dx.$$

We shall prove the following two theorems.

Theorem 2.11.1 *A bounded function* f *on the bounded interval* [a,b] *is Riemann-integrable if and only if it is continuous a.e.*

Theorem 2.11.2 *If a bounded function* f *over a bounded interval* [a,b] *is Riemann-integrable, then it is also Lebesgue-integrable and the two integrals agree—that is,*

$$\int_a^b f(x)\, dx = \int_{[a,b]} f(x)\, dx. \tag{2.11.2}$$

Proof of Theorem 2.11.1. Suppose f is Riemann-integrable. For each positive integer k, let

$$E_k = \left\{x;\ \overline{\lim_{\delta\to 0}} \sup_{\substack{|y-x|<\delta \\ |z-x|<\delta}} |f(y) - f(z)| > \frac{1}{k}\right\}.$$

The set E_0 where f is not continuous coincides with $\bigcup_{k=1}^{\infty} E_k$. If f is not continuous a.e., then $\mu^*(E_0) > 0$. Therefore also

$$\mu^*(E_k) = \alpha > 0 \qquad \text{for some } k.$$

k is now fixed.

Take any partition $\Pi: a = x_0 < x_1 < \ldots < x_{n-1} < x_n = b$ and let $J = \{x_1,\ldots,x_n\}$. The intervals $(x_0,x_1),\ldots,(x_{n-1},x_n)$ form a covering of $E_k - J$ and $\mu^*(E_k - J) = \mu^*(E_k) = \alpha$. Denote by J' the subset of J consisting of all the points x_i for which

$$(x_{i-1},x_i) \cap E_k \neq \varnothing \qquad \text{—that is,} \qquad (x_{i-1},x_i) \cap (E_k - J) \neq \varnothing.$$

By the monotonicity of μ^*,

$$\sum_{x_i \in J'} (x_i - x_{i-1}) \geq \alpha.$$

Since $M_i - m_i > 1/k$ if $(x_{i-1},x_i) \cap E_k \neq \varnothing$, we then have

$$S_\Pi - s_\Pi = \sum_{x_i \in J} (M_i - m_i)(x_i - x_{i-1}) \geq \sum_{x_i \in J'} (M_i - m_i)(x_i - x_{i-1}) > \frac{\alpha}{k}.$$

Hence (2.11.1) is not satisfied. We have thus derived a contradiction to our assumption that f is Riemann-integrable.

Suppose now that f is continuous a.e., and let $|f(x)| \leq M$ for $a \leq x \leq b$. For any $\varepsilon > 0$, cover the set E_0 of discontinuities of f by an open set E with

$$\mu(E) < \frac{\varepsilon}{4M}. \tag{2.11.3}$$

On the closed set $F = [a,b] - E$, f is continuous and, therefore, uniformly continuous. Hence there exists a $\delta > 0$ such that

$$|f(x) - f(y)| < \frac{\varepsilon}{2(b-a)} \qquad \text{if } x \in F,\ |x - y| < \delta. \tag{2.11.4}$$

Let Π_k ($k = 1,2$) be partitions with mesh $<\delta$, given by sets $J_k = \{x_1^k,\ldots,x_{n_k-1}^k,x_{n_k}^k = b\}$, and let Π be the partition given by the set $J = J_1 \cup J_2 = \{x_1,\ldots,x_{n-1},x_n = b\}$. We shall compare sums

$$S_k = \sum_{x_i^k \in J_k} f(y_i^k)(x_i^k - x_{i-1}^k) \qquad (k = 1,2),$$

$$S = \sum_{x_j \in J} f(z_j)(x_j - x_{j-1}),$$

where $x_0^k = x_0 = a$, $y_i^k \in [x_{i-1}^k, x_i^k]$, and $z_j \in F \cap [x_{j-1}, x_j]$ if the latter set is not empty. If

$$F \cap [x_{j-1}, x_j] = \varnothing, \tag{2.11.5}$$

we require only that $z_j \in [x_{j-1}, x_j]$.

We can clearly write

$$S_1 = \sum_{x_j \in J} f(y_j)(x_j - x_{j-1})$$

where $y_j = y_i^1$ if $[x_{j-1}, x_j] \subset [x_{i-1}^1\ x_i^1]$. Then

$$|S - S_1| \le \sum_{x_j \in J} |f(z_j) - f(y_j)|(x_j - x_{j-1}).$$

Denote by J' the subset of J consisting of those points x_j for which (2.11.5) holds, and let J'' be the set $J - J'$. Recalling that $|z_j - y_j| < \delta$ and using (2.11.3), (2.11.4), we get

$$\begin{aligned} |S - S_1| &\le \sum_{x_j \in J'} |f(z_j) - f(y_j)|(x_j - x_{j-1}) \\ &\quad + \sum_{x_j \in J''} |f(z_j) - f(y_j)|(x_j - x_{j-1}) \\ &\le 2M\mu(E) + (b - a)\frac{\varepsilon}{2(b - a)} < \varepsilon. \end{aligned}$$

A similar bound holds for $S - S_2$. Hence,

$$|S_1 - S_2| \le 2\varepsilon \qquad \text{if } |\Pi_1| < \delta,\ |\Pi_2| < \delta. \tag{2.11.6}$$

Choosing the y_i^1, y_j^2 such that

$$|S_1 - S_{\Pi_1}| < \varepsilon, \qquad |S_2 - S_{\Pi_2}| < \varepsilon,$$

we get

$$|S_{\Pi_1} - S_{\Pi_2}| < 4\varepsilon.$$

It follows that $\lim S_\Pi$ exists as $|\Pi| \to 0$. Similarly $\lim s_\Pi$ exists as $|\Pi| \to 0$. Next take $\Pi_1 = \Pi_2$ and choose the y_i^1 such that $|S_1 - S_{\Pi_1}| < \varepsilon$ and the y_i^2 such that $|S_2 - s_{\Pi_1}| < \varepsilon$. We then obtain

$$|S_{\Pi_1} - s_{\Pi_1}| < 4\varepsilon \qquad \text{if } |\Pi_1| < \delta.$$

This proves (2.11.1). Hence f is Riemann-integrable.

Proof of Theorem 2.11.2. We shall first prove that f is measurable. Denote by $\tilde{E}_1$ the set of points where f is continuous and let $E_1 = \tilde{E}_1 - \{a,b\}$. The set $E_2 = [a,b] - E_1$ has measure zero. Denote by f_i $(i = 1,2)$ the restriction of f to E_i. For any open interval I on the real line,

$$f^{-1}(I) = f_1^{-1}(I) \cup f_2^{-1}(I). \tag{2.11.7}$$

For any $x \in E_1$ with $f(x) \in I$ there is an open interval J_x about x with $f(J_x) \subset I$. (Here we needed the fact that $x \neq a$, $x \neq b$.) Denote by J the set $\bigcup J_x$ when x varies over the set $E_1 \cap f^{-1}(I)$. J is an open set and $f_1^{-1}(I) = E_1 \cap J$. Since E_1 and J are measurable sets, also $f_1^{-1}(I)$ is a measurable set. As for the set $f_2^{-1}(I)$, it is a subset of the set E_2 of measure zero. Hence it is a a measurable set. From (2.11.7) it then follows that $f^{-1}(I)$ is a measurable set. Hence f is a measurable function.

Since f is a bounded, measurable function, it is Lebesgue-integrable (by Corollary 2.10.3). It remains to show that (2.11.2) is valid.

We shall associate with each Darboux lower sum s_Π a step function σ_Π defined by

$$\sigma_\Pi(x) = m_i \qquad \text{if } x_{i-1} < x \leq x_i.$$

Since $f(x) \geq \sigma_\Pi(x)$, $\int_{[a,b]} f(x)\,dx \geq \int_{[a,b]} \sigma_\Pi(x)\,dx$. However,

$$\int_{[a,b]} \sigma_\Pi(x)\,dx = \sum_i m_i \mu\{[x_{i-1},x_i]\} = \sum_i m_i(x_i - x_{i-1}) = s_\Pi.$$

Hence

$$\int_{[a,b]} f(x)\,dx \geq \lim_{|\Pi| \to 0} s_\Pi = \int_a^b f(x)\,dx.$$

The reverse inequality is proved similarly, employing upper Darboux sums. The proof of Theorem 2.11.2 is thereby completed.

For unbounded functions or for unbounded intervals of integration, one defines the Riemann integral as an improper integral. Thus, for example, if $f(x)$ is a bounded Riemann-integrable function in any bounded interval, one defines

$$\int_{-\infty}^{\infty} f(x)\,dx = \lim_{\substack{A \to -\infty \\ B \to \infty}} \int_A^B f(x)\,dx.$$

A function f may then be Riemann-integrable without its absolute value $|f|$ being Riemann-integrable. This situation cannot occur for Lebesgue-integrable functions.

PROBLEMS

2.11.1. Prove the following theorem of Darboux: Let $f(x)$ be a bounded function in a bounded interval $[a,b]$. Then there is a number S such that for any sequence Π_m of partitions of $[a,b]$ such that $|\Pi_m| \to 0$, $S_{\Pi_m} \to S$.

2.11.2. Prove that a bounded monotone function can have at most a countable number of points of discontinuity. (Hence, it is Riemann-integrable on any bounded interval.)

2.11.3. The function $f(x) = (\sin x)/x$ is Riemann-integrable (but not Lebesgue-integrable, by Problem 2.7.5) over the interval $(1,\infty)$.

2.11.4. Construct a sequence $\{f_n\}$ of Riemann-integrable functions on the interval $[0,1]$, such that $|f_n(x)| \le 1$ for all $0 \le x \le 1$, $n = 1,2,\ldots$, and $\lim f_n(x) = f(x)$ exists everywhere, but $f(x)$ is not Riemann-integrable on $[0,1]$.

2.11.5. Let f, f_n $(n = 1,2,\ldots)$ be bounded Riemann-integrable functions on the interval $[0,1]$. Assume that there is a constant K such that $|f_n(x)| \le K$ for $0 \le x \le 1$, $n = 1,2,\ldots$. Show that if $\lim f_n(x) = f(x)$ a.e., then

$$\lim_{n\to\infty} \int_0^1 f_n(x)\,dx = \int_0^1 f(x)\,dx.$$

2.11.6. Let f be a bounded Riemann integrable function on bounded intervals of the real line. Assume that $|f|$ is Riemann-integrable on $(-\infty,\infty)$—that is, $\lim_{n\to\infty} \int_{-n}^{n} |f(x)|\,dx < \infty$. Prove that f is Lebesgue-integrable on $(-\infty,\infty)$ and

$$\int_{(-\infty,\infty)} f(x)\,dx = \int_{-\infty}^{\infty} f(x)\,dx.$$

2.11.7. Let R be a rectangle given by $a \le x \le b$, $c \le y \le d$, and let f be a bounded function on R. Let Π_1: $a = x_0 < x_1 < \ldots < x_n = b$ be a partition of (a,b) and let Π_2: $c = y_0 < y_1 < \ldots < y_m = d$ be a partition of (c,d). Let

$$M_{ij} = \sup f(x,y), \qquad m_{ij} = \inf f(x,y)$$

when

$$x_{i-1} \le x \le x_i, \qquad y_{j-1} \le y \le y_j.$$

Set $\Pi = (\Pi_1,\Pi_2)$, $|\Pi| = \max(|\Pi_1|,|\Pi_2|)$ and form the sums

$$S_\Pi = \sum_{i=1}^{n} \sum_{j=1}^{m} M_{ij}(x_i - x_{i-1})(y_j - y_{j-1}),$$

$$s_\Pi = \sum_{i=1}^{n} \sum_{j=1}^{m} m_{ij}(x_i - x_{i-1})(y_j - y_{j-1}).$$

If $\lim S_\Pi$ and $\lim s_\Pi$ exist as $|\Pi| \to 0$, and if they are equal, then we say that f is *Darboux-integrable* on R. We then define the *integral of f over R* to be $\lim_{|\Pi|\to 0} S_\Pi$. Similarly one defines the Riemann integral of f over R as the limit of the Riemann sums $\sum f(\xi_i, \eta_j)(x_i - x_{i-1})(y_j - y_{j-1})$, when $|\Pi| \to 0$. It is well known that the Riemann integral exists if and only if the Darboux integral exists, and then the two integrals coincide.

Prove: If f is measurable, then the Lebesgue integral of f over R is equal to the Riemann integral of f over R.

2.11.8. Analogously to the Darboux integral $\int_b^a f(x)\,dx$, one defines the Riemann-Stieltjes integral as follows: Consider *upper Darboux-Stieltjes* sums

$$S_\Pi = \sum_{i=1}^{n} M_i[g(x_i) - g(x_{i-1})]$$

and *lower Darboux-Stieltjes* sums

$$s_\Pi = \sum m_i[g(x_i) - g(x_{i-1})].$$

If lim S_Π and lim s_Π exist, as $|\Pi| \to 0$, and if they are equal, then we say that f is *Darboux-Stieltjes integrable* (with respect to g) and the limit is denoted by $\int_a^b f(x)\,dg(x)$. This limit is called the *Darboux-Stieltjes integral* of f with respect to g, from a to b. The *Riemann-Stieltjes integral* is defined as the limit of $\sum_{i=1}^n f(y_i)[g(x_i) - g(x_i)]$ when $|\Pi| \to 0$; it is equal to the Darboux-Stieltjes integral. In the above sum, the y_i are any points in $[x_{i-1},x_i]$. Prove that if f is continuous and g of bounded variation, then the Riemann-Stieltjes integral exists.

2.11.9. Prove that if f is continuous in $[a,b]$ and g is monotone increasing and continuous from the right, then

$$\int_a^b f(x)\,dg(x) = \int_{(a,b]} f(x)\,dg(x),$$

where $dg = d\mu_g$, and μ_g is the Lebesgue-Stieltjes measure induced by g.

2.11.10. Let f be continuous on $[a,b]$ and $g(x)$ be of bounded variation; then

$$\left|\int_a^b f(x)\,dg(x)\right| \le V(g) \sup_{a \le x \le b} |f(x)|,$$

where $V(g)$ is the total variation of g over $[a,b]$.

2.11.11. If $f(t)$ is Lebesgue-integrable over $(-\infty,\infty)$ and if $-\infty < a < b < \infty$, then for any real number h,

$$\int_{[a,b]} f_h(x)\,dx = \int_{[a+h,\,b+h]} f(x)\,dx, \qquad \text{where } f_h(x) = f(x+h).$$

2.12 THE RADON-NIKODYM THEOREM

In this section we consider measure spaces $(X,\mathfrak{A},\mu)$, where μ is a signed measure.

Definition 2.12.1 Let $(X,\mathfrak{A},\mu)$ be a measure space and let ν be a signed measure with the domain $\mathfrak{A}$. We say that ν is *absolutely continuous* with respect to μ, and write $\nu \ll \mu$, if $\nu(E) = 0$ for any measurable set E for which $|\mu|(E) = 0$.

Notice that if μ is a measure and if ν is absolutely continuous with respect to μ according to Definition 2.8.1, then ν is also absolutely continuous with respect to μ according to Definition 2.12.1. We shall prove (Theorem 2.12.2) that if $|\nu|$ is finite on every set where μ is finite, then the two concepts of absolute continuity are equivalent.

Lemma 2.12.1 *For any two signed measures* μ *and* ν, *the following three conditions are equivalent:*

(a) $\nu \ll \mu$;
(b) $\nu^+ \ll \mu$ *and* $\nu^- \ll \mu$;
(c) $|\nu| \ll \mu$.

Proof. Let $X = A \cup B$ be a Hahn decomposition with respect to ν. Suppose (a) holds and let $|\mu|(E) = 0$. Then $|\mu|(E \cap A) = 0$, $|\mu|(E \cap B) = 0$. Hence $\nu^+(E) = \nu(E \cap A) = 0, \nu^-(E) = \nu(E \cap B) = 0$. This proves (b). Next, (b) implies (c), since $|\nu| = \nu^+ + \nu^-$. Finally, (c) implies (a), since $0 \le |\nu(E)| \le |\nu|(E)$.

Theorem 2.12.2 *Let* μ *be a signed measure and let* ν *be a signed measure such that* $|\nu|(E) < \infty$ *for any measurable set* E *for which* $|\mu|(E) < \infty$. *Suppose that* $\nu \ll \mu$. *Then for any* $\varepsilon > 0$ *there is a* $\delta > 0$ *such that* $|\nu|(E) < \varepsilon$ *for any measurable set* E *for which* $|\mu|(E) < \delta$.

Proof. If the assertion is false, then there is an $\varepsilon > 0$ and a sequence $\{E_n\}$ for measurable sets such that $|\mu|(E_n) < 1/2^n$ and $|\nu|(E_n) \ge \varepsilon$ for all $n \ge 1$. Let $E = \overline{\lim_n} E_n$. Then

$$|\mu|(E) \le \sum_{i=n}^{\infty} |\mu|(E_i) \le \frac{1}{2^{n-1}} \qquad \text{for any } n \ge 1.$$

Hence $|\mu|(E) = 0$. On the other hand $|\nu|(\bigcup_{i=1}^{\infty} E_i) < \infty$ (since $|\mu|(\bigcup_{i=1}^{\infty} E_i) \le \sum_{i=1}^{\infty} |\mu|(E_i) \le 1$) and, therefore, by Theorem 1.2.2,

$$|\nu|(E) \ge \overline{\lim_{n\to\infty}} |\nu|(E_n) \ge \varepsilon.$$

This contradicts the assumption $\nu \ll \mu$.

From Theorem 2.12.2 and the fact (see Theorem 2.7.4) that $\int_E f\, d\mu = 0$ if $\mu(E) = 0$, we immediately obtain the following generalization of that part of Theorem 2.8.4 which is concerned with absolute continuity.

Corollary 2.12.3 *If* f *is a measurable function, integrable on every measurable set* E *of finite measure* (μ *is assumed to be a measure*), *then the set* *function* $\lambda(E)$ *defined by* (2.8.1) *is absolutely continuous in the sense of Definition* 2.8.1.

We now state the theorem of *Radon-Nikodym.*

Theorem 2.12.4 *Let* $(X,\mathcal{A},\mu)$ *be a* σ*-finite measure space with* μ *a measure, and let* ν *be a* σ*-finite signed measure on* $\mathcal{A}$*, absolutely continuous with respect to* μ*. Then there exists a real-valued measurable function* f *on* X *such that*

$$\nu(E) = \int_E f\,d\mu \tag{2.12.1}$$

for every measurable set E *for which* $|\nu|(E) < \infty$*. If* g *is another function such that* $\nu(E) = \int_E g\,d\mu$ *for any measurable set* E *for which* $|\nu|(E) < \infty$*, then* $f = g$ *a.e. with respect to* μ.

Note that f is not asserted to be integrable. In fact, it is clear that f is integrable if and only if $|\nu|(X) < \infty$. Thus if $\nu \ll \mu$ and $|\nu|(X) < \infty$, then ν is the indefinite integral of some μ-integrable function f.

Note also that if f is any measurable function, then, by Theorem 2.7.4, the set function $\nu(E)$, defined for all measurable sets E for which $\int_E |f|d\mu < \infty$, satisfies $|\nu|(E) = 0$ if $|\mu|(E) = 0$. If, further, f is integrable on any measurable set E with $|\mu|(E) < \infty$, then (by Corollary 2.12.3) $\nu(E)$ is absolutely continuous in the sense of Definition 2.8.1.

Definition The function f occurring in (2.12.1) is called the *Radon-Nikodym derivative* of ν with respect to μ, and one writes

$$f = \frac{d\nu}{d\mu}, \qquad \text{or } d\nu = f\,d\mu. \tag{2.12.2}$$

Proof. We can write X as a countable disjoint union of sets X_j, each having a finite μ-measure [that is, $\mu(X_j) < \infty$]. We can also write each set X_j as a countable disjoint union of sets Y_{jh}, each having a finite ν-measure [that is, $|\nu|(Y_{jh}) < \infty$]. Write the double sequence $\{Y_{jh}\}$ as a sequence $\{Z_n\}$. Suppose we can prove the existence part of the theorem for each subspace Z_n. Then on each Z_n we have a μ-integrable function f_n such that

$$\nu(E \cap Z_n) = \int_{E\cap Z_n} f_n\,d\mu, \qquad E \in \mathcal{A}.$$

Define $f(x) = f_n(x)$ if $x \in Z_n$. f is clearly a measurable function. If $|\nu|(E) < \infty$, then

$$\sum_n \int_{E\cap Z_n} |f|\,d\mu = \sum_n |\nu|(E \cap Z_n) = |\nu|(E) < \infty.$$

It follows (by Problem 2.10.7 with $f_n = f\chi_{E \cap Z_n}$) that f is μ-integrable on E and (by Theorem 2.8.4) that (2.12.1) holds. As for uniqueness, if g is a function as in the last part of Theorem 2.12.4, then $\int_F g\, d\mu = \int_F f\, d\mu$ for any measurable subset of each set Z_n. Hence, by Theorem 2.7.6, $g = f$ a.e. in Z_n. Therefore also $g = f$ a.e. in X. We have thus shown that it suffices to prove the existence part of the theorem in each subspace Z_n, where μ and ν are both finite. Thus, without loss of generality we may assume that $\mu(X) < \infty$, $|\nu|(X) < \infty$.

Let $X = A \cup B$ be a Hahn decomposition for ν. The argument above shows that it suffices to prove the existence part of the theorem in each of the subspaces A and B separately. Since ν is a measure on A and $-\nu$ is a measure on B, we may assume, without loss of generality, that ν is already a measure on X.

We now consider the case where μ and ν are finite measures, and prove the existence of f. Denote by $\mathscr{D}$ the class of all nonnegative functions $\hat{f}$, integrable with respect to μ, such that

$$\int_E \hat{f}\, d\mu \le \nu(E) \qquad \text{for all } E \in \mathfrak{A}.$$

Let

$$\alpha = \sup_{\hat{f} \in \mathscr{D}} \int \hat{f}\, d\mu. \tag{2.12.3}$$

Then there is a sequence $\{f_n\}$ in $\mathscr{D}$ such that

$$\lim_{n\to\infty} \int f_n\, d\mu = \alpha.$$

Let $g_n = \max(f_1, \ldots, f_n)$. We write each measurable set E as a disjoint union $\bigcup_{j=1}^{n} E_j$ of measurable sets E_j defined as follows: $x \in E_1$ if $f_1(x) \ge f_j(x)$ for $2 \le j \le n$; $x \in E_2$ if $x \notin E_1$ and if $f_2(x) \ge f_j(x)$ for $3 \le j \le n$; and so on. It follows that $g_n(x) = f_j(x)$ for $x \in E_j$. Hence

$$\int_E g_n\, d\mu = \sum_{j=1}^{n} \int_{E_j} f_j\, d\mu \le \sum_{j=1}^{n} \nu(E_j) = \nu(E).$$

Let $f_0(x) = \sup_n f_n(x)$. Then $f_0(x) = \lim g_n(x)$, and $\{g_n\}$ is a monotone-increasing sequence of nonnegative integrable functions. The Lebesgue monotone convergence theorem then implies that $f_0(x)$ is integrable, $\int f_0\, d\mu = \alpha$, and $\int_E f_0\, d\mu \le \nu(E)$ for any $E \in \mathfrak{A}$ (so that $f_0 \in \mathscr{D}$). Since f_0 is integrable, it is real-valued a.e. Let f be a real-valued function that is equal a.e. to f_0. Then f also belongs to $\mathscr{D}$, and

$$\int f\, d\mu = \alpha. \tag{2.12.4}$$

Consider the set function $\lambda(E) = \nu(E) - \int_E f\,d\mu$. If we show that $\lambda \equiv 0$, then we have completed the proof of the theorem. Note that $\lambda \ll \mu$, and λ is a measure.

Suppose $\lambda \not\equiv 0$ and let $X = A_m \cup B_m$ be a Hahn decomposition for the signed measure $\lambda - (1/m)\mu$ $(m = 1,2,\ldots)$. Write

$$A_0 = \bigcup_{m=1}^{\infty} A_m, \qquad B_0 = \bigcap_{m=1}^{\infty} B_m.$$

Since $B_0 \subset B_m$,

$$0 \le \lambda(B_0) \le \frac{1}{m}\,\mu(B_0) \to 0 \text{ if } m \to \infty.$$

Hence $\lambda(B_0) = 0$. Since $\lambda \not\equiv 0$ and $A_0 \cap B_0 = \varnothing$, $A_0 \cup B_0 = X$, we must have $\lambda(A_0) > 0$. By absolute continuity of λ with respect to μ, we also have $\mu(A_0) > 0$. Hence $\mu(A_m) > 0$ for at least one m. Denoting $1/m$ by ε and A_m by A, we then have [by the positivity of A_m with respect to $\lambda - (1/m)\mu$]:

$$\varepsilon\mu(E \cap A) \le \lambda(E \cap A) = \nu(E \cap A) - \int_{E \cap A} f\,d\mu$$

for any measurable set E.

Consider the function $g = f + \varepsilon\chi_A$. It satisfies

$$\int_E g\,d\mu = \int_E f\,d\mu + \varepsilon\mu(E \cap A) \le \int_{E-A} f\,d\mu + \nu(E \cap A)$$
$$\le \nu(E - A) + \nu(E \cap A) = \nu(E)$$

for any measurable set E. Hence $g \in \mathcal{D}$. However, by (2.12.4),

$$\int g\,d\mu = \int f\,d\mu + \varepsilon\mu(A) > \alpha,$$

thereby contradicting (2.12.3). We have thus proved that $\lambda \equiv 0$.

Let $(X,\mathfrak{A},\mu)$ be a measure space, with μ a signed measure, and let $\mu = \mu^+ - \mu^-$ be the Jordan decomposition of μ. A measurable function f is said to be *integrable* (*on* E) if it is integrable (on E) with respect to the measure $|\mu|$. The *integral of* f (*over* E) is then defined by

$$\int f\,d\mu = \int f\,d\mu^+ - \int f\,d\mu^- \qquad \left(\int_E f\,d\mu = \int_E f\,d\mu^+ - \int_E f\,d\mu^-\right). \tag{2.12.5}$$

PROBLEMS

2.12.1. Prove that if f is integrable with respect to μ then it is also integrable with respect to μ^+ and μ^-. [Therefore the integrals in (2.12.5) are well defined.]

2.12.2. The Radon-Nikodym theorem remains true in case μ is a σ-finite signed measure.

2.12.3. If ν and μ are σ-finite signed measures and $\nu \ll \mu$, then the set $\{x; (d\nu/d\mu)(x) = 0\}$ has ν-measure zero.

2.12.4. Let ν and μ be σ-finite signed measures and let $\nu \ll \mu$. For any ν-integrable function φ, the function $\varphi(d\nu/d\mu)$ is μ-integrable and

$$\int \varphi \, d\nu = \int \varphi \frac{d\nu}{d\mu} \, d\mu. \tag{2.12.6}$$

[*Hint:* Consider first the case where μ and ν are measures and φ is a simple function.]

2.12.5. Let λ, μ be σ-finite measures and let ν be a σ-finite signed measure. Assume that $\nu \ll \mu$, $\mu \ll \lambda$. Then

$$\frac{d\nu}{d\lambda} = \frac{d\nu}{d\mu} \frac{d\mu}{d\lambda} \qquad \text{a.e. with respect to } \lambda.$$

2.12.6. Let μ be a σ-finite measure and let ν_1, ν_2 be σ-finite signed measures such that $\nu_1 \ll \mu$, $\nu_2 \ll \mu$. Then $\nu_1 + \nu_2 \ll \mu$ and

$$\frac{d(\nu_1 + \nu_2)}{d\mu} = \frac{d\nu_1}{d\mu} + \frac{d\nu_2}{d\mu}.$$

2.12.7. Let μ and ν be σ-finite measures and assume that $\mu - \nu$ is a measure. If $\nu \ll \mu - \nu$, then the set of points where $d\nu/d\mu = 1$ has zero μ-measure.

2.13 THE LEBESGUE DECOMPOSITION

Let μ and ν be two signed measures defined in the measure space $(X, \mathfrak{A})$. We say that μ and ν are *mutually singular*, and write $\mu \perp \nu$, if there exist measurable sets A and B such that

$$X = A \cup B, \qquad A \cap B = \varnothing,$$

and

$$|\mu|(A) = 0, \qquad |\nu|(B) = 0.$$

We also say that μ is *singular relative* to ν (or that ν is singular relative to μ).

PROBLEMS

2.13.1. For each signed measure μ, $\mu^+ \perp \mu^-$, $\mu^+ \ll |\mu|$, $\mu^- \ll |\mu|$.

2.13.2. If ν_1 and ν_2 are singular relative to μ, then $\nu_1 + \nu_2$ is also singular relative to μ. [*Hint:* If $X = A_i \cup B_i$ is a decomposition relative to

ν_i and μ, then $X = [(A_1 \cap A_2) \cup (A_1 \cap B_2) \cup (A_2 \cap B_1)] \cup (B_1 \cap B_2)$ is a decomposition relative to ν and μ.]

2.13.3. If $\nu \ll \mu$ and $\nu \perp \mu$, then $\nu = 0$.

Theorem 2.13.1 *Let μ and ν be σ-finite signed measures in a measure space $(X,\mathfrak{A})$. Then there exist σ-finite signed measures ν_0 and ν_1 such that*

$$\nu = \nu_0 + \nu_1, \qquad \nu_0 \perp \mu \quad and \quad \nu_1 \ll \mu. \tag{2.13.1}$$

The pair (ν_0,ν_1) is uniquely determined.

The decomposition of ν, given by (2.13.1), is called the *Lebesgue decomposition* of ν relative to μ.

Proof. As in the proof of Theorem 2.12.4, we first reduce the general situation to the special one where μ and ν are finite. We may further assume that μ is a measure, since $\nu_0 \perp |\mu|$ and $\nu_1 \ll |\mu|$ imply that $\nu_0 \perp \mu$ and $\nu_1 \ll \mu$. Finally, we may assume that ν is a measure, for otherwise we can treat ν^+ and ν^- separately.

Since μ and ν are measures, ν is absolutely continuous with respect to $\mu + \nu$. By the Radon-Nikodym theorem, there exists an integrable function f such that

$$\nu(E) = \int_E f\, d\mu + \int_E f\, d\nu \tag{2.13.2}$$

for any measurable set E. Since $0 \le \nu(E) \le \mu(E) + \nu(E)$, we have (by Problem 2.7.3) $0 \le f \le 1$ a.e. with respect to the measure $\mu + \nu$. Hence $0 \le f \le 1$ a.e. with respect to the measure ν. Let

$$A = \{x; f(x) = 1\}, \qquad B = X - A.$$

Then

$$\nu(A) = \int_A d\mu + \int_A d\nu = \mu(A) + \nu(A).$$

Since $\nu(A) < \infty$, it follows that $\mu(A) = 0$. Therefore, if we define

$$\nu_0(E) = \nu(E \cap A), \qquad \nu_1(E) = \nu(E \cap B) \qquad (E \in \mathfrak{A}),$$

then $\nu_0 \perp \mu$. It is also clear that $\nu_0 + \nu_1 = \nu$.

It remains to prove that $\nu_1 \ll \mu$. Suppose $\mu(E) = 0$. Then (2.13.2) gives

$$\int_{E\cap B} d\nu = \int_{E\cap B} f\, d\nu$$

—that is, $\int_{E\cap B} (1 - f)\, d\nu = 0$. Since, on $E \cap B$, $1 - f > 0$ a.e. with respect to ν, we must have $\nu(E \cap B) = 0$—that is, $\nu_1(E) = 0$.

To prove uniqueness, suppose $v = \bar{v}_0 + \bar{v}_1$ is another decomposition of v, with $\bar{v}_0 \perp \mu$, $\bar{v}_1 \ll \mu$. Then, the signed measure $\lambda = v_0 - \bar{v}_0 = \bar{v}_1 - v_1$ is both singular (by Problem 2.13.2) and absolutely continuous with respect to μ. Hence, by Problem 2.13.3, $\lambda = 0$. Thus $v_0 = \bar{v}_0$ and $v_1 = \bar{v}_1$.

PROBLEM

2.13.4. Let A be a sequence $\{x_m\}$ in the Euclidean space R^n, and let $\{p_m\}$ be a sequence of positive numbers. Denote by v the σ-finite measure (cf. Problem 1.2.5) given by

$$v(E) = \sum_{x_m \in E} p_m \qquad (E \text{ any subset of } R^n).$$

Find the Lebesgue decomposition of v with respect to the Lebesgue measure μ of R^n.

2.14 THE LEBESGUE INTEGRAL ON THE REAL LINE

In this section we restrict our interest to the real line, and find a connection between integrals and derivatives. This connection, stated in Theorem 2.14.1, generalizes the Fundamental Theorem of Calculus. It will have implications for some of the abstract theorems we have previously proved, such as the Radon-Nikodym theorem.

We shall denote the Lebesgue integral of a function $f(t)$ over an interval (c,d) by $\int_c^d f(t)\,dt$. We also define $\int_d^c f(t)\,dt = -\int_c^d f(t)\,dt$, $\int_c^c f(t)\,dt = 0$. We shall take (a,b) to be a fixed bounded interval.

Definition If f is Lebesgue-integrable on (a,b), then the function

$$g(x) = \int_a^x f(t)\,dt + c \qquad (c \text{ any constant}) \tag{2.14.1}$$

is called the *indefinite integral* of f.

Note the similarity between the definitions of the indefinite integral given here and in Section 2.8.

By Problem 2.8.2, $g(x)$ is an absolutely continuous function. We shall prove the following important result.

Theorem 2.14.1 *If* f *is a Lebesgue-integrable function on* (a,b), *then its indefinite integral* g(x) *is differentiable a.e., and* g′(x) = f(x) *a.e.*

The assertion $g'(x) = f(x)$ a.e. can also be stated as follows:

$$\lim_{h \to 0} \frac{1}{h} \int_x^{x+h} f(t)\, dt = f(x) \qquad \text{a.e.}$$

We shall need a few lemmas.

Lemma 2.14.2 *Let* f(x) *be a Lebesgue-integrable function on* (a,b). *If* $\int_a^x$ f(t) dt = 0 *a.e., then* f(x) = 0 *a.e.*

Proof. The assumptions on f imply that $\int_E f(x)\, dx = 0$ for any interval E. Since the set function $\int_E f(x)\, dx$ is completely additive, it vanishes on any open set E. Hence it vanishes also on every closed set E of (a,b). Suppose now that the assertion is false. Then there is a set F of positive measure such that either $f > 0$ on F or $f < 0$ on F. It is enough to consider the first case. Since $\mu[(a,b) - F] < b - a$, there is (by Problem 1.9.7) an open set E containing $(a,b) - F$ such that $\mu(E) < b - a$. The set $G = (a,b) - E$ is a closed subset of F and $\mu(G) > 0$. Since $f > 0$ on G, we get (by Theorem 2.7.5) $\int_G f(x)\, dx > 0$, a contradiction.

Lemma 2.14.3 *If* f(x) *is bounded and Lebesgue-integrable on* (a,b), *then the assertion of Theorem* 2.14.1 *is valid.*

Proof. Write

$$g(x) = \int_a^x f^+(t)\, dt - \int_a^x f^-(t)\, dt. \tag{2.14.2}$$

Each of the indefinite integrals is a monotone-increasing function. But it is known (the proof, which is rather involved, is omitted) that monotone functions are differentiable a.e. Hence $g'(x)$ exists a.e. Let $|f(x)| \le M$ for all $a < x < b$. Then, if $0 < h < b - x$,

$$\left| \frac{g(x+h) - g(x)}{h} \right| = \left| \frac{1}{h} \int_x^{x+h} f(t)\, dt \right| \le M.$$

Since also

$$\lim_{h \to 0} \frac{g(x+h) - g(x)}{h} = g'(x) \qquad \text{a.e.,}$$

the Lebesgue-bounded convergence theorem shows that $g'(x)$ is integrable and

$$\int_a^x \frac{g(t+h) - g(t)}{h}\, dt \to \int_a^x g'(t)\, dt. \tag{2.14.3}$$

The integral on the left in (2.14.3) is equal to (here we use Problem 2.11.11)

$$\frac{1}{h}\int_{a+h}^{x+h} g(t)\,dt - \frac{1}{h}\int_{a}^{x} g(t)\,dt = \frac{1}{h}\int_{x}^{x+h} g(t)\,dt - \frac{1}{h}\int_{a}^{a+h} g(t)\,dt,$$

and this tends to $g(x) - g(a)$ as $h \to 0$. Hence

$$\int_a^x g'(t)\,dt = g(x) - g(a)$$

—that is,

$$\int_a^x [g'(t) - f(t)]\,dt = 0$$

for all x. Lemma 2.14.2 then gives $g'(x) = f(x)$ a.e.

Lemma 2.14.4 *If* $\varphi(x)$ *is a continuous nondecreasing function in* $[a,b]$, *then* $\varphi'(x)$ *is integrable, and*

$$\int_a^b \varphi'(x)\,dx \leq \varphi(b) - \varphi(a). \tag{2.14.4}$$

Proof. Define $\varphi(x) = \varphi(b)$ if $x > b$ and $\varphi(x) = \varphi(a)$ if $x < a$. Since $[\varphi(x+h) - \varphi(x)]/h \geq 0$ and $[\varphi(x+h) - \varphi(x)]/h \to \varphi'(x)$ a.e. as $h \to 0$, Fatou's lemma gives

$$\varliminf_{h\to 0} \int_a^b \frac{\varphi(x+h) - \varphi(x)}{h}\,dx \geq \int_a^b \varphi'(x)\,dx;$$

if the left-hand side is finite, then $\varphi'(x)$ is integrable. The integral on the left can be evaluated in the same way as the integral on the left in (2.14.3). We find that, as $h \to 0$, this integral converges to $\varphi(b) - \varphi(a)$. This proves (2.14.4).

We can now complete the proof of Theorem 2.14.1. We may suppose that $f(x) \geq 0$ a.e., for otherwise we can treat each of the indefinite integrals in (2.14.2) separately. Since $g(x)$ is monotone, $g'(x)$ exists a.e. Define $f_n(x) = \min\{f(x), n\}$. Then $f(x) - f_n(x) \geq 0$. It follows that

$$\int_a^x [f(t) - f_n(t)]\,dt$$

is nondecreasing. Hence its derivative (which exists a.e.) is ≥ 0. This gives

$$\frac{d}{dx}\int_a^x f(t)\,dt \geq \frac{d}{dx}\int_a^x f_n(t)\,dt.$$

By Lemma 2.14.3, the expression on the right is equal to $f_n(x)$ a.e. Hence

$g'(x) \geq f_n(x)$ a.e. It follows that $g'(x) \geq f(x)$ a.e. Therefore,

$$\int_a^b g'(x)\,dx \geq \int_a^b f(x)\,dx = g(b) - g(a).$$

Using Lemma 2.14.4, we get the reverse inequality. Consequently,

$$\int_a^b [g'(x) - f(x)]\,dx = 0.$$

Since, however, $g'(x) \geq f(x)$ a.e., we get $g'(x) = f(x)$ a.e.

Recall (see Problem 2.8.2) that the *indefinite integral is an absolutely continuous function.*

The converse is also true, namely, *if $g(x)$ is an absolutely continuous function in (a,b) then there exists a Lebesgue-integrable function $f(x)$ such that*

$$g(x) = g(a) + \int_a^x f(t)\,dt.$$

We shall briefly indicate the proof. From Problems 2.8.3 and 2.8.4 we know that $g(x)$ can be written as a difference of two monotone-increasing, continuous functions, say $g(x) = g_1(x) - g_2(x)$. Lemma 2.14.4 shows that $g'_1(x)$ and $g'_2(x)$ are integrable. Hence $g'(x)$ is integrable. Consider the function

$$\varphi(x) = g(x) - \int_a^x g'(t)\,dt.$$

φ is absolutely continuous and $\varphi'(x) = 0$ a.e. But it can be shown (we omit the proof, which is rather involved) that such a function is necessarily a constant.

PROBLEMS

2.14.1. To every function f, monotone-increasing and continuous from the right on (a,b), there corresponds a Lebesgue-Stieltjes measure, which we denote by μ_f. Every function f of bounded variation and continuous from the right can be written in the form $f = f_1 - f_2$, where f_1 and f_2 are monotone-increasing and continuous from the right. We write $\mu_f = \mu_{f_1} - \mu_{f_2}$. Prove that the function $f(x)$ is absolutely continuous if and only if the signed measure μ_f is absolutely continuous with respect to the Lebesgue measure (which we shall denote by) μ_x.

2.14.2. Let $f(x)$ be absolutely continuous on (a,b). Prove that $d\mu_f/d\mu_x = f'(x)$ a.e., and that for any μ_f-integrable function φ,

$$\int_{(a,b)} \varphi\,df = \int_a^b \varphi f'\,dx.$$

Here we use the notation: $df = d\mu_f$.

2.14.3. Let $f(x)$ be Lebesgue-integrable on (a,b) and let $g(x)$ be absolutely continuous on (a,b). Then the following *integration-by-parts* formula holds:

$$\int_a^b fg\,dx = F(b)g(b) - F(a)g(a) - \int_a^b Fg'\,dx.$$

where $F(x)$ is any indefinite integral of $f(x)$.

2.14.4. If $f(x)$ and $g(x)$ are absolutely continuous functions on (a,b), then

$$\int_{(a,b)} f\,dg + \int_{(a,b)} g\,df = f(b)g(b) - f(a)g(a).$$

(This is actually true if f and g are just assumed to be continuous and of bounded variation.)

2.14.5. A function $f(x)$ is called *singular* if $f'(x) = 0$ a.e. Show that every function g of bounded variation in (a,b) can be written as a sum $g = g_1 + g_2$, where g_1 is absolutely continuous and g_2 is singular. Show also that this decomposition is unique, except for an additive constant.

2.14.6. Let g, g_1, g_2 be as in Problem 2.14.5. Show that in the decomposition $\mu_g = \mu_{g_1} + \mu_{g_2}$, μ_{g_1} is absolutely continuous with respect to μ_x and μ_{g_2} is singular relative to μ_x.

2.14.7. If $f(x)$ is a continuous function in (a,b) and if $f'(x)$ exists everywhere in (a,b) and is bounded, then $\int_a^x f'(t)\,dt = f(x) - f(a)$ for any $a < x < b$.

2.14.8. For any measurable set E on the real line, write $E(x_0,h) = E \cap [x_0 - h, x_0 + h]$. Denote by μ the Lebesgue measure. A point x_0 on the real line is called a *point of density θ* with respect to E if $\mu[E(x_0,h)]/2h \to \theta$ as $h \to 0$. Prove that almost every point of E is a point of density 1, and almost every point of $R^1 - E$ is a point of density 0. [*Hint:* Suppose E is contained in a bounded interval $[\alpha,\beta]$ and let $g(x) = \int_{\alpha-1}^x \chi_E(t)\,dt$. Then $g'(x) = \chi_E(x)$ a.e. Note now that $g(x+h) - g(x-h) = \mu[E(x,h)]$.]

2.15 PRODUCT OF MEASURES

Let X and Y be two spaces. The *Cartesian product* $X \times Y$ is the set of all ordered pairs (x,y), where $x \in X$, $y \in Y$. Any subset of $X \times Y$ of the form $A \times B$, where $A \subset X$, $B \subset Y$, is called a *generalized rectangle* (or, briefly, a *rectangle*) with *sides* A and B.

Let $\mathfrak{A}$ be a σ-algebra of subsets of X and let $\mathscr{B}$ be a σ-algebra of subsets of Y. We shall denote by $\mathfrak{A} \times \mathscr{B}$ the σ-algebra generated by all the rectangles $A \times B$ with $A \in \mathfrak{A}$, $B \in \mathscr{B}$. We call $\mathfrak{A} \times \mathscr{B}$ the *Cartesian product* of $\mathfrak{A}$ and $\mathscr{B}$.

A rectangle $A \times B$ with $A \subset \mathcal{A}$, $B \in \mathcal{B}$ will be called a *rectangle of* $\mathcal{A} \times \mathcal{B}$.

If $E \subset X \times Y$ and x is any point of X, we call the set $E_x = \{y; (x,y) \in E\}$ an *X-section* of E. Similarly, $E^y = \{x; (x,y) \in E\}$ is called the *Y-section* of E.

Lemma 2.15.1 *Every X-section of a set in $\mathcal{A} \times \mathcal{B}$ is a set of $\mathcal{B}$, and every Y-section of a set in $\mathcal{A} \times \mathcal{B}$ is a set of $\mathcal{A}$.*

Proof. Denote by $\mathcal{D}$ the class of all sets in $\mathcal{A} \times \mathcal{B}$ with the property that every X-section is in $\mathcal{B}$. $\mathcal{D}$ clearly contains all the rectangles of $\mathcal{A} \times \mathcal{B}$. It is also a σ-algebra of subsets of $X \times Y$. Hence it must coincide with $\mathcal{A} \times \mathcal{B}$. The proof for Y-sections is similar.

Lemma 2.15.2 *Let $(X,\mathcal{A},\mu)$ and $(Y,\mathcal{B},\nu)$ be measure spaces. Denote by $\mathcal{F}$ the class of all finite disjoint unions of rectangles of $\mathcal{A} \times \mathcal{B}$ that are contained in a fixed subset of $X \times Y$. Then $\mathcal{F}$ is a ring.*

Proof. $\mathcal{F}$ is clearly closed under the formation of finite disjoint unions. One easily verifies the relations

$$\bigcap_{j=1}^{m} (A_j \times B_j) = \left(\bigcap_{j=1}^{m} A_j\right) \times \left(\bigcap_{j=1}^{m} B_j\right), \tag{2.15.1}$$

$$(A_1 \times B_1) - (A_2 \times B_2) = [(A_1 \cap A_2) \times (B_1 - B_2)] \cup (A_1 - A_2) \times B_1]. \tag{2.15.2}$$

We shall use these relations to prove that if E and F belong to $\mathcal{F}$, then also $E - F$ belongs to $\mathcal{F}$. This will complete the proof that $\mathcal{F}$ is a ring.

Write E and F as disjoint unions of rectangles of $\mathcal{A} \times \mathcal{B}$: $E = \bigcup_{i=1}^{n} E_i$, $F = \bigcup_{j=1}^{m} F_j$. Then

$$E - F = \bigcup_{i=1}^{n} \bigcap_{j=1}^{m} (E_i - F_j).$$

By (2.15.2), for each i and j, $E_i - F_j = D_{j1} \cup D_{j2}$, where D_{j1} and D_{j2} are disjoint rectangles of $\mathcal{A} \times \mathcal{B}$ (depending on i). We shall use the relation

$$\bigcap_{j=1}^{m} (D_{j1} \cup D_{j2}) = \bigcup_{k_s} [D_{1k_1} \cap \ldots \cap D_{mk_m}], \qquad \text{where } k_s = 1,2;\ s = 1,\ldots,m.$$

Noting that the sets in the union, on the right, are mutually disjoint, and using (2.15.1), we see that each set $G_i = \bigcap_j (E_i - F_j)$ is in $\mathcal{F}$. Since $E - F$ is a disjoint union of the G_i, it also belongs to $\mathcal{F}$.

Lemma 2.15.3 *If* μ *and* ν *are* σ*-finite measures, then every set in* $X \times Y$ *can be covered by a countable disjoint union of rectangles of* $\mathfrak{A} \times \mathcal{B}$ *having sides of finite measure.*

Proof. It is sufficient to cover $X \times Y$. Let $X = \bigcup_n X_n$ be a countable disjoint union of sets in $\mathfrak{A}$ with $\mu(X_n) < \infty$. Similarly, let $Y = \bigcup_m Y_m$ be a countable disjoint union of sets in $\mathcal{B}$ with $\nu(Y_m) < \infty$. Then $X \times Y = \bigcup_{n,m} (X_n \times Y_m)$ gives the desired covering of $X \times Y$.

Definition A class $\mathcal{M}$ of sets is called *monotone*, if for any monotone sequence $\{E_n\}$ of sets in $\mathcal{M}$ we have: $\lim E_n$ is in $\mathcal{M}$.

Lemma 2.15.4 *A monotone ring* $\mathcal{R}$ *is a* σ*-ring.*

Proof. We have to show that $\mathcal{R}$ is closed under the formation of countable unions. Thus, let $\{E_m\}$ be a sequence in $\mathcal{R}$. Since $\mathcal{R}$ is a ring, the finite unions $F_n = \bigcup_{m=1}^{n} E_m$ belong to $\mathcal{R}$. Since $\mathcal{R}$ is monotone, $\bigcup_{m=1}^{\infty} E_m = \lim_n F_n$ is also in $\mathcal{R}$.

Lemma 2.15.5 *If a monotone class* $\mathcal{M}$ *contains a ring* $\mathcal{R}$*, then it contains also the* σ*-ring* $\mathcal{S}(\mathcal{R})$ *generated by* $\mathcal{R}$*.*

Proof. The class of all subsets of X is a monotone class and the intersection of any collection of monotone classes is a monotone class. Hence for any class $\mathcal{K}$ of sets there exists the smallest monotone class $\mathcal{M}_0(\mathcal{K})$ containing $\mathcal{K}$. Let $\mathcal{M}_0 = \mathcal{M}_0(\mathcal{R})$. If we prove that $\mathcal{M}_0$ is a ring, then, by Lemma 2.15.4, $\mathcal{M}_0$ is a σ-ring and hence $\mathcal{M} \supset \mathcal{M}_0 \supset \mathcal{S}(\mathcal{R})$. To prove the $\mathcal{M}_0$ is a ring, we introduce for each set F the class $\mathcal{K}_F$ of all sets E for which $E - F$, $F - E$, and $E \cup F$ are all in $\mathcal{M}_0$. Note that $E \in \mathcal{K}_F$ if and only if $F \in \mathcal{K}_E$. $\mathcal{K}_F$ is a monotone class, since, for any monotone sequence $\{E_n\}$ in $\mathcal{K}_F$,

$$\lim_n E_n - F = \lim_n (E_n - F) \in \mathcal{M}_0,$$

$$F - \lim_n E_n = \lim_n (F - E_n) \in \mathcal{M}_0,$$

$$F \cup \left(\lim_n E_n\right) = \lim_n (F \cup E_n) \in \mathcal{M}_0.$$

If $F \in \mathcal{R}$, then $E \in \mathcal{R}$ implies $E \in \mathcal{K}_F$. Hence $\mathcal{R} \subset \mathcal{K}_F$. Therefore $\mathcal{M}_0 \subset \mathcal{K}_F$. This implies that if $E \in \mathcal{M}_0$, $F \in \mathcal{R}$, then $F \in \mathcal{K}_E$. Thus $\mathcal{K}_E \supset \mathcal{R}$. From the definition of $\mathcal{M}_0$ it follows that $\mathcal{K}_E \supset \mathcal{M}_0$. Since this is true for all $E \in \mathcal{M}_0$, $\mathcal{M}_0$ is a ring.

If $\mathcal{D}$ is a class of sets and A is any set, we denote the class of all sets $E \cap A$, where $E \in \mathcal{D}$, by $\mathcal{D} \cap A$.

Lemma 2.15.6 *Let $\mathcal{D}$ be any class of sets and* A *any subset of* X. *Then*

$$\mathcal{S}(\mathcal{D}) \cap A = \mathcal{S}(\mathcal{D} \cap A).$$

Proof. Since $\mathcal{S}(\mathcal{D}) \cap A$ is a σ-ring containing $\mathcal{D} \cap A$, it contains also $\mathcal{S}(\mathcal{D} \cap A)$. To prove the converse, denote by $\mathcal{K}$ the class of all sets of the form $B \cup (C - A)$, where $B \in \mathcal{S}(\mathcal{D} \cap A)$ and $C \in \mathcal{S}(\mathcal{D})$. Writing $E = (E \cap A) \cup (E - A)$ for any $E \in \mathcal{D}$, we see that $\mathcal{D} \subset \mathcal{K}$. Since, as easily verified, $\mathcal{K}$ is a σ-ring, we find that $\mathcal{S}(\mathcal{D}) \subset \mathcal{K}$. Hence $\mathcal{S}(\mathcal{D}) \cap A \subset \mathcal{K} \cap A$. Noting, however, that $\mathcal{K} \cap A = \mathcal{S}(\mathcal{D} \cap A)$, we get $\mathcal{S}(\mathcal{D}) \cap A \subset \mathcal{S}(\mathcal{D} \cap A)$.

Theorem 2.15.7 *Let* $(X,\mathfrak{A},\mu)$ *and* $(Y,\mathcal{B},\nu)$ *be* σ*-finite measure spaces and let* E *be any set in* $\mathfrak{A} \times \mathcal{B}$. *Define functions* f *and* g *by* $f(x) = \nu(E_x)$ *for all* $x \in X$ *and* $g(y) = \mu(E^y)$ *for all* $y \in Y$. *Then* f *and* g *are measurable functions, and*

$$\int f\, d\mu = \int g\, d\nu. \tag{2.15.3}$$

Proof. Note that, by Lemma 2.15.1, f and g are well defined. We shall break up the proof into several steps.

(a) Denote by $\mathcal{D}$ the class of all the sets E in $\mathfrak{A} \times \mathcal{B}$ for which the assertion of the theorem is true. It is clear that if $\mathcal{D}$ contains a sequence of mutually disjoint sets, then $\mathcal{D}$ contains also their union.

(b) $\mathcal{D}$ contains all the rectangles $E = A \times B$ of $\mathfrak{A} \times \mathcal{B}$ with sides having finite measure. Indeed, in this case $f = \nu(B)\chi_A$, $g = \mu(A)\chi_B$. It follows that f and g are measurable and

$$\int f\, d\mu = \mu(A)\nu(B) = \int g\, d\nu.$$

(c) Let $E_0 = A_0 \times B_0$ be a rectangle of $\mathfrak{A} \times \mathcal{B}$ with $\mu(A_0) < \infty$, $\nu(B_0) < \infty$. Let $\{E_n\}$ be a monotone sequence of sets in $\mathfrak{A} \times \mathcal{B}$ such that $E_n \subset E_0$ if $n \geq 1$. We claim that if $E_n \in \mathcal{D}$ for all $n \geq 1$, then also $E = \lim E_n$ is in $\mathcal{D}$. Since $\mathfrak{A} \times \mathcal{B}$ is a σ-algebra, E belongs to $\mathfrak{A} \times \mathcal{B}$. The sequence of X-sections $E_{n,x}$ of E_n (for each fixed x) clearly converges monotonically to the corresponding X-section E_x of E. Similarly, $\{E_n^y\}$ converges to E^y. Denote by f_n and g_n the functions f and g corresponding to E_n. Then $\{f_n\}$ and $\{g_n\}$ converge monotonically to f and g, respectively, and

$$0 \leq f_n \leq \nu(B_0), \qquad 0 \leq g_n \leq \mu(A_0).$$

The Lebesgue-bounded convergence theorem implies that f and g are integrable and

$$\int f \, d\mu = \lim_n \int f_n \, d\mu = \lim_n \int g_n \, dv = \int g \, dv.$$

Hence E belongs to $\mathscr{D}$.

(d) Let E be a rectangle in $\mathfrak{A} \times \mathscr{B}$ with sides having finite measure. Denote by $\mathscr{D}_E$ the class of all subsets of $\mathscr{D}$ that are contained in E. Denote by $\mathscr{F}_E$ the class of all finite disjoint unions of rectangles of $\mathfrak{A} \times \mathscr{B}$ that are subsets of E. By Lemma 2.15.2, $\mathscr{F}_E$ is a ring. By (c) and (a), (b), $\mathscr{D}_E$ is a monotone class and $\mathscr{D}_E \supset \mathscr{F}_E$. Hence, by Lemma 2.15.5, $\mathscr{D}_E$ contains the σ-ring generated by $\mathscr{F}_E$. Denote by $\mathscr{K}$ the class of all rectangles of $\mathfrak{A} \times \mathscr{B}$. Then $\mathscr{K} \cap E = \{F \in \mathscr{K}; F \subset E\}$. Consequently, $\mathscr{S}(\mathscr{F}_E) \supset \mathscr{S}(\mathscr{K} \cap E)$. By Lemma 2.15.6, the latter class coincides with $\mathscr{S}(\mathscr{K}) \cap E$—that is, with $(\mathfrak{A} \times \mathscr{B}) \cap E$. Thus, $\mathscr{D} \supset \mathscr{D}_E \supset (\mathfrak{A} \times \mathscr{B}) \cap E$.

(e) Let F be any set of $\mathfrak{A} \times \mathscr{B}$. By Lemma 2.15.3, F is contained in a countable disjoint union $\bigcup_{n=1}^{\infty} E_n$ of rectangles E_n of $\mathfrak{A} \times \mathscr{B}$ having sides of finite measure. By (d), each set $F \cap E_n$ is in $\mathscr{D}$. Hence, by (a), the disjoint union $\bigcup_n (F \cap E_n)$ is also in $\mathscr{D}$—that is, $F \in \mathscr{D}$. This completes the proof.

Theorem 2.15.8 *Let* $(X,\mathfrak{A},\mu)$ *and* $(Y,\mathscr{B},v)$ *be σ-finite measure spaces. Then the set function λ defined for every set* E *in* $\mathfrak{A} \times \mathscr{B}$ *by*

$$\lambda(E) = \int v(E_x) \, d\mu(x) = \int \mu(E^y) \, dv(y) \tag{2.15.4}$$

is a σ-finite measure on $\mathfrak{A} \times \mathscr{B}$, *having the property that, for any rectangle* $A \times B$ *of* $\mathfrak{A} \times \mathscr{B}$

$$\lambda(A \times B) = \mu(A) \cdot v(B). \tag{2.15.5}$$

This last property determines λ uniquely.

Definition The measure λ is called the *product* of the measures μ and v, and we write $\lambda = \mu \times v$. The measure space $(X \times Y, \mathfrak{A} \times \mathscr{B}, \mu \times v)$ is called the *Cartesian product* of the measure spaces $(X,\mathfrak{A},\mu)$ and $(Y,\mathscr{B},v)$.

Proof. Since $v(E_x)$ is a nonnegative measurable function, its integral $\lambda(E)$ is either a nonnegative real number or $+\infty$. The complete additivity of λ follows from the complete additivity of v and from the Lebesgue monotone convergence theorem (compare Problem 2.10.7). Since $X \times Y$ can be covered by a countable class of rectangles of $\mathfrak{A} \times \mathscr{B}$ whose sides have finite measure, λ is σ-finite. It thus remains to prove uniqueness.

Suppose $\bar{\lambda}$ is another measure satisfying (2.15.5). By Lemma 2.15.3, we can write $X \times Y$ as a countable disjoint union of rectangles E_n of $\mathfrak{A} \times \mathcal{B}$ with sides having finite measure. Denote by $\mathcal{D}_n$ the class of all sets in $(\mathfrak{A} \times \mathcal{B}) \cap E_n$ on which λ and $\bar{\lambda}$ agree. Denote by $\mathcal{K}_n$ the class of all finite disjoint unions of rectangles of $\mathfrak{A} \times \mathcal{B}$ contained in E_n. Clearly $\mathcal{K}_n \subset \mathcal{D}_n$ and $\mathcal{D}_n$ is a monotone class (by Theorem 1.2.1 (iii)). Since, by Lemma 2.15.2, $\mathcal{K}_n$ is a ring, Lemma 2.15.5 implies that $\mathcal{D}_n$ contains the class $\mathcal{K} \cap E_n$, where $\mathcal{K}$ is the class of all the rectangles of $\mathfrak{A} \times \mathcal{B}$. It follows (using Lemma 2.15.6) that $\mathcal{D}_n = (\mathfrak{A} \times \mathcal{B}) \cap E_n$. Writing any set F in $\mathfrak{A} \times \mathcal{B}$ in the form $F = \bigcup (F \cap E_n)$ and using the last result, it follows that $\lambda(F) = \bar{\lambda}(F)$. Thus $\lambda \equiv \bar{\lambda}$.

The rectangles of $\mathfrak{A} \times \mathcal{B}$ are also called *measurable rectangles*. The reason for this name is that if a rectangle $A \times B$ is a set in $\mathfrak{A} \times \mathcal{B}$, then (by Lemma 2.15.1), $A \in \mathfrak{A}$, $B \in \mathcal{B}$.

PROBLEMS

2.15.0. Prove that $\mathcal{B}_m \times \mathcal{B}_k = \mathcal{B}_{m+k}$.

2.15.1. A Borel set E in R^n is called a G_δ *set* if it has the form $E = \bigcap_{k=1}^{\infty} A_k$, where the A_k are open sets. From Problems 1.9.7, 1.9.8 we have that every bounded Lebesgue set F in R^n has the form $F = E - N$, where E is a G_δ set and $\mu(N) = 0$. Verify (2.15.1) for $m = \infty$, thereby concluding that if A and B are G_δ sets in R^n and R^k, respectively, then $A \times B$ is a G_δ set in R^{n+k}.

2.15.2. Let N be a set in R^n having Lebesgue measure 0. Then for any set B in R^k, $N \times B$ is a Lebesgue set in R^{n+k} having Lebesgue measure 0. [*Hint:* Consider first the case where B is bounded.]

2.15.3. Denote by $\mathcal{L}^n$ and $\mathcal{L}^k$ the classes of the Lebesgue sets in R^n and R^k, respectively. Then any bounded rectangle of $\mathcal{L}^n \times \mathcal{L}^k$ has the form $E - N$, where $E = E_1 \times E_2$ is a G_δ set in R^{n+k}, N is a set of Lebesgue measure 0, and E_1, E_2 are G_δ sets of R^n and R^k, respectively. It follows that $\mathcal{L}^n \times \mathcal{L}^k$ is contained in the σ-algebra $\mathfrak{A}$ of all sets $E - N$, where E is any Borel set in R^{n+k} and N is any set in R^{n+k} having Lebesgue measure 0. Note (compare Problem 2.15.1) that $\mathfrak{A}$ coincides with the class $\mathcal{L}^{n+k}$ of all Lebesgue sets in R^{n+k}.

2.15.4. Every open set in R^{n+k} is in $\mathcal{L}^n \times \mathcal{L}^k$. Hence $\mathcal{L}^n \times \mathcal{L}^k$ contains all the Borel sets of R^{n+k}.

2.15.5. Prove that $\mathcal{L}^n \times \mathcal{L}^k \neq \mathcal{L}^{n+k}$.

2.15.6. Denote by $[R^m, \mathcal{L}^m, (dx)^m]$ the Lebesgue measure space in R^m. Prove that the measure $(dx)^{n+k}$ and the product $(dx)^n \times (dx)^k$ coincide on rectangles $A \times B$ with bounded sides, when (i) A and B are intervals; (ii) A and B are open sets; (iii) A and B are G_δ sets, and, finally, (iv) A and B are any Lebesgue sets.

2.15.7. Prove that the measure space $(R^{n+k}, \mathcal{L}^{n+k}, (dx)^{n+k})$ is the completion of the Cartesian product $[(R^n, \mathcal{L}^n, (dx)^n) \times (R^k, \mathcal{L}^k, (dx)^k]$. Prove also that the latter product is an extension of the restriction of $(R^{n+k},\mathscr{L}^{n+k},(dx)^{n+k})$ to the Borel σ-algebra of R^{n+k}.

2.15.8. If $\mu_1 \ll \mu_2$, $\nu_1 \ll \nu_2$, then $\mu_1 \times \nu_1 \ll \mu_2 \times \nu_2$; the μ_i, ν_i are finite measures.

2.15.9. A function φ in a subset Ω of R^n is called a *special simple function* if there is a finite number of intervals I_{a_k,b_k} and real numbers c_k such that $\varphi = \sum c_k \chi_k$, where χ_k is the characteristic function of I_{a_k,b_k}. Prove that if f is Lebesgue-integrable in a bounded open set Ω of R^n, then there exists a sequence $\{\varphi_m\}$ of special simple functions such that $\int_\Omega |f - \varphi_m|\, dx \to 0$ as $m \to \infty$, and $\lim \varphi_m = f$ a.e. in Ω. [*Hint:* Compare Problem 2.6.5, and use Problem 1.9.7.]

2.15.10. If f is Lebesgue-integrable on a bounded open set Ω of R^n, then there exists a sequence $\{\varphi_m\}$ of continuous functions, of the form $\sum_{i=1}^{p} \psi_{i1}(x_1)\psi_{i2}(x_2)\ldots\psi_{in}(x_n)$, with ψ_{ij} continuous on the real line, such that $\int_\Omega |f - \varphi_m|\, dx \to 0$ as $m \to \infty$, and $\lim \varphi_m = f$ a.e. in Ω.

2.15.11. If f is an a.e. real-valued measurable function on a bounded open set Ω of R^n, then for any $\varepsilon > 0$, $\delta > 0$ there exists a continuous function $g(x)$ in $\overline{\Omega}$ such that $|g(x) - f(x)| < \varepsilon$ for all x in a subset E of Ω, and $\mu(\Omega - E) < \delta$.

2.16 FUBINI'S THEOREM

We maintain the notation of Section 2.15.

If f is a function defined on a subset E of $X \times Y$, then, for each x in X, we call the function

$$f_x(y) = f(x,y) \qquad \text{for } y \in E_x$$

an *X-section* of f. Similarly, a *Y-section* of f is defined by

$$f^y(x) = f(x,y) \qquad \text{for } x \in E^y.$$

Lemma 2.16.1 *Every* X*-section* (Y*-section*) *of a measurable function in the measure space* $(X \times Y, \mathfrak{A} \times \mathscr{B}, \mu \times \nu)$ *is a measurable function in the measure space* $(Y,\mathscr{B},\nu)$ $[(X,\mathfrak{A},\mu)]$.

Proof. Let f be a measurable function. For any open interval I on the real line,

$$\begin{aligned} f_x^{-1}(I) &= \{y; f_x(y) \in I\} = \{y; f(x,y) \in I\} \\ &= \{y; (x,y) \in f^{-1}(I)\} = [f^{-1}(I)]_x. \end{aligned}$$

Since $f^{-1}(I)$ is in $\mathfrak{A} \times \mathscr{B}$, the set on the right is in $\mathscr{B}$ (by Lemma 2.15.1). Thus f_x is measurable. The proof for f^y is similar.

For any measurable function h on $X \times Y$, we write the integral $\int h\, d\lambda$ ($\lambda = \mu \times \nu$) also in the form

$$\int h(x,y)\, d\lambda(x,y) \qquad \text{or} \qquad \int h(x,y)\, d(\mu \times \nu)(x,y).$$

We call it the *double integral* of h. Note that here h is assumed either to be integrable or, if it is not integrable, to be nonnegative (and then $\int h\, d\lambda = \infty$ by definition).

By Lemma 2.16.1, the X-sections h_x are measurable functions. We set

$$f(x) = \int h_x(y)\, d\nu(y) \tag{2.16.1}$$

if h_x is integrable, or if it is nonnegative. If $f(x)$ is either integrable or nonnegative and measurable, then we write

$$\int f\, d\mu = \iint h(x,y)\, d\nu(y)\, d\mu(x) = \int d\mu(x) \int h(x,y)\, d\nu(y)$$

and call it an *iterated integral* of h. Similarly we define the iterated integral

$$\int g\, d\nu = \iint h(x,y)\, d\mu(x)\, d\nu(y) = \int d\nu(y) \int h(x,y)\, d\mu(x),$$

where

$$g(y) = \int h^y(x)\, d\mu(x). \tag{2.16.2}$$

The connection between the double integral and the iterated integrals is given by the following theorem, known as *Fubini's theorem*.

Theorem 2.16.2 *If* h *is a nonnegative, measurable function on* X × Y, *then the functions* f(x) *and* g(y) *defined by* (2.16.1) *and* (2.16.2) *are measurable, and*

$$\int h\, d(\mu \times \nu) = \iint h\, d\mu\, d\nu = \iint h\, d\nu\, d\mu. \tag{2.16.3}$$

Proof. If h is a characteristic function of a set E, then

$$\int h(x,y)\, d\nu(y) = \nu(E_x), \qquad \int h(x,y)\, d\mu(x) = \mu(E^y),$$

and the assertion follows from Theorem 2.15.7. By linearity, the assertion follows also when h is a nonnegative simple function. By Theorem 2.2.5, there exists an increasing sequence $\{h_n\}$ of nonnegative simple functions that converges to h everywhere. By the Lebesgue monotone convergence theorem we then have

$$\lim_{n\to\infty} \int h_n \, d\lambda = \int h \, d\lambda. \tag{2.16.4}$$

Let $f_n(x) = \int h_n(x,y) \, d\nu(y)$. Then $\{f_n\}$ is a nonnegative increasing sequence of measurable functions and, again by the Lebesgue monotone convergence theorem,

$$\lim_{n\to\infty} f_n(x) = \int h(x,y) \, d\nu(y) = f(x).$$

Hence $f(x)$ is measurable and nonnegative, and

$$\lim_{n\to\infty} \int f_n \, d\mu = \int f \, d\mu.$$

Combining this with (2.16.4), and recalling that (2.16.3) holds for each h_n, we get

$$\int h \, d\lambda = \int f \, d\mu.$$

Similarly one proves that g is nonnegative and measurable, and $\int h \, d\lambda = \int g \, d\nu$.

The following theorem, which follows almost immediately from Theorem 2.16.2, is also usually referred to as *Fubini's theorem*.

Theorem 2.16.3 *If* h *is an integrable function on* X × Y, *then almost every section of* h *is integrable, the functions* f(x) *and* g(y) *defined in* (2.16.1) *and* (2.16.2) *are integrable, and* (2.16.3) *holds.*

Proof. We may suppose that $h \geq 0$, for otherwise we write $h = h^+ - h^-$ and consider each of the functions h^+, h^- separately. The integrability of f and g follows from (2.16.3) and the fact that $\int h \, d\lambda < \infty$. Since f and g are then finite-valued a.e., it follows that almost all sections of h are integrable.

We give another version, perhaps the most useful one, of *Fubini's theorem.*

Theorem 2.16.4 *If* h *is a measurable function on* X × Y *and if either* $\iint |h| \, d\mu \, d\nu < \infty$ *or* $\iint |h| \, d\nu \, d\mu < \infty$, *then* h *is integrable on* X × Y *and the assertions of Theorem* 2.16.3 *are valid.*

Proof. By Theorem 2.16.2, applied to $|h|$, it follows that $\int |h| \, d\mu$, $\int |h| \, d\nu$ are measurable functions, so that the integrals,

$$\iint |h| \, d\mu \, d\nu, \qquad \iint |h| \, d\nu \, d\mu$$

are indeed defined. Furthermore,

$$\int |h| \, d(\mu \times \nu) = \iint |h| \, d\mu \, d\nu = \iint |h| \, d\nu \, d\mu.$$

Hence $\int |h| \, d(\mu \times \nu) < \infty$. We conclude that $|h|$ is integrable on $X \times Y$. It follows that h is also integrable. Now apply Theorem 2.16.3.

Theorem 2.16.2 for h a characteristic function of a measurable set E, or Theorem 2.15.8, gives

$$\lambda(E) = \int \nu(E_x) \, d\mu(x) = \int \mu(E^y) \, d\nu(y). \tag{2.16.5}$$

Corollary 2.16.5 *A measurable set* E *of* X $\times$ Y *has measure zero if and only if almost every* X*-section (or almost every* Y*-section) of* E *has measure zero.*

Indeed, if $\nu(E_x) = 0$ a.e., then $\lambda(E) = 0$ by (2.16.5). Conversely, if $\lambda(E) = 0$, then $\int \nu(E_x) \, d\mu(x) = 0$ and $\nu(E_x) \geq 0$ a.e. imply that $\nu(E_x) = 0$ a.e.

The *Cartesian product* of n measure spaces $(X_j, \mathcal{A}_j, \mu_j)$ can be defined by induction. It obeys the associate law. The product

$$\mu = \mu_1 \times \cdots \times \mu_n = \prod_{i=1}^{n} \mu_i$$

is uniquely determined by the requirement that for any n-rectangle $A_1 \times \cdots \times A_n$ of $\mathcal{A}_1 \times \cdots \times \mathcal{A}_n$,

$$\mu(A_1 \times \cdots \times A_n) = \prod_{i=1}^{n} \mu_i(A_i).$$

The Lebesgue measure space of R^n coincides with the completion of the product of n measure spaces, each being the Lebesgue measure space of R^1. Fubini's theorem also extends to the case of a product of n measure spaces.

PROBLEMS

2.16.1. If E and F are measurable sets of $X \times Y$ such that $\nu(E_x) = \nu(F_x)$ for almost all $x \in X$, then $\lambda(E) = \lambda(F)$ $(\lambda = \mu \times \nu)$.

2.16.2. Let $f(x)$ and $g(y)$ be integrable functions on X and Y, respectively. Then the function $h(x,y) = f(x)g(y)$ is integrable on $X \times Y$, and

$$\int h \, d(\mu \times \nu) = \int f \, d\mu \int g \, d\nu.$$

2.16.3. Show that

$$\int_0^1 \int_0^1 f \, dx \, dy \neq \int_0^1 \int_0^1 f \, dy \, dx \qquad \text{if } f(x,y) = \frac{x^2 - y^2}{(x^2 + y^2)^2}.$$

2.16.4. If $f(x,y)$ is a nonnegative Borel-measurable function in R^{n+k} ($x \in R^n$, $y \in R^k$), then $\int f(x,y)\,dx$, $\int f(x,y)\,dy$ are Borel-measurable and

$$\int f(x,y)\,dx\,dy = \int dy \int f(x,y)\,dx = \int dx \int f(x,y)\,dy.$$

Here $dx\,dy$ is the product of the Lebesgue measures dx and dy.

2.16.5. Let E be a planar domain bounded by two continuous curves $y = \varphi_1(x)$, $y = \varphi_2(x)$ for $a \le x \le b$, where $\varphi_1(x) < \varphi_2(x)$ if $a < x < b$. Prove that if f is a Lebesgue-integrable, Borel-measurable function on E, then

$$\int f(x,y)\,d\mu = \int_a^b \int_{\varphi_1(x)}^{\varphi_2(x)} f(x,y)\,dy\,dx, \tag{2.16.6}$$

where μ is the Lebesgue measure in R^2.

2.16.6. The formula (2.16.6) remains true (with the same proof) if x varies in a bounded domain of R^m instead of the interval $[a,b]$ of R^1. Using this formula, compute the volume of the unit ball in R^n.

2.16.7. Consider the transformation $Tx = Ax + k$ in R^n, where A is a nonsingular $n \times n$ matrix and x,k are column n-vectors. Denote by λ the Lebesgue measure in R^n. Prove the relation

$$\lambda[T(I_{a,b})] = |\det A|\lambda(I_{a,b}) \tag{2.16.7}$$

for $n = 2$. [*Hint:* use (2.16.6).]

2.16.8. Prove the relation (2.16.7) for any $n \ge 2$. [*Hint:* Proceed by induction, and assume that the assertion is true for $n - 1$. Write $A = B_1 \cdots B_k A_0$, where the B_j are elementary matrices and A_0 is the unit matrix. Use Problem 2.16.1 and the inductive assumption to prove that $\lambda[B_j(E)] = |\det B_j|\lambda(E)$ for any interval E; hence, by Problem 1.9.11, also for any Lebesgue set E.]

2.16.9. Let Ω be a bounded open set in R^n. Let $Tx = Ax + k$ be a linear map of R^n into itself with $\det A \ne 0$. Denote by Ω' the image of Ω under the map T. Prove: If $f(y)$ is Lebesgue-integrable for $y \in \Omega'$, then $f(Tx)$ is Lebesgue-integrable as a function of x in Ω, and

$$\int_{\Omega'} f(y)\,dy = \int_{\Omega} f(Tx)\,|\det A|\,dx.$$

[*Hint:* Use Problem 1.9.11.]

2.16.10. Let G be a bounded open set in R^n and let Ω be an open set in R^n. Let $f(x,y)$ be a function of $(x,y) \in \Omega \times G$, and assume that all its derivatives with respect to x exist and are bounded measurable functions in $\Omega \times G$. Let $g(y)$ be a Lebesgue-integrable function in G. Prove that the function

$$h(x) = \int_G f(x,y)g(y)\,dy$$

has continuous derivatives of all orders in Ω, and

$$D_x^m h(x) = \int_G D_x^m f(x,y) \cdot g(y)\, dy.$$

Here D_x^m denotes any partial derivative of order m. Note that the assertion for $m = 1$, $n = 1$ is similar to the assertion in Problem 2.10.5.

2.16.11. Let Ω be a bounded open set in R^n. Let $f(x)$ and $g(x)$ be functions having continuous derivatives in $x \in \Omega$, of all orders $\leq m$. Assume further that $g(x) = 0$ outside a compact subset of Ω. Prove that for any partial derivative D_x^m of order m

$$\int_\Omega f(x) D_x^m g(x)\, dx = (-1)^m \int_\Omega D_x^m f(x) \cdot g(x)\, dx.$$

[*Hint:* Suppose $m = 1$, $D_x^1 = \partial/\partial x_1$. Replace Ω by an interval containing it and use Fubini's theorem with $X = R^1$, $Y = R^{n-1}$, and Problem 2.14.3.]

CHAPTER 3

METRIC SPACES

3.1 TOPOLOGICAL AND METRIC SPACES

Let X be a set. We call X also a *space* and we call its elements also *points*. Let $\mathscr{K}$ be a class of subsets of X having the following properties:

(i) $\varnothing$ and X belong to $\mathscr{K}$.
(ii) The intersection of any finite number of sets of $\mathscr{K}$ is in $\mathscr{K}$.
(iii) The union of any class of sets of $\mathscr{K}$ is in $\mathscr{K}$.

We then say that $\mathscr{K}$ defines a *topology* on X, and we call the pair $(X,\mathscr{K})$ a *topological space* and the sets of $\mathscr{K}$ *open sets*. For brevity, we often denote the topological space by X. A set B is called a *closed set* if its complement $X - B$ is an open set.

A topological space is called a *Hausdorff topological space* if it satisfies the following property:

(iv) For any two distinct points x and y there exist disjoint open sets A and B such that $x \in A$ and $y \in B$.

Note that (iv) implies the following:

(v) Sets consisting of single points are closed.

A Hausdorff topological space is called *normal* if it satisfies the following property:

(vi) For any pair of disjoint closed sets E and F there exist disjoint open sets A and B such that $E \subset A$ and $F \subset B$.

Note that (v) and (vi) together imply (iv).

Let X be a topological space. A *neighborhood* of a set B (point x) is any open set A containing B (x).

It is easily verified that the intersection of any class of closed sets is a closed set, and that a finite union of closed sets is a closed set. Given a set A, we define the *closure* of A to be the smallest closed set containing A. It coincides with the intersection of all the closed sets containing A. The closure of A is denoted by $\bar{A}$.

The *interior* of a set A is the set consisting of all the points of A that have neighborhoods contained in A. It is denoted by int A.

It is easily verified that $\overline{A \cup B} = \bar{A} \cup \bar{B}$, $\bar{\bar{A}} = \bar{A}$. Furthermore, $x \in \bar{A}$ if and only if every neighborhood of x intersects A.

If A is closed and B is open, then $A - B$ is closed, since $X - (A - B) = (X - A) \cup B$ is open.

Let Y be a subset of X. With the aid of the topology $\mathscr{K}$ on X we can define a topology on Y as follows: a subset B of Y is in $\mathscr{K}_Y$ (that is, is open) if and only if there exists a set D in $\mathscr{K}$ such that $B = D \cap Y$. We call $(Y,\mathscr{K}_Y)$ (or, simply, Y) *a topological subspace* of X, and we say that it has the *topology induced* by $(X,\mathscr{K})$ (or, simply, by X).

An *open covering* of X is a class of open sets whose union contains X. A topological space X is called *compact* if every open covering of X contains a finite subclass that forms a covering of X. A subset Y of X is called *compact* if it is a compact space in the induced topology. This is the case if and only if every open covering of Y by sets of X contains a finite subclass that forms a covering of Y.

In a topological space one defines concepts such as convergence, points of accumulation, and so on. Since, however, we shall be interested primarily in topological spaces with a special structure, namely, metric spaces, we shall proceed directly to consider these notions in a metric space.

Recall that a space X is called a *metric space* with *metric* ρ if to any two points x,y there corresponds a real number $\rho(x,y)$ such that

(i) $\rho(x,y) \geq 0$, and $\rho(x,y) = 0$ if and only if $x = y$,
(ii) $\rho(x,y) = \rho(y,x)$ (symmetry),
(iii) $\rho(x,z) \leq \rho(x,y) + \rho(y,z)$ (the triangle inequality).

We have defined in Section 1.7 the concept of an open set. It is easily seen that the class of open sets defines a topology. Moreover, *metric spaces are normal*. Indeed, to prove (iv) take $A = \{z;\, \rho(z,x) < \frac{1}{2}\rho(x,y)\}$, $B = \{z;\, \rho(z,y) < \frac{1}{2}\rho(x,y)\}$. To prove (vi), we take

$$A = \{x;\, \rho(x,E) < \rho(x,F)\}, \qquad B = \{x;\, \rho(x,F) < \rho(x,E)\},$$

where, by definition,

$$\rho(C,D) = \inf_{\substack{x \in C \\ y \in D}} \rho(x,y), \qquad \rho(x,D) = \rho(\{x\},D).$$

It is easily seen that A and B are open sets, and, obviously (see Problem 1.7.6), $E \subset A, F \subset B, A \cap B = \emptyset$.

A subset Y of a metric space X is called *bounded* if $\sup_{x,y \in Y} \rho(x,y) < \infty$. Note that a set is bounded if and only if it is contained in some ball.

If Y is a subset of a metric space X, then we can make Y into a metric space (called a *metric subspace* of X) by taking its metric to be the restriction of the metric ρ of X to Y. We call this metric the one *induced* by the metric of X.

A subset Y of a metric space X is called *sequentially compact* if every sequence $\{y_n\}$ in Y has a subsequence $\{y_{n_k}\}$ that converges to a point in Y—that is, $\rho(y_{n_k},y) \to 0$ as $n_k \to \infty$, for some $y \in Y$.

A subset Y of X is called *dense* if $\bar{Y} = X$. Y is called *nowhere dense* if $\bar{Y}$ has no interior points.

A space X is called *separable* if it contains a countable dense set.

We finally recall that a metric space is called *complete* if any Cauchy sequence $\{x_n\}$ [that is, $\rho(x_n,x_m) \to 0$ as $m,n \to \infty$] is convergent.

We shall derive later on various relations between some of the concepts defined above. In particular, it will be shown that a set Y is compact if and only if it is sequentially compact.

Example 1. R^n, the n-dimensional Euclidean space, is a metric space with

$$\rho(x,y) = \left\{ \sum_{i=1}^{n} (x_i - y_i)^2 \right\}^{1/2}, \tag{3.1.1}$$

where $x = (x_1,\dots,x_n)$, $y = (y_1,\dots,y_n)$. R^n is not compact. The Heine-Borel theorem asserts that bounded closed subsets of R^n are compact sets. R^n is complete. It is also separable, since the points $x = (x_1,\dots,x_n)$ with rational components x_i $(1 \le i \le n)$ form a dense sequence.

Example 2. The real (complex) space l^∞. Its elements are sequences $x = (\xi_1,\xi_2,\dots,\xi_n,\dots)$ with real (complex) components ξ_n, such that $\sup_n |\xi_n| < \infty$. If $y = (\eta_1,\eta_2,\dots,\eta_n,\dots)$ is another point of l^∞, we define

$$\rho(x,y) = \sup_n |\xi_n - \eta_n|.$$

A sequence $\{x_m\}$ in l^∞, with $x_m = (\xi_{m1},\dots,\xi_{mn},\dots)$, is convergent to $y = (y_1,\dots,y_n,\dots)$ if and only if $\lim_m x_{mn} = y_n$ uniformly with respect to n.

Example 3. The real (complex) space l^1. Its elements are sequences $x = (\xi_1,\xi_2,\dots,\xi_n,\dots)$ with real (complex) components ξ_n such that $\sum |\xi_n| < \infty$.

If $y = (\eta_1,\eta_2,\ldots,\eta_n,\ldots)$ is another point of l^1, we define

$$\rho(x,y) = \sum_{n=1}^{\infty} |\xi_n - \eta_n|.$$

Note that l^∞ and l^1 are special cases of the spaces $L^\infty(X,\mu)$ and $L^1(X,\mu)$ introduced in Problems 2.10.16 and 2.8.1, respectively. Indeed, they are obtained by taking X to consist of all the positive integers and $\mu(E)$, for any subset E of X, to be the number of positive integers contained in E.

Example 4. The space c is the metric subspace of l^∞ consisting of all sequences $(\xi_1,\xi_2,\ldots,\xi_n,\ldots)$ for which $\lim_n \xi_n$ exists. The space c_0 is the metric subspace of c consisting of all the sequences for which $\lim_n \xi_n = 0$.

Example 5. The space s consists of all the sequences $x = (\xi_1,\xi_2,\ldots\xi_n,\ldots)$ with real components. Let $y = (\eta_1,\eta_2,\ldots,\eta_n\ldots)$ be another sequence. Then we define

$$\rho(x,y) = \sum_{n=1}^{\infty} \frac{1}{2^n} \frac{|\xi_n - \eta_n|}{1 + |\xi_n - \eta_n|}. \tag{3.1.2}$$

To prove the triangle inequality, it suffices to show that for any real numbers a,b,

$$\frac{|a+b|}{1 + |a+b|} \le \frac{|a|}{1+|a|} + \frac{|b|}{1+|b|}. \tag{3.1.3}$$

Suppose a and b have the same sign, say both are positive. Then

$$\frac{a+b}{1+a+b} = \frac{a}{1+a+b} + \frac{b}{1+a+b} < \frac{a}{1+a} + \frac{b}{1+b}.$$

Suppose next that a and b have different signs. We may assume that $|a| \ge |b|$, so that $|a+b| \le |a|$. Since the function $x/(1+x)$ is monotone increasing for $x > 0$, we get

$$\frac{|a+b|}{1+|a+b|} \le \frac{|a|}{1+|a|} \le \frac{|a|}{1+|a|} + \frac{|b|}{1+|b|}.$$

Example 6. Let $-\infty < a < b < \infty$. $C[a,b]$ is the space of all continuous functions on $[a,b]$ (real-valued for the *real space* $C[a,b]$, and complex-valued for the *complex space* $C[a,b]$.) If f,g are two continuous functions on $[a,b]$, then we define

$$\rho(f,g) = \max_{a \le t \le b} |f(t) - g(t)|. \tag{3.1.4}$$

We call this metric the *maximum metric*. Convergence of a sequence in $C[a,b]$ (with the maximum metric) means that the sequence of functions is uniformly convergent on $[a,b]$. For this reason we call the metric (3.1.4) also the *uniform metric.*

Definition Two metrics ρ and $\hat{\rho}$ on a space X are called *equivalent* if there exist positive constants α and β such that

$$\alpha\hat{\rho}(x,y) \leq \rho(x,y) \leq \beta\hat{\rho}(x,y)$$

for all x,y in X.

For example, the metrics

$$\hat{\rho}(x,y) = \sup_{1\leq i\leq n} |x_i - y_i|, \qquad \bar{\rho}(x,y) = \sum_{i=1}^{n} |x_i - y_i|$$

are both equivalent to the Euclidean metric (3.1.1) of R^n.

Example 7. Let $(X_1,\rho_1),\ldots,(X_m,\rho_m)$ be metric spaces. Their *Cartesian product* (X, ρ) is defined as follows: X is the Cartesian product $X_1 \times \cdots \times X_m$ —that is, its elements are n-tuples $x = (x_1,\ldots,x_m)$ with $x_i \in X_i$, and

$$\rho(x,y) = \sum_{i=1}^{m} \rho_i(x_i,y_i) \tag{3.1.5}$$

if $y = (y_1,\ldots,y_m)$. Note that if $c_1,\ldots,c_m$ are any positive numbers, then the metric

$$\rho_c(x,y) = \sum_{i=1}^{m} c_i\,\rho_i(x_i,y_i)$$

is equivalent to ρ.

Example 8. Let $\{(X_i,\rho_i)\}$ be an infinite sequence of metric spaces. Let X be the Cartesian product $X_1 \times X_2 \times \cdots \times X_n \times \cdots$—that is, it consists of points $x = (x_1,x_2,\ldots,x_n,\ldots)$, with $x_i \in X_i$. We cannot define ρ simply by (3.1.5) with $m = \infty$, since the series may not converge. Instead we define

$$\rho(x,y) = \sum_{n=1}^{\infty} \frac{1}{2^n}\frac{\rho_n(x_n,y_n)}{1 + \rho_n(x_n,y_n)}, \tag{3.1.6}$$

where $y = (y_1,y_2,\ldots,y_n,\ldots)$. We call (X,ρ) the *Cartesian product* of the sequence $\{(X_i,\rho_i)\}$. Note that the space s is obtained as a special case, with all the X_i coinciding with R^1.

PROBLEMS

3.1.1. Prove that if (X,ρ) is a metric space, and if

$$\hat{\rho}(x,y) = \frac{\rho(x,y)}{1 + \rho(x,y)},$$

then also $(X,\hat{\rho})$ is a metric space. [*Hint:* Cf. the proof of (3.1.3).]

3.1.2. Let $X,\rho,\hat{\rho}$ be as in Problem 3.1.1. Prove that $\rho(x_n,x)\to 0$ if and only if $\hat{\rho}(x_n,x)\to 0$. Give an example showing that ρ and $\hat{\rho}$ are not equivalent in general.

3.1.3. Prove that the Cartesian product (X,ρ) defined in Example 7 is a complete (separable) space if and only if each space (X_i,ρ_i) is complete (separable).

3.1.4. Prove that a sequence $\{x_m\}$, with $x_m = (x_{m1},\ldots,x_{mn},\ldots)$ in the Cartesian product defined in Example 8, is convergent if and only if, for each n, $\{x_{mn}\}$ is a convergent sequence in (X_n,ρ_n). Prove also that (X,ρ) is complete if and only if each space (X_n,ρ_n) is complete.

3.1.5. Prove that the spaces l^1,l^∞,s,c,c_0, and $C[a,b]$ are complete metric spaces.

3.1.6. Prove that the spaces l^1,s,c,c_0 are separable metric spaces.

3.1.7. Prove that l^∞ is not separable. [*Hint:* Consider the points $x_\xi = (\xi_1,\xi_2,\ldots,\xi_n\ldots)$, where each ξ_n is either 0 or 1. This is a nondenumerable set, and $\rho(x_\xi,x_{\xi'}) = 1$ if $\xi \neq \xi'$. The open balls B_ξ of radius $\frac{1}{3}$ and center ξ satisfy $B_\xi \cap B_{\xi'} = \varnothing$ if $\xi \neq \xi'$. If $\{x_m\}$ is any sequence, then there exist indices ξ such that $x_m \notin B_\xi$ for all m. Another proof: Suppose $A = \{x_m\}$ is a dense sequence, and $x_m = (\xi_1^m,\ldots,\xi_n^m,\ldots)$. Construct a point $z = (z_1,\ldots,z_n,\ldots)$ not in $\bar{A}$.]

3.1.8. Construct, in each of the spaces $l^1,l^\infty,s,c,c_0,C[a,b]$, bounded closed sets that are not sequentially compact.

3.1.9. Let X be a metric space, and let Y be a metric subspace of X. Prove that Y is also a topological subspace of X.

3.1.10. If $\rho(x_n,x)\to 0$, $\rho(y_n,y)\to 0$, then $\rho(x_n,y_n)\to\rho(x,y)$.

3.1.11. For what values of p is $\rho(x,y) = |x-y|^p$ a metric on the real line?

3.1.12. If K is a closed subset of a topological space X, and if L is a closed subset of the topological subspace K, then L is a closed subset of X.

3.2 L^p SPACES

Let $(X,\mathfrak{A},\mu)$ be a measurable space, and let p be any positive number. We denote by $\mathscr{L}^p(X,\mu)$ the class of all measurable functions f such that $|f|^p$ is integrable. We write

$$\|f\|_p = \left\{\int |f|^p \, d\mu\right\}^{1/p}. \tag{3.2.1}$$

Similarly, $\mathscr{L}^\infty(X,\mu)$ is the class of all measurable and essentially bounded functions. We write

$$\|f\|_\infty = \operatorname*{ess\,sup}_X |f|. \tag{3.2.2}$$

We shall derive an important inequality known as *Hölder's inequality.*

Theorem 3.2.1 *Let* p *and* q *be extended real numbers,* $1 \le p \le \infty$, $1 \le q \le \infty$, $1/p + 1/q = 1$. *If* $f \in \mathscr{L}^p(X,\mu)$, $g \in \mathscr{L}^q(X,\mu)$, *then* $fg \in \mathscr{L}^1(X,\mu)$ *and*

$$\|fg\|_1 \le \|f\|_p \|g\|_q. \tag{3.2.3}$$

Proof. The assertions for $p = 1$ and for $p = \infty$ are rather obvious. We therefore shall assume that $1 < p < \infty$. We shall need the elementary inequality

$$ab \le \frac{a^p}{p} + \frac{b^q}{q} \qquad (a \ge 0, b \ge 0). \tag{3.2.4}$$

To prove it, consider the function $h(t) = t^p/p + t^{-q}/q$. It (strictly) decreases from ∞ to 1 in the interval $(0,1]$, and it (strictly) increases from 1 to ∞ in the interval $[1,\infty)$, and $h(1) = 1$. Hence

$$1 \le \frac{t^p}{p} + \frac{t^{-q}}{q} \qquad \text{for all } t > 0.$$

Suppose $a > 0$, $b > 0$. Taking $t = a^{1/q}/b^{1/p}$ and multiplying the resulting inequality by ab, (3.2.4) follows for $a > 0$, $b > 0$. The general case $a \ge 0$, $b \ge 0$ follows immediately by continuity.

We may assume that $\|f\|_p \ne 0$, $\|g\|_q \ne 0$, for otherwise the assertion of the theorem is trivially true. We then take

$$a = \frac{|f(x)|}{\|f\|_p}, \qquad b = \frac{|g(x)|}{\|g\|_q}$$

in (3.2.4). We get

$$\frac{|f(x)g(x)|}{\|f\|_p \|g\|_q} \le \frac{1}{p}\frac{|f(x)|^p}{\|f\|_p^p} + \frac{1}{q}\frac{|g(x)|^q}{\|g\|_q^q}. \tag{3.2.5}$$

Since each of the two terms on the right is in $\mathscr{L}^1(X,\mu)$, the same is true of $|fg|$. Further, by integrating both sides of (3.2.5) we obtain (3.2.3).

We shall use Hölder's inequality to derive *Minkowski's inequality*:

Theorem 3.2.2 *Let* $1 \le p \le \infty$. *If* f *and* g *belong to* $\mathscr{L}^p(X,\mu)$, *then also* $f + g$ *belongs to* $\mathscr{L}^p(X,\mu)$, *and*

$$\|f + g\|_p \le \|f\|_p + \|g\|_p. \tag{3.2.6}$$

Proof. The assertions for $p = 1$ and for $p = \infty$ are rather obvious. We shall therefore assume that $1 < p < \infty$. From

$$|f(x) + g(x)|^p \leq \begin{cases} 2^p |f(x)|^p, & \text{if } |f(x)| \geq |g(x)|, \\ 2^p |g(x)|^p, & \text{if } |g(x)| \geq |f(x)|, \end{cases}$$

it follows that $|f + g|^p \leq 2^p |f|^p + 2^p |g|^p$. Hence $f + g$ belongs to $\mathscr{L}^p(X,\mu)$. Next, using Hölder's inequality, we can write

$$\begin{aligned} \|f + g\|_p{}^p &= \int |f + g|^p \, d\mu \\ &\leq \int |f| |f + g|^{p-1} \, d\mu + \int |g| |f + g|^{p-1} \, d\mu \\ &\leq \left(\int |f|^p \, d\mu\right)^{1/p} \left(\int |f + g|^p \, d\mu\right)^{1/q} + \left(\int |g|^p \, d\mu\right)^{1/p} \left(\int |f + g|^p \, d\mu\right)^{1/q} \\ &= (\|f\|_p + \|g\|_p) \, \|f + g\|_p^{p/q}. \end{aligned}$$

From this (3.2.6) immediately follows.

We say that f is *equivalent* to g if $f = g$ a.e. For any measurable function f, we denote by $\bar{f}$ the class of all the measurable functions equivalent to f. Note that if f and g belong to $\mathscr{L}^p(X,\mu)$, then $g \in \bar{f}$ if and only if $\|f - g\|_p = 0$.

We denote by $L^p(X,\mu)$ the space of all classes $\bar{f}$ of functions f of $\mathscr{L}^p(X,\mu)$ and define

$$\rho(\bar{f},\bar{g}) = \rho(f,g) = \|f - g\|_p. \tag{3.2.7}$$

Theorem 3.2.3 *If* $1 \leq p \leq \infty$, *then* $L^p(X,\mu)$ *is a complete metric space.*

Proof. If $\rho(\bar{f},\bar{g}) = 0$, then $\bar{f} = \bar{g}$. Also, $\rho(\bar{f},\bar{g}) = \rho(\bar{g},\bar{f})$. Finally, the triangle inequality follows from Minkowski's inequality. Thus, $L^p(X,\mu)$ is a metric space [with the metric given by (3.2.7)]. We shall prove completeness in case $1 \leq p < \infty$; the proof for $p = \infty$ is left to the reader. Let $\{\bar{f}_m\}$ be a Cauchy sequence in $L^p(X,\mu)$. Then

$$\int |f_m - f_n|^p \, d\mu \to 0 \qquad \text{if } m,n \to \infty. \tag{3.2.8}$$

The proof of Lemma 2.5.2 shows that $\{f_m\}$ is Cauchy in measure. Hence, by Theorem 2.4.3, there is a subsequence $\{f_{m_k}\}$ that converges almost uniformly to a measurable function f. $\{f_{m_k}\}$ then converges to f also a.e.

Note now that [by (3.2.8)] $\{\int |f_m|^p \, d\mu\}$ is a bounded sequence. Since $\lim |f_{m_k}|^p = |f|^p$ a.e., Fatou's lemma gives

$$\int |f|^p \, d\mu \leq \varliminf_k \int |f_{m_k}|^p \, d\mu < \infty.$$

Hence $|f|^p$ is integrable—that is, $\bar{f} \in L^p(X,\mu)$.

Next, again by Fatou's lemma,

$$\int |f - f_{m_k}|^p \, d\mu \leq \varliminf_{j} \int |f_{m_j} - f_{m_k}|^p \, d\mu \to 0 \qquad \text{if } m_j \to \infty. \qquad (3.2.9)$$

Finally, by (3.2.8) and (3.2.9), for any $\varepsilon > 0$,

$$\rho(f,f_n) \leq \rho(f,f_{m_k}) + \rho(f_{m_k},f_n) \leq \varepsilon$$

if $n \geq m_k \geq \overline{m}(\varepsilon)$. Fixing $m_k \geq \overline{m}(\varepsilon)$, we get $\rho(f,f_n) \leq \varepsilon$ if $n \geq m_k$. This shows that $\rho(\bar{f}_n,\bar{f}) \to 0$ as $n \to \infty$.

The completeness part of Theorem 3.2.3 is sometimes called the *Riesz-Fischer theorem.*

Notation For simplicity we shall refer from now on to the elements of $L^p(X,\mu)$ as functions f (instead of classes $\bar{f}$). Equality between functions f and g of $L^p(X,\mu)$ then means that $f = g$ a.e. If the functions are all extended real-valued (complex-valued), then we speak of the *real space* $L^p(X,\mu)$ [the *complex space* $L^p(X,\mu)$].

Definition The *real* (*complex*) l^p is the space of sequences $\xi = (\xi_1,\xi_2,\ldots,\xi_n,\ldots)$ with real (complex) components, such that

$$\|\xi\|_p = \left(\sum_{n=1}^{\infty} |\xi_n|^p\right)^{1/p} < \infty.$$

Theorems 3.2.1 and 3.2.2 yield the following results as special cases (compare the remark at the end of Example 3 in 3.1):

Corollary 3.2.4 *If* $\xi = (\xi_1,\xi_2,\ldots,\xi_n,\ldots) \in l^p$ *and* $\eta = (\eta_1,\eta_2,\ldots,\eta_n,\ldots) \in l^q$, *where* $1 \leq p \leq \infty$, $1 < q < \infty$, $1/p + 1/q = 1$, *then*

$$\sum_{n=1}^{\infty} |\xi_n \eta_n| \leq \left(\sum_{n=1}^{\infty} |\xi_n|^p\right)^{1/p} \left(\sum_{n=1}^{\infty} |\eta_n|^q\right)^{1/q}. \qquad (3.2.10)$$

Corollary 3.2.5 *If* $\xi = (\xi_1,\xi_2,\ldots,\xi_n,\ldots)$ *and* $\eta = (\eta_1,\eta_2,\ldots,\eta_n,\ldots)$ *belong to* l^p, *where* $1 \leq p \leq \infty$, *then also* $\xi + \eta$ *belongs to* l^p, *and*

$$\left(\sum_{n=1}^{\infty} |\xi_n + \eta_n|^p\right)^{1/p} \leq \left(\sum_{n=1}^{\infty} |\xi_n|^p\right)^{1/p} + \left(\sum_{n=1}^{\infty} |\eta_n|^p\right)^{1/p}. \qquad (3.2.11)$$

Hölder's inequality for $p = q = 2$ is also known as *Schwarz's inequality.* The inequality (3.2.10) for $p = q = 2$ is also known as *Cauchy's inequality.*

PROBLEMS

3.2.1. Prove that any function f in $L^p(X,\mu)$ $(1 \le p < \infty)$ is the limit, in the metric of $L^p(X,\mu)$, of a sequence of simple functions. [*Hint:* For f bounded and nonnegative, take the sequence constructed in Theorem 2.2.5. For f nonnegative, the sequence given by $f_n = \min(f,n)$ is convergent to f in $L^p(X,\mu)$.]

3.2.2. Let μ be the Lebesgue measure on R^n. Prove that if $1 \le p < \infty$, then $L^p(R^n,\mu)$ is separable. [*Hint:* Take the sequence of simple functions $\sum c_i \chi_{E_i}$, where c_i are rational and E_i are intervals with rational end-points. To show that $f = \chi_E\,[\mu(E) < \infty]$ is in the closure, use Problem 1.9.7.]

3.2.3. Show that if $f \in L^p(R^n,\mu)$ (μ = Lebesgue's measure) and $1 \le p < \infty$, then there is a sequence of continuous functions f_m on R^n such that $\|f - f_m\|_p \to 0$.

3.2.4. Prove that l^p is separable if $1 \le p < \infty$.

3.2.5. Prove that l^p is not a metric space if $0 < p < 1$.

3.2.6. Prove that the space $C[a,b]$ with the metric

$$\rho(f,g) = \int_a^b |f(t) - g(t)|\,dt$$

is not a complete metric space.

3.2.7. Prove that equality holds in Minkowski's inequality (3.2.6) if and only if either $f = 0$ a.e., or $g = 0$ a.e., or $f(x) = \lambda g(x)$ a.e., where λ is a positive constant. [*Hint:* If equality holds in Hölder's inequality, then $|f(x)|^p = \mu |g(x)|^q$ a.e., where μ is a constant.]

3.2.8. Let $f_1,\dots,f_m,g_1,\dots,g_m$ belong to $L^p(X,\mu)$, $1 \le p \le \infty$. Prove that

$$\left\{\sum_{i=1}^m \int |f_i + g_i|^p\,d\mu\right\}^{1/p} \le \left\{\sum_{i=1}^m \int |f_i|^p\,d\mu\right\}^{1/p} + \left\{\sum_{i=1}^m \int |g_i|^p\,d\mu\right\}^{1/p}.$$

[*Hint:* Introduce a measure space (A,ν), where $A = \{1,2,\dots,m\}$, $\nu\{i\} = 1$, and apply Minkowski's inequality in the product of the spaces (X,μ) and (A,ν).]

3.2.9. If $f_n \to f$ in $L^p(X,\mu)$ (that is, if $\|f_n - f\|_p \to 0$) and if $g_n \to g$ in $L^q(X,\mu)$, where $1 \le p \le \infty$, $1 \le q \le \infty$, $1/p + 1/q = 1$, then $f_n g_n \to fg$ in $L^1(X,\mu)$.

3.3 COMPLETION OF METRIC SPACES; $H^{m,p}$ SPACES

Let (X,ρ) and $(\hat{X},\hat{\rho})$ be two metric spaces. A map σ from X into $\hat{X}$ is called *isometric* if $\hat{\rho}(\sigma x,\sigma y) = \rho(x,y)$ for all x,y in X. We call σ also an *imbedding* of (X,ρ) into $(\hat{X},\hat{\rho})$. If σ is onto, then we say that σ is an *isomorphic map* and the two metric spaces are said to be *isomorphic* to each other.

The following theorem shows that any metric space can be imbedded in a complete metric space such that its image is dense.

Theorem 3.3.1 *Let* (X,ρ) *be a metric space. Then, there exists a metric space* $(\hat{X},\hat{\rho})$ *with the following properties:*

(i) $(\hat{X},\hat{\rho})$ *is complete.*

(ii) *There is an imbedding* σ *from* X *into* $\hat{X}$.

(iii) σ(X) *is dense in* $\hat{X}$.

If $(\hat{\hat{X}},\hat{\hat{\rho}})$ *is another metric space with these three properties, then it is isomorphic to* $(\hat{X},\hat{\rho})$.

We call $(\hat{X},\hat{\rho})$ the *completion* of the metric space (X,ρ).

Proof. We divide the proof into several steps.

(a) Two Cauchy sequences $\{x_n\}$ and $\{y_n\}$ in X will be called *equivalent* if $\rho(x_n,y_n) \to 0$ as $n \to \infty$. This relation is reflexive, symmetric, and transitive. We can therefore form classes $\tilde{x}$ of Cauchy sequences, each class containing only mutually equivalent Cauchy sequences, such that sequences of different classes are not equivalent to each other. We take the space $\hat{X}$ to consist of these classes $\tilde{x}$, and we define

$$\hat{\rho}(\tilde{x},\tilde{y}) = \lim_{n\to\infty} \rho(x_n,y_n) \tag{3.3.1}$$

where $\{x_n\} \in \tilde{x}$, $\{y_n\} \in \tilde{y}$.

Since

$$\rho(x_n,y_n) \le \rho(x_n,x_m) + \rho(x_m,y_m) + \rho(y_m,y_n),$$

we have $\rho(x_n,y_n) - \rho(x_m,y_m) \le \rho(x_n,x_m) + (\rho(y_m,y_n)$. Writing also the same inequality but with m and n interchanged, we see that

$$|\rho(x_n,y_n) - \rho(x_m,y_m)| \le \rho(x_n,x_m) + \rho(y_m,y_n).$$

It follows that the limit in (3.3.1) exists.

(b) We shall show that the definition of $\hat{\rho}$ is unambiguous—that is, if $\{\bar{x}_n\} \in \tilde{x}$, $\{\bar{y}_n\} \in \tilde{y}$, then

$$\lim_n \rho(x_n,y_n) = \lim_n \rho(\bar{x}_n,\bar{y}_n). \tag{3.3.2}$$

Since

$$\rho(x_n,y_n) \le \rho(x_n,\bar{x}_n) + \rho(\bar{x}_n,\bar{y}_n) + \rho(\bar{y}_n,y_n),$$

we get

$$\lim_n \rho(x_n,y_n) \le \lim \rho(\bar{x}_n,\bar{y}_n).$$

The reverse inequality is obtained in a similar way.

(c) $\hat{\rho}$ is a metric. Indeed, clearly $\hat{\rho}(\tilde{x},\tilde{y}) \geq 0$. Next, if $\hat{\rho}(\tilde{x},\tilde{y}) = 0$, then there are Cauchy sequences $\{x_n\}$ in $\tilde{x}$ and $\{y_n\}$ in $\tilde{y}$ such that $\lim \rho(x_n,y_n) = 0$. But this implies that $\tilde{x} = \tilde{y}$. The symmetry of $\hat{\rho}$ is obvious. Thus it remains to prove the triangle inequality. Let $\{x_n\} \in \tilde{x}$, $\{y_n\} \in \tilde{y}$, $\{z_n\} \in \tilde{z}$. Then

$$\hat{\rho}(\tilde{x},\tilde{z}) = \lim_n \rho(x_n,z_n) \leq \lim_n \rho(x_n,y_n) + \lim_n \rho(y_n,z_n) = \hat{\rho}(\tilde{x},\tilde{y}) + \hat{\rho}(\tilde{y},\tilde{z})$$

(d) Denote by $\tilde{x}^*$ the class containing the *constant* Cauchy sequence $\{x,x,\ldots,x,\ldots\}$, where $x \in X$. Then $\hat{\rho}(\tilde{x}^*,\tilde{y}^*) = \rho(x,y)$. Hence the map $\sigma: x \to \tilde{x}^*$ is an isometry of X into $\hat{X}$.

(e) We shall prove that $(\hat{X},\hat{\rho})$ is a complete space. Take a Cauchy sequence $\{\tilde{x}_n\}$ in $\hat{X}$. In each class $\tilde{x}_n$, choose a Cauchy sequence $\{x_{n,1},x_{n,2},\ldots, x_{n,m},\ldots\}$. Then, for each positive integer n there is a positive integer k_n such that

$$\rho(x_{n,m},x_{n,k_n}) \leq \frac{1}{n} \qquad \text{if } m \geq k_n.$$

Consider the sequence

$$\{x_{1,k_1},x_{2,k_2},\ldots,x_{n,k_n},\ldots\}. \tag{3.3.3}$$

We claim that this is a Cauchy sequence in X. Indeed, if $\tilde{x}^*_{n,k_n}$ is the class of the constant sequence of x_{n,k_n}, then

$$\hat{\rho}(\tilde{x}_n,\tilde{x}^*_{n,k_n}) = \lim_{j\to\infty} \rho(x_{n,j},x_{n,k_n}) \leq \frac{1}{n}. \tag{3.3.4}$$

Hence,

$$\begin{aligned}\rho(x_{n,k_n},x_{m,k_m}) &= \hat{\rho}(\tilde{x}^*_{n,k_n},\tilde{x}^*_{m,k_m}) \\ &\leq \hat{\rho}(\tilde{x}^*_{n,k_n},\tilde{x}_n) + \hat{\rho}(\tilde{x}_n,\tilde{x}_m) + \hat{\rho}(\tilde{x}_m,\tilde{x}^*_{m,k_m}) \\ &\leq \hat{\rho}(\tilde{x}_n,\tilde{x}_m) + \frac{1}{n} + \frac{1}{m}.\end{aligned}$$

It follows that $\rho(x_{n,k_n},x_{m,k_m}) \to 0$ if $m,n \to \infty$.

(f) Denote by $\tilde{x}$ the class containing the Cauchy sequence of (3.3.3). Then

$$\hat{\rho}(\tilde{x},\tilde{x}_n) \leq \hat{\rho}(\tilde{x},\tilde{x}^*_{n,k_n}) + \hat{\rho}(\tilde{x}^*_{n,k_n},\tilde{x}_n) \leq \hat{\rho}(\tilde{x},\tilde{x}^*_{n,k_n}) + \frac{1}{n}$$

by (3.3.4). Since also, for any $\varepsilon > 0$,

$$\hat{\rho}(\tilde{x},\tilde{x}^*_{n,k_n}) = \lim_j \rho(x_{j,k_j},x_{n,k_n}) < \varepsilon$$

if n is sufficiently large, we conclude that

$$\lim_n \hat{\rho}(\tilde{x},\tilde{x}_n) = 0.$$

Thus the given Cauchy sequence $\{\tilde{x}_n\}$ converges to $\tilde{x}$. This proves that $(\hat{X},\hat{\rho})$ is complete.

(g) We next show that $\sigma(X)$ is dense in $\hat{X}$. Let $\tilde{x} \in \hat{X}$. Take $\{x_n\} \in \tilde{x}$. Then each class $\tilde{x}_n^*$, containing the constant sequence $\{x_n,x_n,\ldots,x_n,\ldots\}$, is in $\sigma(X)$, and $\hat{\rho}(\tilde{x},\tilde{x}_n^*) = \lim_m \rho(x_n,x_m) \to 0$ if $n \to \infty$.

(h) Suppose, finally, that $(\hat{\hat{X}},\hat{\hat{\rho}})$ is another metric space satisfying (i)–(iii). Denote the corresponding imbedding from X into $\hat{\hat{X}}$ by $\bar{\sigma}$. For any $\hat{\hat{x}} \in \hat{\hat{X}}$, let $\{x_n\}$ be a sequence in X such that $\bar{\sigma}(x_n) \to \hat{\hat{x}}$. Since $\bar{\sigma}$ is an isometry, $\{x_n\}$ is a Cauchy sequence. Denote by $\tilde{x}$ a class containing it. We then define a correspondence γ by $\gamma(\hat{\hat{x}}) = \tilde{x}$. Note that if $\hat{\hat{y}} \in \hat{\hat{X}}$ and $\bar{\sigma}(y_n) \to \hat{\hat{y}}$, $\{y_n\} \in \tilde{y}$, then, by Problem 3.1.10,

$$\hat{\hat{\rho}}(\hat{\hat{x}},\hat{\hat{y}}) = \lim_n \hat{\hat{\rho}}(\bar{\sigma}x_n,\bar{\sigma}y_n) = \lim_n \rho(x_n,y_n) = \hat{\rho}(\tilde{x},\tilde{y}).$$

Thus γ is an isometry. γ is also an isomorphy. Indeed, every $\tilde{x} \in \hat{X}$ contains a Cauchy sequence $\{x_n\}$. The sequence $\{\bar{\sigma}x_n\}$ is a Cauchy sequence in $\hat{\hat{X}}$. It we denote by $\hat{\hat{x}}$ its limit in $\hat{\hat{X}}$, then clearly $\gamma(\hat{\hat{x}}) = \tilde{x}$. This completes the proof of the theorem.

Theorem 3.3.1 shows that metric spaces can be completed. However, the elements of the completed space are more complicated entities than those of the original space. In many instances when this completion needs to be carried out, the problem arises of giving a simple characterization of the elements of the completed space. We shall give one important example.

Let Ω be a bounded open set in R^n. We denote by $C^m(\Omega)$ the set of functions $u(x)$ that are continuous in Ω together with all their first m derivatives. The subset consisting of the functions that have a compact support (that is, that vanish outside a compact subset of Ω) is denoted by $C_0^m(\Omega)$. The corresponding spaces of infinitely differentiable functions are denoted by $C^\infty(\Omega)$ and $C_0^\infty(\Omega)$. We also write $C(\Omega) = C^0(\Omega)$ and $C_0(\Omega) = C_0^0(\Omega)$.

We write $D^\alpha = D_1^{\alpha_1} \cdots D_n^{\alpha_n}$, where $D_j = \partial/\partial x_j$, and let

$$\|u\|_{m,p} = \|u\|_{m,p}^\Omega = \left\{ \sum_{|\alpha| \le m} \int_\Omega |D^\alpha u|^p \, dx \right\}^{1/p} \qquad (1 \le p < \infty), \qquad (3.3.5)$$

where the integral is taken in the sense of Lebesgue, and the summation is taken over all the indices $\alpha = (\alpha_1,\ldots,\alpha_n)$ with $|\alpha| = \alpha_1 + \cdots + \alpha_n \le m$. We also write $\|u\|_p = \|u\|_p^\Omega = \|u\|_{0,p}^\Omega$; the latter is defined for any measurable set Ω in R^n and for any u measurable on Ω.

PROBLEMS

3.3.1. Let A be a compact subset of R^n and let B be an open set containing A. Prove that there exists a C^∞ function h on R^n such that $h = 1$ on A, $h = 0$ outside B, and $0 \le h \le 1$ in $B - A$. [*Hint:* Let

$$\rho(x) = \begin{cases} 0, & \text{if } |x| \ge 1, \\ c \exp\left\{\dfrac{1}{|x|^2 - 1}\right\}, & \text{if } |x| < 1, \end{cases} \tag{3.3.6}$$

where c is a constant such that $\int_{R^n} \rho(x)\,dx = 1$, and $|x| = (\sum_{i=1}^{n} x_i^2)^{1/2}$. Take

$$h(x) = \varepsilon^{-n} \int_G \rho\left(\frac{x-y}{\varepsilon}\right) dy,$$

where G is a bounded open set, $A \subset G \subset \bar{G} \subset B$, and $\varepsilon > 0$ sufficiently small. Use Problems 2.16.9, 2.16.10.]

3.3.2. We denote by $L^p(\Omega)$ $(1 \le p < \infty)$ the space $L^p(X,\mu)$, where (X,μ) is the Lebesgue measure space corresponding to an open set Ω in R^n. Let $u \in L^p(\Omega)$, and define

$$(J_\varepsilon u)(x) = \varepsilon^{-n} \int_\Omega \rho\left(\frac{x-y}{\varepsilon}\right) u(y)\,dy \qquad (\varepsilon > 0). \tag{3.3.7}$$

We call $J_\varepsilon u$ a *mollifier* of u. Prove that

(i) $J_\varepsilon u$ is in $C^\infty(R^n)$.

(ii) If u vanishes outside a subset A of Ω, then $J_\varepsilon u$ vanishes outside an ε-neighborhood of A [that is, outside the set $\{x; \rho(x,A) < \varepsilon\}$].

(iii) If K is a closed subset of Ω with $\rho(K, R^n - \Omega) \ge \delta > 0$, then

$$(J_\varepsilon u)(x) = \varepsilon^{-n} \int_{|y-x|<\varepsilon} \rho\left(\frac{x-y}{\varepsilon}\right) u(y)\,dy = \int_{|z|<1} \rho(z) u(x - \varepsilon z)\,dz$$

for any $x \in K$, provided $\varepsilon < \delta$.

3.3.3. If $u \in L^p(\Omega)$ and K, ε are as in Problem 3.3.2, then

$$\|J_\varepsilon u\|_{L^p(K)} \le \|u\|_{L^p(\Omega)}.$$

Here we have used the notation: $\|w\|_{L^p(G)} = \|w\|_p^G$. [*Hint*: Use Hölder's inequality.]

3.3.4. If $u \in L^p(\Omega)$ and K, ε are as in Problem 3.3.2, then

$$\|J_\varepsilon u - u\|_{L^p(K)} \to 0 \qquad \text{as } \varepsilon \to 0.$$

[*Hint:* Prove it first for continuous functions. Then approximate u by continuous functions, using Problem 3.2.3.]

3.3.5. $C_0^\infty(\Omega)$ is dense in $L^p(\Omega)$.

3.3.6. Let $f \in L^p(\Omega)$. If $\int_\Omega f\varphi\, dx = 0$ for all $\varphi \in C_0^\infty(\Omega)$, then $f = 0$ a.e. [*Hint:* Prove that $\int_\Omega f\chi_E\, dx = 0$, E any compact subset of Ω.]

Denote by $\hat{C}^m(\Omega)$ the subset of $C^m(\Omega)$ consisting of those functions u for which $\|u\|_{m,p} < \infty$. Then $\hat{C}^m(\Omega)$ is a metric space with the metric

$$\rho(u,v) = \|u - v\|_{m,p}. \tag{3.3.8}$$

Note that the triangle inequality follows from Problem 3.2.8.

Let $\{u_m\}$ be a Cauchy sequence in $\hat{C}^m(\Omega)$ with respect to the metric (3.3.8)—that is,

$$\sum_{|\alpha| \le m} \int_\Omega |D^\alpha u_m - D^\alpha u_k|^p\, dx \to 0 \qquad \text{as } m,k \to \infty. \tag{3.3.9}$$

By Theorem 3.2.3 there exist functions u^α $(0 \le |\alpha| \le m)$ in $L^p(\Omega)$ such that

$$\sum_{|\alpha| \le m} \int_\Omega |D^\alpha u_m - u^\alpha|^p dx \to 0 \qquad \text{as } m \to \infty. \tag{3.3.10}$$

Any two equivalent Cauchy sequences give the same u^α [as elements of $L^p(\Omega)$]. Thus, each class of equivalent Cauchy sequences gives a unique vector $\{u^\alpha;\ 0 \le |\alpha| \le m\}$ with components in $L^p(\Omega)$.

Lemma 3.3.2 *For any* $\varphi \in C^\infty(\Omega)$,

$$\int_\Omega u^\alpha \varphi\, dx = (-1)^{|\alpha|} \int u^0 D^\alpha \varphi\, dx. \tag{3.3.11}$$

Proof. Integration by parts gives

$$\int_\Omega D^\alpha u_m \cdot \varphi\, dx = (-1)^{|\alpha|} \int u_m D^\alpha \varphi\, dx.$$

Now take $m \to \infty$.

Corollary 3.3.3 *If* $u^0 = 0$, *then* $u^\alpha = 0$ *for all* $0 \le |\alpha| \le m$.

Indeed, if $u^0 = 0$, then, by (3.3.11), $\int_\Omega u^\alpha \varphi\, dx = 0$ for all $\varphi \in C_0^\infty(\Omega)$. Now use Problem 3.3.6.

The corollary shows that the u^α for $0 < |\alpha| \le m$ are determined uniquely by u^0.

We denote the completion of $\hat{C}^m(\Omega)$ with respect to the metric (3.3.8) by $H^{m,p}(\Omega)$. We can then state:

Theorem 3.3.4 *The elements of* $H^{m,p}(\Omega)$ *can be identified with functions of* $L^p(\Omega)$. *A function* u *in* $L^p(\Omega)$ *belongs to* $H^{m,p}(\Omega)$ *if and only if there exists* *a Cauchy sequence* $\{u_m\}$ *in the metric* (3.3.8) *of functions that belong to* $\hat{C}^m(\Omega)$, *such that* $\int_\Omega |u_m - u|^p\, dx \to 0$ *as* $m \to \infty$.

PROBLEMS

3.3.7. If $u \in H^{m,p}(\Omega)$ and $v \in C_0^m(\Omega)$, then uv belongs to $H^{m,p}(\Omega)$.

3.3.8. The components u^α occurring in (3.3.10) are called the αth *strong derivatives* of u. Prove that if $v \in C^m(\Omega)$ and $u \in H^{m,p}(\Omega)$, then, for any open set Ω_0 with $\overline{\Omega}_0 \subset \Omega$,

$$D^\alpha(uv) = \sum_{\beta \le \alpha} \binom{\alpha}{\beta} D^\beta u \cdot D^{\alpha-\beta} v \qquad \text{as elements of } L^p(\Omega_0),$$

where $D^\alpha(uv)$, $D^\beta u$ are taken as strong derivatives, and $\beta \le \alpha$ means that $\beta_i \le \alpha_i$ for all i.

3.4 COMPLETE METRIC SPACES

Theorem 3.4.1 *Let* $\{F_n\}$ *be a monotone-decreasing sequence of nonempty closed sets, in a complete metric space* (X,ρ). *If the sequence of the diameters* $d(F_n)$ *converges to zero, then there exists one and only one point that belongs to* $\bigcap_{n=1}^{\infty} F_n$.

This theorem is an extension to metric spaces of the nested sequence theorem on the real line, with F_n being closed intervals.

Proof. There cannot be two distinct points x and y that belong to $\bigcap_n F_n$, for otherwise $\rho(x,y) \le d(F_n) \to 0$ as $n \to \infty$—that is, $\rho(x,y) = 0$. To prove that $\bigcap_n F_n$ is nonempty, choose in each F_m a point x_n. Then $\rho(x_n, x_m) \le d(F_m) \to 0$ *if* $n \ge m \to \infty$. Thus $\{x_n\}$ is a Cauchy sequence. Denote by x its limit. Since each set F_n is a closed set that contains all the x_m with $m \ge n$, we have $x \in F_n$. Hence $x \in \bigcap_n F_n$.

Definition A metric space is said to be *of the first category* if it can be written as a countable union of sets that are nowhere dense. If a metric space is not of the first category, then we say it is *of the second category.*

The space of rational numbers on the real line with the Euclidean metric is clearly a space of the first category. On the other hand, the real line is a space of the second category. In fact, the last assertion is a special case of the following general theorem, known as the *Baire category theorem.*

***Theorem* 3.4.2** *A complete metric space is a space of the second category.*

Proof. Suppose $X = \bigcup_{j=1}^{\infty} X_j$, where X_j are nowhere dense—that is, the sets $\bar{X}_j$ have no interior points. Fix a ball $B(x_0,1)$. Since $\bar{X}_1$ does not contain $B(x_0,1)$, there is a point x_1 in $B(x_0,1)$ such that $x_1 \notin \bar{X}_1$. But then there is also a ball $B(x_1,r_1)$ such that $\bar{B}(x_1,r_1) \subset B(x_0,1)$ and $\bar{B}(x_1,r_1) \cap \bar{X}_1 = \varnothing$. We may assume that $r_1 < \frac{1}{2}$. Similarly, there is a point x_2 such that $\bar{B}(x_2,r_2) \subset B(x_1,r_1)$ and $\bar{B}(x_2,r_2) \cap \bar{X}_2 = \varnothing$. We may assume that $r_2 < \frac{1}{3}$. Proceeding similarly step by step, we get a sequence of decreasing closed balls $\bar{B}(x_n,r_n)$ such that $\bar{B}(x_n,r_n) \cap \bar{X}_j = \varnothing$ if $1 \le j \le n$, and $r_n \to 0$. By Theorem 3.4.1, there is a point x that belongs to all the closed balls $\bar{B}(x_n,r_n)$. But then $x \notin \bar{X}_j$ for all $j \ge 1$. Hence $x \notin \bigcup_j X_j = X$, a contradiction.

Let Ω be a bounded open set of R^n. We denote by $C^m(\bar{\Omega})$ the spaces of functions $u(x)$ all of whose first m derivatives are uniformly continuous in Ω (and thus can be extended, by continuity, to $\bar{\Omega}$). If $u \in C^m(\bar{\Omega})$ then the number

$$|u|_{m,\Omega} = |u|_m = \sum_{|\alpha| \le m} \sup_{\Omega} |D^\alpha u| \tag{3.4.1}$$

is finite.

PROBLEMS

3.4.1. Prove that $C^m(\bar{\Omega})$ is a complete metric space with the metric

$$\rho(u,v) = |u - v|_m. \tag{3.4.2}$$

When $m = 0$, we call this metric the *maximum metric*, or the *uniform metric.*

3.4.2. Prove that the metrics (3.3.8) and (3.4.2) on $C^m(\bar{\Omega})$ are not equivalent.

3.4.3. Give an example of a sequence of monotone-decreasing closed sets F_n in a complete metric space such that $d(F_n) = \infty$ and $\bigcap_n F_n = \varnothing$.

3.4.4. Give an example of a sequence of monotone-decreasing closed sets F_n in a complete metric space such that $d(F_1) < \infty$, $\lim_n d(F_n) > 0$, and $\bigcap_n F_n = \varnothing$. [*Hint:* Let $\{y_j\}$ be any bounded sequence with no points of accumulation (compare Problem 3.1.7). Write it as a double sequence $\{x_{mn}\}$ and take $F_n = \{x_{mj}; 1 \le m < \infty, n \le j < \infty\}$.]

3.4.5. A set Y in a metric space X is said to be *of the first category in* X if it is contained in a countable union of nowhere dense sets of X. If Y is not of the first category in X, then it is said to be *of the second category in* X. The real line with the Euclidean metric is a space of the second category. Prove, however, that, as a subset of the Euclidean plane, the real line is a set of the first category.

3.4.6. A countable union of subsets (of X) of the first category in X is again a set of the first category in X.

3.4.7. Let $f(x)$ be a real-valued function on the real line. Prove that there is a nonempty interval (a,b) and a positive number c such that for any $x \in (a,b)$ there is a sequence $\{x_n\}$ such that $x_n \to x$ and $|f(x_n)| \le c$.

3.5 COMPACT METRIC SPACES

A subset Y of a subset Z of a space X is called *dense in* Z if $\overline{Y} \supset Z$. If Z has a countable subset that is dense (in Z) then we say that Z is *separable*.

Theorem 3.5.1 *A sequentially compact subset of a metric space is a closed set.*

Proof. Let K be a sequentially compact subset of a metric space X. We have to show that if $\{x_n\}$ is any sequence in K that converges to a point y in X, then $y \in K$. By assumption, $\{x_n\}$ has a subsequence $\{x_{n_k}\}$ that converges to a point x in K. Since, clearly, $y = x$, we conclude that $y \in K$.

Theorem 3.5.2 *A sequentially compact subset of a metric space is bounded.*

Proof. If a sequentially compact subset K of a metric space X is not bounded, then it contains a sequence $\{x_n\}$ such that $\rho(x_n,z) \to \infty$ as $n \to \infty$, where z is any given fixed point. But then, for any point y and for any subsequence $\{x_{n_k}\}$ of $\{x_n\}$, $\rho(x_{n_k},y) \to \infty$ as $k \to \infty$. Thus $\{x_n\}$ does not have a convergent subsequence.

Theorem 3.5.3 *A sequentially compact subset of a metric space is separable.*

Proof. Let K be a compact subset of a metric space X. Take any point x_0 and let $d_0 = \sup_{x \in K} \rho(x_0,x)$. By Theorem 3.5.2, $d_0 < \infty$. Now choose

x_1 in K such that $\rho(x_0,x_1) \geq d_0/2$. Proceeding by induction, we choose x_{n+1} in K such that $\min_{1 \leq j \leq n} \rho(x_j,x_{n+1}) \geq d_n/2$, where $d_n = \sup_{x \in K} \min_{1 \leq j \leq n} \rho(x_j,x)$. Clearly $d_0 \geq d_1 \geq \cdots$. If $d_n \geq \delta > 0$ for all n, then no subsequence of $\{x_n\}$ is a Cauchy sequence. This, however, contradicts the assumption that K is sequentially compact. Hence $d_n \to 0$ as $n \to \infty$. But this implies that $\{x_n\}$ is dense in K—that is, K is separable.

Theorem 3.5.4 *A subset of a metric space is compact if and only if it is sequentially compact.*

Proof. Suppose K is a compact subset of a metric space X. If it is not sequentially compact, then there exists a sequence $\{x_n\}$ in K having no points of accumulation in K. This sequence contains an infinite subsequence of mutually distinct elements. Denote this subsequence by $\{y_n\}$. Since each point y_m is not a limit point of the sequence $\{y_n\}$, and since $y_n \neq y_m$ if $n \neq m$, there is a ball $B(y_m,r_m)$ such that $B(y_m,r_m) \cap K$ does not contain any of the points y_n with $n \neq m$. The sequence $\{B(y_m,r_m) \cap K\}$, together with the set $K - \{y_n\}$, form an open covering of K. Since no finite subclass of these open sets can cover K, we have obtained a contradiction to the assumption that K is compact.

Suppose conversely that K is sequentially compact. We shall show first that any countable open (in X) covering of K has a finite subclass of open sets that cover K. Let $\{E_j\}$ be the open sets of the covering. Suppose that for any n, $\bigcup_{j=1}^{n} E_j$ does not cover K. Then there exists a sequence $\{x_n\}$ in K such that $x_n \notin \bigcup_{j=1}^{n} E_j$. Let $\{x_{n_k}\}$ be a subsequence of $\{x_n\}$ that converges to a point y in K. Since $X - \bigcup_{j=1}^{n} E_j$ is a closed set containing all the points x_m with $m \geq n$, y belongs to this set. But then $y \in (X - \bigcup_{j=1}^{\infty} E_j)$—that is, $y \notin \bigcup_{j=1}^{\infty} E_j$. Thus $y \notin K$, which is impossible.

Now take any covering of K by a class $\{E_\alpha\}$ of open sets in X. By Theorem 3.5.3, there is a dense sequence $\{y_m\}$ in K. The sequence of balls

$$B\left(y_m, \frac{1}{n}\right), \qquad \text{where } m,n = 1,2,\ldots, \tag{3.5.1}$$

then has the following property: for any open set G of X there is a subclass of the balls from (3.5.1), say $\{B(y_{m'},1/n')\}$ such that $G \cap K$ is the union of the sets $K \cap B(y_{m'},1/n')$. Indeed, for any $x \in G \cap K$ there is a ball $B(x,r)$ in G. Take $y_{\bar{m}}$ such that $\rho(y_{\bar{m}},x) < 1/\bar{n}$, where $2/\bar{n} < r$, $\bar{n}$ integer. Then $x \in B(y_{\bar{m}},1/\bar{n}) \subset G$. Now denote by $B(y_{m'},1/n')$ the distinct balls $B(y_{\bar{m}},1/\bar{n})$ obtained when x varies in $G \cap K$.

Let $B_1, B_2, \ldots$ be the sequence of all the balls $B(y_{m'}, 1/n')$ that occur in the above way when G varies over the sets E_α. Each B_j is contained in some set E_{α_j}. The sets $\{E_{\alpha_j}\}$ then form an open covering on K. We can now apply the result of the last paragraph.

Combining Theorems 3.5.1–3.5.4, we get:

Corollary 3.5.5 *A compact subset of a metric space is closed, bounded, and separable.*

Definition A subset K of a metric space X is called *totally bounded* if for any $\varepsilon > 0$ there is a finite number of balls $B(x_i, \varepsilon)$, with $x_i \in X$, that cover K. These balls are said to form an *ε-covering*.

Theorem 3.5.6 *Let* X *be a complete metric space. A closed subset* K *is compact if and only if it is totally bounded.*

Proof. Suppose K is compact. For any $\varepsilon > 0$, the class of balls $\{B(x,\varepsilon); x \in K\}$ forms an open covering of K. Hence there is a finite subclass, say $\{B(x_i,\varepsilon); i = 1,\ldots,m\}$, that covers K. Thus K is totally bounded.

Conversely, let K be closed and totally bounded. We shall prove that K is sequentially compact. Let $\{y_n\}$ be a sequence of points in K. Take a finite 1-covering of K: $\{B(x_1,1),\ldots,B(x_m,1)\}$. At least one of the balls $B(x_i,1)$ contains an infinite subsequence $\{y_{n,1}\}$ of $\{y_n\}$. Next, take a $\frac{1}{2}$-covering of K by a finite number of balls, and extract a subsequence $\{y_{n,2}\}$ from $\{y_{n,1}\}$ that is completely contained in one of the balls of radius $\frac{1}{2}$. Proceeding in this way step by step, we extract in the mth step a subsequence $\{y_{m,n}\}$, of $\{y_{m,n-1}\}$, which is contained in a ball of radius $1/n$. The diagonal sequence $\{y_{m,m}\}$ is then a Cauchy sequence. Since X is complete, $\lim_m y_{m,m}$ exists in X. This limit also belongs to K, since K is closed. Thus K is sequentially compact.

PROBLEMS

3.5.1. A subset Y of a metric space X is separable if there exists a sequence of points in X whose closure contains Y.

3.5.2. A subset K of a metric space X is totally bounded if for any $\varepsilon > 0$ there is a finite covering of K by open balls with centers in X.

3.5.3. Let Y, Z, U be subsets of a metric space X. If Y is dense in Z and Z is dense in U, then Y is dense in U.

3.5.4. A subset F of a compact metric space is compact if and only if it is closed.

3.5.5. A subset Y of a metric space is totally bounded if and only if its closure $\overline{Y}$ is totally bounded.

3.5.6. The intersection of any number of compact subsets of a metric space is a compact set.

3.5.7. A subset of a separable metric space is a separable space.

3.5.8. Show that a metric space is compact if and only if it has the following property: for every collection of closed subsets $\{F_\alpha\}$, if any finite subcollection has a nonempty intersection, then the whole collection has a nonempty intersection.

3.5.9. Prove that a closed set K of points $x = (x_1,x_2,\ldots)$ in l^p $(1 \le p < \infty)$ is compact if (i) $\sum_{n=1}^{\infty} |x_n|^p \le C$ for all $x \in K$, where C is a constant, and (ii) for any $\varepsilon > 0$ there is an n_0 such that $\sum_{n=n_0}^{\infty} |x_n|^p \le \varepsilon$ for all $x \in K$.

3.5.10. The Cartesian product of a countable number of compact metric spaces is a compact metric space.

3.6 CONTINUOUS FUNCTIONS ON METRIC SPACES

A function f (real- or complex-valued) defined on a subset Y of a metric space is said to be *continuous at a point* y_0 of Y if for any $\varepsilon > 0$ there is a $\delta > 0$ such that $|f(x) - f(y_0)| < \varepsilon$ for all $x \in Y \cap B(y_0,\delta)$. If f is continuous at each point of Y, then we say that f is *continuous on* Y. This definition is equivalent to the one given in 2.1. A function f is said to be *uniformly continuous* on Y if for any $\varepsilon > 0$ there is a $\delta > 0$ such that $|f(y) - f(x)| < \varepsilon$ whenever y and x belong to Y and $\rho(y,x) < \delta$.

One easily verifies that f is continuous at y_0 if and only if for any sequence $\{x_n\}$ in Y such that $\rho(x_n,y_0) \to 0$ as $n \to \infty$, $f(x_n) \to f(y_0)$ as $n \to \infty$.

Theorem 3.6.1 *If* f *is a continuous function on a compact subset* Y *of a metric space* X, *then* f *is uniformly continuous on* Y.

Proof. If the assertion is false, then there is an $\varepsilon > 0$ and sequences $\{y_n\}$, $\{x_n\}$ in Y such that

$$|f(y_n) - f(x_n)| \ge \varepsilon \tag{3.6.1}$$

and $\rho(y_n,x_n) \to 0$. There exist subsequences $\{y_{n_k}\}$ and $\{x_{n_k}\}$ that converge to points y and x, respectively, belonging to Y. Since

$$\rho(y,x) \le \rho(y,y_{n_k}) + \rho(y_{n_k},x_{n_k}) + \rho(x_{n_k},x) \to 0 \qquad \text{as } k \to \infty,$$

we have $\rho(y,x) = 0$—that is, $y = x$. Hence

$$|f(y_{n_k}) - f(x_{n_k})| \le |f(y_{n_k}) - f(y)| + |f(x) - f(x_{n_k})| \to 0$$

as $k \to \infty$, which contradicts (3.6.1).

Theorem 3.6.2 *A real-valued continuous function on a compact subset* Y *of a metric space* X *is bounded and has both maximum and minimum.*

Proof. Let $M = \sup_Y f$. Then there exists a sequence $\{y_n\}$ in Y such that $f(y_n) \to M$. Let $\{y_{n_k}\}$ be a convergent subsequence of $\{y_n\}$, and denote its limit by y. Then $f(y_{n_k}) \to f(y)$ as $n \to \infty$. This shows that $M < \infty$ and that $f(y) = M$. Thus f has a maximum on Y. The proof that f has a (finite) minimum is similar.

The following result is usually called *the Tietze extension theorem.*

Theorem 3.6.3 *Let* f *be a bounded, real-valued, continuous function on a closed set* Y *of a metric space* X. *Then there is a continuous function* F *defined on* X *with* F(x) = f(x) *for* x ∈ Y, *and*

$$\inf_X F = \inf_Y f, \qquad \sup_X F = \sup_Y f. \tag{3.6.2}$$

Proof. Suppose first that $f \ge 0$. Consider the function

$$M_x(r) = \sup_{Y \cap B(x,r)} f.$$

For each $x \in X$, $M_x(r)$ is bounded and monotone increasing in r. Hence it is integrable on finite r-intervals. We define F by $F(x) = f(x)$ if $x \in Y$, and

$$F(x) = \frac{1}{\delta(x)} \int_{\delta(x)}^{2\delta(x)} M_x(r)\,dr \tag{3.6.3}$$

if $x \notin Y$, where $\delta(x) = \rho(x, Y)$. Note that $\delta(x) > 0$ if $x \notin Y$, since Y is closed. We first prove that F is continuous at any point $y \in Y$. Since, for $x \notin Y$,

$$\min_{Y \cap B(y,3\delta)} f \le F(x) \le \max_{Y \cap B(y,3\delta)} f, \qquad \text{where } \delta = \rho(x,y),$$

and since $3\delta \to 0$ as $x \to y$ $(x \notin Y)$, it follows that, for any $\varepsilon > 0$, $|F(x) - f(y)| < \varepsilon$ if $x \notin Y$ and $\delta(x) \le \delta_0$, where δ_0 is sufficiently small. On the other hand, $|F(x) - f(y)| = |f(x) - f(y)| < \varepsilon$ if $x \in Y$ and $\rho(x,y) \le \delta_1$, where δ_1 is sufficiently small. Thus $F(x)$ is continuous at y.

Consider next the continuity of F at a point $z \notin Y$. Let x be any point in X with $\rho(x,z) < \frac{1}{3}\rho(z,Y)$. Let $h = \rho(x,z)$. Then $2h < \rho(x,Y)$. Since

$|\delta(x) - \delta(z)| \le h$, $M_z(r) \ge M_x(r-h)$, and $M_z(r) \ge 0$, we have

$$
\begin{aligned}
&F(x) - F(z) \\
&\quad = \frac{1}{\delta(x)} \int_{\delta(x)}^{2\delta(x)} M_x(r)\,dr - \frac{1}{\delta(z)} \int_{\delta(z)}^{2\delta(z)} M_z(r)\,dr \\
&\quad \le \frac{1}{\delta(x)} \int_{\delta(x)}^{2\delta(x)} M_x(r)\,dr - \frac{1}{\delta(x)+h} \int_{\delta(x)+h}^{2\delta(x)-2h} M_x(r-h)\,dr \\
&\quad = \frac{1}{\delta(x)} \int_{\delta(x)}^{2\delta(x)-3h} M_x(r)\,dr + \frac{1}{\delta(x)} \int_{2\delta(x)-3h}^{2\delta(x)} M_x(r)\,dr \\
&\qquad - \frac{1}{\delta(x)+h} \int_{\delta(x)}^{2\delta(x)-3h} M_x(s)\,ds \\
&\quad = \frac{h}{\delta(x)[\delta(x)+h]} \int_{\delta(x)}^{2\delta(x)-3h} M_x(r)\,dr + \frac{1}{\delta(x)} \int_{2\delta(x)-3h}^{2\delta(x)} M_x(r)\,dr \le \frac{4hM}{\delta(x)},
\end{aligned}
$$

where $M = \sup_Y f$. A similar inequality holds with x and z interchanged. Hence $F(x) \to F(z)$ as $\rho(x,z) \to 0$.

It is now easy to complete the proof of the theorem. Let $c = \inf_Y f$ and consider the function $\tilde{f} = f - c$. Then $0 \le \tilde{f} \le C - c$, where $C = \sup_Y f$. The continuous extension $\tilde{F}$ given by (3.6.3), with f replaced by $\tilde{f}$, satisfies $0 \le \tilde{F}(x) \le C - c$. Hence the function $F = \tilde{F} + c$ is a continuous extension of f into X and it satisfies $c \le F(x) \le C$—that is, it satisfies (3.6.2).

Definition A family $\{f_\alpha\}$ of functions defined on a set Z is said to be *uniformly bounded* if there is a constant C such that $|f_\alpha(x)| \le C$ for all α and for all $x \in Z$.

Definition A family $\{f_\alpha\}$ of continuous functions defined on a subset Z of a metric space (X,ρ) is said to be *equicontinuous* if for any $\varepsilon > 0$ there exists a $\delta > 0$ such that

$$|f_\alpha(x) - f_\alpha(y)| < \varepsilon \qquad \text{if } \rho(x,y) < \delta,$$

for all x,y in Z and for all α.

We can now state the following result known as the lemma of *Ascoli-Arzela*.

Theorem 3.6.4 *Let $\mathscr{K}$ be a family of functions, uniformly bounded and equicontinuous on a compact metric space* X. *Then any sequence* $\{f_n\}$ *of functions of $\mathscr{K}$ has a subsequence that is uniformly convergent in* X *to a continuous function.*

Proof. Let $\{x_m\}$ be a dense sequence in X. Since the sequence $\{f_n(x_1)\}$ is bounded, there is a subsequence $\{f_{n,1}(x_1)\}$ that is convergent. Next, the sequence $\{f_{n,1}(x_2)\}$ is bounded. Hence it has a subsequence $\{f_{n,2}(x_2)\}$ that is convergent. We proceed in this way step by step. In the kth step, we extract a convergent subsequence $\{f_{n,k}(x_k)\}$ of the bounded sequence $\{f_{n,k-1}(x_k)\}$. Consider now the diagonal sequence $\{f_{n,n}\}$ of the double sequence $\{f_{n,k}\}$, and write $g_n = f_{n,n}$. Then $\{g_n(x_k)\}$ is convergent for every k, since, except for the first k terms, it is a subsequence of $\{f_{n,k}(x_k)\}$.

We shall prove that $\{g_n\}$ is uniformly convergent in X. Since the family $\{g_n\}$ is equicontinuous, for any $\varepsilon > 0$ there is a $\delta > 0$ such that

$$|g_n(x) - g_n(y)| < \varepsilon \qquad \text{whenever } \rho(x,y) < \delta,$$

for all x,y in X and $1 \leq n < \infty$. For any $x \in X$ we now write

$$|g_n(x) - g_m(x)| \leq |g_n(x) - g_n(x_k)| + |g_n(x_k) - g_m(x_k)| + |g_m(x_k) - g_m(x)|,$$

where x_k is such that $\rho(x,x_k) < \delta$. Then

$$|g_n(x) - g_m(x)| < 2\varepsilon + |g_n(x_k) - g_m(x_k)|. \tag{3.6.4}$$

We now claim that there is a finite number of the points x_k, say $x_1,\ldots,x_h$, such that for any $x \in X$ there is a point x_k with $1 \leq k \leq h$ such that $\rho(x,x_k) < \delta$. Indeed, take a finite $(\delta/2)$-covering of X by balls $B_1,\ldots,B_p$ and choose in each ball B_j a point x_{α_j} from the sequence $\{x_m\}$. Then the claim above holds with $h = \max(\alpha_1,\ldots,\alpha_p)$.

For each k, $1 \leq k \leq h$, there is a positive integer n_k such that

$$|g_m(x_k) - g_n(x_k)| < \varepsilon \qquad \text{if } m \geq n \geq n_k.$$

Using this in (3.6.4), we get $|g_n(x) - g_m(x)| < 3\varepsilon$ if $m \geq n \geq \bar{n}$, where $\bar{n} = \max(n_1,\ldots,n_h)$. Thus $\{g_n\}$ is uniformly convergent. Denote by $f(x)$ the uniform limit of $\{g_n(x)\}$. Then, for any $\varepsilon > 0$,

$$\begin{aligned}|f(x) - f(y)| &\leq |f(x) - g_n(x)| + |g_n(x) - g_n(y)| + |g_n(y) - f(y)| \\ &< 2\varepsilon + |g_n(x) - g_n(y)|\end{aligned}$$

if n is sufficiently large. We now fix n. The uniform continuity of g_n then implies that $|g_n(x) - g_n(y)| < \varepsilon$ if $\rho(x,y) < \delta$. Hence $|f(x) - f(y)| < 3\varepsilon$ if $\rho(x,y) < \delta$. Thus, $f(x)$ is continuous, and the proof of the theorem is complete.

For any compact metric space X, we denote by $C(X)$ the space of all functions $f(x)$ that are continuous on X. We introduce in this space the *uniform metric*

$$\rho(f,g) = \max_{x \in X} |f(x) - g(x)|. \tag{3.6.5}$$

$C(X)$ is then a metric space. It is easily seen that $C(X)$ is also complete.

Definition A subset K of a metric space is called *relatively compact* (or *conditionally compact*) if its closure $\overline{K}$ is compact.

We can now state a criterion for relative compactness of subsets of $C(X)$.

Theorem 3.6.5 *Let X be a compact metric space. A family $\mathscr{K}$ of functions of C(X) is relatively compact in C(X) if and only if $\mathscr{K}$ is both uniformly bounded and equicontinuous.*

Proof. The lemma of Ascoli-Arzela shows that if $\mathscr{K}$ is uniformly bounded and equicontinuous, then any sequence of functions $\{f_n\}$ in $\mathscr{K}$ has a subsequence that converges in the metric of $C(X)$ to an element of $C(X)$. From this it easily follows that the closure $\overline{\mathscr{K}}$ of $\mathscr{K}$ is sequentially compact and, hence, compact. Thus $\mathscr{K}$ is relatively compact.

Suppose conversely that $\mathscr{K}$ is relatively compact. By Theorem 3.5.2, $\mathscr{K}$ is a bounded set in $C(X)$. But this is equivalent to $\mathscr{K}$ being a uniformly bounded family of functions. To prove that the family $\mathscr{K}$ is equicontinuous, take an ε-covering of $\mathscr{K}$ by a finite number of balls whose centers are functions $f_1,\ldots,f_h$. Then, for any $f \in \mathscr{K}$ there is an f_j, with $1 \le j \le h$, such that $\rho(f,f_j) < \varepsilon$. That means that

$$|f(x) - f_j(x)| < \varepsilon \qquad \text{for all } x \in X. \tag{3.6.6}$$

Since each of the functions $f_1,\ldots,f_h$ is uniformly continuous (by Theorem 3.6.1), there is a $\delta > 0$ such that

$$|f_j(y) - f_j(z)| < \varepsilon \qquad \text{if } \rho(y,z) < \delta \tag{3.6.7}$$

for all y,z in X and for all $1 \le j \le h$. Combining (3.6.6), (3.6.7), we get

$$|f(y) - f(z)| < 3\varepsilon \qquad \text{if } \rho(y,z) < \delta.$$

This proves the equicontinuity of the family $\mathscr{K}$.

PROBLEMS

3.6.1. Let Ω be an open set in R^n, and let $\{\Omega_m\}$ be an increasing sequence of bounded open sets such that $\overline{\Omega}_m \subset \Omega_{m+1}, \Omega = \bigcup_{m=1}^{\infty} \Omega_m$. Prove that $C(\Omega)$ is a complete metric space with the metric

$$\rho(f,g) = \sum_{m=1}^{\infty} \frac{1}{2^m} \frac{\max_{\overline{\Omega}_m} |f - g|}{1 + \max_{\overline{\Omega}_m} |f - g|}. \tag{3.6.8}$$

Prove also that $\rho(f_m, f) \to 0$ if and only if $\{f_m\}$ converges to f uniformly on compact subsets of Ω.

3.6.2. Let Ω be an open set in the complex plane. Denote by $H(\Omega)$ the space of all the complex analytic functions that are holomorphic in Ω. Let $\{\Omega_m\}$ be as in Problem 3.6.1, and define $\rho(f,g)$, for f and g in $H(\Omega)$, by (3.6.8). Prove that $H(\Omega)$ is a complete metric space.

3.6.3. A family $\mathscr{F}$ of complex analytic functions holomorphic in Ω that are uniformly bounded on every compact subset of Ω is a relatively compact subset of $H(\Omega)$. [*Hint:* Use Cauchy's formula and the lemma of Ascoli-Arzela.]

3.6.4. A function f that is upper semicontinuous (for definition, see Problem 2.1.11) on a compact metric space is bounded from above and has a maximum.

3.6.5. Let Z be a set of functions u in $L^p(\Omega)$ $(1 \le p < \infty)$, where Ω is a bounded open set in R^n, and extend each u into $R^n - \Omega$ by 0. Prove that Z is relatively compact in $L^p(\Omega)$ if the following conditions are satisfied: (i) $\rho(J_\varepsilon u, u) \equiv \|J_\varepsilon u - u\|_{L^p(R^n)} \to 0$ as $\varepsilon \to 0$, uniformly with respect to $u \in Z$, and (ii) $\|u\|_{L^p(\Omega)} \le C$ for all $u \in Z$. Here $J_\varepsilon u$ is the function defined in (3.3.7). [*Hint:* It suffices to show that Z has a finite δ-covering for any $\delta > 0$. Take $\varepsilon_0 > 0$ such that $\|J_{\varepsilon_0} u - u\|_{L^p(R^n)} < \delta/2$ for all $u \in Z$. It suffices to find a finite $(\delta/2)$-covering of $W = \{J_{\varepsilon_0} u; u \in Z\}$. Show that W is uniformly bounded and equicontinuous and use Theorems 3.5.4, 3.5.6, and 3.6.4.]

3.6.6. Let Z be a set of functions u in $L^p(\Omega)$, $1 \le p < \infty$, where Ω is a bounded open set in R^n, and extend each u into $R^n - \Omega$ by 0. Prove that if Z is relatively compact, then the conditions (i) and (ii) of Problem 3.6.5 are satisfied.

3.6.7. In every infinite metric space there is an infinite sequence such that no point of accumulation of the sequence is a point of the sequence.

3.6.8. If a compact metric space X is such that bounded closed subsets of $C(X)$ are compact, then X consists of a finite number of points. [*Hint:* Use Problem 3.6.7 and the Tietze extension theorem.]

3.6.9. If F is a Lebesgue-measurable bounded set in R^n, then for any $\varepsilon > 0$ there exists a closed set $G \subset F$ such that $\mu(G) > \mu(F) - \varepsilon$. [*Hint:* Use Problem 1.9.8.]

3.6.10. Prove *Lusin's theorem:* let $f(x)$ be an a.e. real-valued Lebesgue-measurable function on a Lebesgue-measurable and bounded set $E \subset R^n$. Then, for any $\delta > 0$, there is a continuous function $\varphi(x)$ in R^n such that $\mu\{x; f(x) \neq \varphi(x)\} < \delta$. Furthermore, if $|f(x)| \leq C$ a.e., then $|\varphi(x)| \leq C$. [*Hint:* Use Problem 2.15.11, Egoroff's theorem, Problem 3.6.9, and the Tietze extension theorem.]

3.7 THE STONE-WEIERSTRASS THEOREM

Let X be a compact metric space. In the real metric space $C(X)$ we introduce the following operations for any f,g in $C(X)$ and λ a real number:

$$(f+g)(x) = f(x) + g(x),$$
$$(fg)(x) = f(x)g(x),$$
$$(\lambda f)(x) = \lambda f(x).$$

A subset $\mathfrak{A}$ of $C(X)$ is called an *algebra* if for any f,g in $\mathfrak{A}$ and λ real we have: $f+g \in \mathfrak{A}$, $fg \in \mathfrak{A}$ and $\lambda f \in \mathfrak{A}$.

One easily verifies that the intersection of any number of algebras is again an algebra. In particular, the intersection of all the algebras containing a given set $\mathscr{B}$ of $C(X)$ is an algebra. We call it the *algebra generated* by $\mathscr{B}$.

It is also easily verified that if $\mathfrak{A}$ is an algebra, then also its closure $\overline{\mathfrak{A}}$ [in $C(X)$] is an algebra.

A set $\mathscr{B}$ in $C(X)$ is said to *distinguish between the points of X* if for any two distinct points x and y in X there is a function $f \in \mathscr{B}$ such that $f(x) \neq f(y)$.

For example, in $C[a,b]$ $(-\infty < a < b < \infty)$ the algebra generated by the constant function 1 and x is the set P of all polynomials. P distinguishes between the points of $[a,b]$. The classical *Weierstrass theorem* states that every continuous function $f(x)$ on $[a,b]$ is the uniform limit of polynomials. That means that $\bar{P} = C[a,b]$. The following theorem, called the *Stone-Weierstrass theorem*, gives a far-reaching generalization of the Weierstrass theorem.

Theorem 3.7.1 *Let* X *be a compact metric space and let* $\mathfrak{A}$ *be a closed algebra in the metric space* C(X) *of continuous real-valued functions. If* $\mathfrak{A}$ *contains the constant function* 1 *and if it also distinguishes between the points of* X, *then* $\mathfrak{A} = \mathrm{C(X)}$.

The proof depends upon the following lemma, which is a special case of the Weierstrass theorem.

Lemma 3.7.2 *Let* λ *be a real variable. For any* $\varepsilon > 0$ *there exists a polynomial* $\mathrm{p}(\lambda)$ *such that* $||\lambda| - \mathrm{p}(\lambda)| < \varepsilon$ *for* $-1 \leq \lambda \leq 1$.

Proof. One can easily check that the function $(\frac{1}{4}\varepsilon^2 + t)^{1/2}$ has a uniformly convergent Taylor series (about $t = \frac{1}{2}$) for $0 \leq t \leq 1$. Hence there is a polynomial $q(t)$ such that

$$|(\tfrac{1}{4}\varepsilon^2 + t)^{1/2} - q(t)| < \frac{\varepsilon}{2} \qquad \text{if } 0 \leq t \leq 1.$$

Now, if $-1 \leq \lambda \leq 1$,

$$||\lambda| - q(\lambda^2)| \leq ||\lambda| - (\tfrac{1}{4}\varepsilon^2 + \lambda^2)^{1/2}| + |(\tfrac{1}{4}\varepsilon^2 + \lambda^2)^{1/2} - q(\lambda^2)| < \varepsilon.$$

Thus, the assertion follows with $p(\lambda) = q(\lambda^2)$.

Proof of Theorem 3.7.1. We first prove that if $f \in \mathfrak{A}$, then $|f| \in \mathfrak{A}$. Suppose first that $\max |f| \leq 1$. Let $p(\lambda) = \sum p_j \lambda^j$ be the polynomial occurring in Lemma 3.7.2, and let $p(f) = \sum p_j f^j$. Then $p(f) \in \mathfrak{A}$ and $||f(x)| - p[f(x)]| < \varepsilon$ for all $x \in X$. This shows that $|f|$ belongs to $\overline{\mathfrak{A}}$—that is, to $\mathfrak{A}$.If $\max |f| > 1$, then$|\mu f| \in \mathfrak{A}$, where $\mu = 1/(\max |f|)$. Hence also$|f| \in \mathfrak{A}$.

The function $\min(f,g)$ defined by

$$[\min(f,g)](x) = \min[f(x),g(x)]$$

can be written in the form

$$\min(f,g) = \tfrac{1}{2}(f+g) - \tfrac{1}{2}|f-g|.$$

Hence, if f and g belong to $\mathfrak{A}$, then also $\min(f,g)$ belongs to $\mathfrak{A}$. Similarly, the function $\max(f,g)$ defined by

$$[\max(f,g)](x) = \max[f(x),g(x)]$$

belongs to $\mathfrak{A}$, since it can be written in the form $\frac{1}{2}(f+g) + \frac{1}{2}|f-g|$.

We shall now approximate any function f of $C(X)$ by functions in $\mathfrak{A}$. Let $x,y \in X$ with $x \neq y$. By assumption, there is a function $h \in \mathfrak{A}$ with $h(x) \neq h(y)$. But then one can find real numbers λ,μ such that the function

$$f_{xy} = \lambda h + \mu \cdot 1$$

satisfies: $f_{xy}(x) = f(x), f_{xy}(y) = f(y)$. Note that $f_{xy} \in \mathfrak{A}$.

Let ε be any positive number. Then there is a ball B_y with center y such that $f_{xy}(z) < f(z) + \varepsilon$ for all $z \in B_y$. Since X is compact, we can cover it by a finite number of such balls; call them $B_{y_1},\ldots,B_{y_m}$. Let

$$f_x = \min(f_{xy_1},\ldots,f_{xy_m}).$$

By that which was proved above, $f_x \in \mathfrak{A}$. Note that $f_x(x) = f(x)$ and $f_x(z) < f(z) + \varepsilon$ for all $z \in X$.

There is a ball D_x with center x such that $f_x(z) > f(z) - \varepsilon$ for all $z \in D_x$. Since X is compact, we can cover it by a finite number of such balls; call them $D_{x_1},\ldots,D_{x_k}$. The function

$$g = \max(f_{x_1},\ldots,f_{x_k})$$

belongs to $\mathfrak{A}$, and it satisfies $g(z) > f(z) - \varepsilon$ for all $z \in X$. Since it also satisfies $g(z) < f(z) + \varepsilon$, we have $\rho(g,f) < \varepsilon$. Since ε is arbitrary, we conclude that $f \in \overline{\mathfrak{A}} = \mathfrak{A}$. This completes the proof.

PROBLEMS

3.7.1. A continuous function f (real- or complex-valued) on a compact set G of R^n can be approximated uniformly on G by a sequence of polynomials.

3.7.2. For every continuously differentiable function $f(t)$ on $0 \le t \le 1$ there exists a sequence of polynomials $p_m(t)$ such that

$$\max_{0 \le t \le 1} |p_m(t) - f(t)| + \max_{0 \le t \le 1} \left| \frac{d}{dt} p_m(t) - \frac{d}{dt} f(t) \right| \to 0 \qquad \text{if } m \to \infty.$$

3.7.3. The algebra generated by 1 and a subset $\mathscr{D} \subset C(X)$ coincides with the class of all polynomials with real coefficients whose variables are elements of $\mathscr{D}$. This algebra is separable if $\mathscr{D}$ is separable.

3.7.4. Let $\{x_n\}$ be a dense sequence in a compact metric space X. Let $\chi_{nm}(x) = 1/m - \rho(x,x_n)$ if $\rho(x,x_n) \le 1/m$ and $\chi_{nm}(x) = 0$ if $\rho(x,x_n) > 1/m$, when $m = 1,2,\ldots$. Let $\mathscr{B} = \{\chi_{nm}$, where $1 \le n < \infty, 1 \le m < \infty\}$. Prove that the algebra $\mathfrak{A}$ generated by $\mathscr{B}$ is separable, and that it distinguishes between the points of X.

3.7.5. If X is a compact metric space, then $C(X)$ is separable.

3.7.6. Let A and B be any disjoint closed subsets of a metric space. Then there exists a continuous bounded function f on X such that $f = 1$ on A, $f = 0$ on B, and $0 \le f \le 1$ on X. In particular it follows that $C(X)$, for X compact, distinguishes between the points of X. [*Hint:* Use the Tietze extension theorem.]

3.7.7. Show, by an example, that the Stone-Weierstrass theorem is generally false for the metric space $C(X)$ of complex-valued functions. [*Hint:* Consider the algebra generated by 1 and z in any domain of the complex plane.]

3.8 A FIXED-POINT THEOREM AND APPLICATIONS

Let (X,ρ) and (Y,σ) be metric spaces. A function f from a subset $X_0 \subset X$ into Y is said to be *continuous at a point* $y \in X_0$ if for any $\varepsilon > 0$ there is a $\delta > 0$ such that for all $x \in X_0$ with $\rho(x,y) < \delta$, the inequality $\sigma[f(x),f(y)] < \varepsilon$ holds. If f is continuous at each point y of X_0 then we say that f is *continuous on* X_0, and if, *furthermore*, δ can be taken to be independent of y, then we say that f is *uniformly continuous* on X_0.

The following result is very useful.

Theorem 3.8.1 *Let* X *and* Y *be metric spaces and let* Y *be complete. If* f *is a uniformly continuous function from a dense subset* X_0 *of* X *into* Y, *then* f *has a unique continuous extension* $\tilde{f}$ *from* X *into* Y. *This extension is uniformly continuous on* X.

Proof. For any $x \in (X - X_0)$, take a sequence $\{x_n\} \subset X_0$ such that $\lim x_n = x$. Define $\tilde{f}(x) = \lim f(x_n)$. It is easily seen that the limit exists and is independent of the choice of the sequence. The uniform continuity of $\tilde{f}$ follows readily from the uniform continuity of f. Finally it is clear that there can be no other continuous extension of f.

A map T from a subset X_0 of a metric space (X,ρ) into X is called a *contraction* (on X_0) if there is a number θ, $0 < \theta < 1$, such that

$$\rho(Tx,Ty) \leq \theta\rho(x,y) \tag{3.8.1}$$

for all x,y in X_0. Here, and in what follows, we often write Tx instead of $T(x)$.

Theorem 3.8.2 *Let* T *be a map of a complete metric space* X *into itself. If* T *is a contraction, then there exists a unique point* z *such that* Tz = z.

A point z for which $Tz = z$ is called a *fixed point of T*. Theorem 3.8.2 is a *fixed-point theorem*. Fixed-point theorems are very important in analysis. They are used to prove the existence of solutions of differential equations, integral equations, and so on. We shall give some such applications later on.

Proof. Fix a point x_0 in X and define inductively $x_{n+1} = T(x_n)$ for $n = 0,1,2,\ldots$. Then

$$\begin{aligned}\rho(x_{n+1},x_n) &= \rho(Tx_n,Tx_{n-1}) \leq \theta\rho(x_n,x_{n-1})\\ &\leq \theta^2\rho(x_{n-1},x_{n-2}) \leq \cdots \leq \theta^n\rho(x_1,x_0),\end{aligned}$$

Hence

$$\begin{aligned}\rho(x_{n+p},x_n) &\leq \rho(x_{n+p},x_{n+p-1}) + \cdots + \rho(x_{n+1},x_n)\\ &\leq (\theta^{n+p-1} + \cdots + \theta^n)\rho(x_1,x_0) \leq K\theta^n,\end{aligned}$$

where $K = \rho(x_1,x_0)/(1-\theta)$. We conclude that $\{x_n\}$ is a Cauchy sequence. Since X is complete, there is a point $z \in X$ such that $\rho(x_n,z) \to 0$ as $n \to \infty$. We claim that $Tz = z$—that is, $\rho(z,Tz) = 0$. Indeed,

$$\begin{aligned}\rho(z,Tz) &\leq \rho(z,x_n) + \rho(x_n,Tz) = \rho(z,x_n) + \rho(Tx_{n-1},Tz)\\ &\leq \rho(z,x_n) + \theta\rho(x_{n-1},z) \to 0 \quad \text{if } n \to \infty.\end{aligned}$$

It remains to show that z is the unique fixed point of T. Suppose then that y is also a fixed point of T. Then

$$\rho(y,z) = \rho(Ty,Tz) \leq \theta\rho(y,z).$$

Since $\theta \neq 1$, we conclude that $\rho(y,z) = 0$—that is, $y = z$.

We restate Theorem 3.8.2 in a form that is often more convenient:

Theorem 3.8.3 *Let* Y *be a closed subset of a complete metric space* X. *If* T *is a map from* Y *into itself, and if* T *is a contraction* (*on* Y), *then there exists a unique point* z *in* Y *such that* Tz = z.

In applying this theorem one has to verify two facts about T: (i) T maps Y into itself, and (ii) T is a contraction on Y.

Let $f(x,y)$ be a real-valued function defined on an open set Ω of the Euclidean plane R^2. We say that $f(x,y)$ is uniformly *Lipschitz continuous with respect to y* in Ω if there is a constant K such that

$$|f(x,y_1) - f(x,y_2)| \leq K|y_1 - y_2| \tag{3.8.2}$$

for all (x,y_1) and (x,y_2) in Ω.

Consider the ordinary differential equation

$$\frac{dy}{dx} = f(x,y) \tag{3.8.3}$$

with the initial condition

$$y(x_0) = y_0, \tag{3.8.4}$$

where (x_0,y_0) is a fixed point in Ω. Denote by I_δ the interval $x_0 - \delta \leq x \leq x_0 + \delta$. By a *solution of the differential system* (3.8.3), (3.8.4) in I_δ we mean a continuously differentiable function $y(x)$ defined on I_δ such that $(x,y(x)) \in \Omega$ for all $x \in I_\delta$, $y'(x) = f(x,y(x))$ for all $x \in I_\delta$, and $y(x_0) = y_0$.

Theorem 3.8.4 *Assume that* f(x,y) *is a continuous bounded function in* Ω, *and uniformly Lipschitz continuous with respect to* y. *Then there exists a unique solution of* (3.8.3), (3.8.4) *in some interval* I_δ.

Proof. Observe that $y(x)$ is a solution of (3.8.3), (3.8.4) on I_δ if and only if

$$y(x) = y_0 + \int_{x_0}^{x} f(t,y(t))\,dt \qquad \text{for all } x \in I_\delta. \tag{3.8.5}$$

Let M be a bound on $|f|$ in Ω. We take δ such that the rectangle

$$R = \{(x,y);\ x_0 - \delta \leq x \leq x_0 + \delta,\ y_0 - M\delta \leq y \leq y_0 + M\delta\}$$

lies in Ω and such that $K\delta < 1$, where K is the constant occurring in (3.8.2). In the space $C(I_\delta)$ we consider the subset Y consisting of those continuous functions $\varphi(t)$ for which

$$\varphi(x_0) = y_0 \qquad \text{and} \qquad |\varphi(x) - y_0| \le M\delta. \tag{3.8.6}$$

Y is a closed subset of $C(I_\delta)$. We define a transformation T from Y into $C(I_\delta)$ as follows:

$$(T\varphi)(x) = y_0 + \int_{x_0}^{x} f(t,\varphi(t))\, dt \qquad \text{for } x \in I_\delta. \tag{3.8.7}$$

T is well defined, since the points $(t,\varphi(t))$ lie in R.

If $y(x)$ is a fixed point of T, then $y(x)$ satisfies (3.8.5) and is thus a solution of (3.8.3), (3.8.4). Conversely, every solution of (3.8.3), (3.8.4) in I_δ is a solution of (3.8.5) and it thus satisfies (3.8.6)—that is, it belongs to Y. It is therefore a fixed point of T. Thus what we have to show is that the transformation T has a unique fixed point in Y.

T maps Y into itself, since $(T\varphi)(x_0) = y_0$ and since

$$|(T\varphi)(x) - y_0| \le \left| \int_{x_0}^{x} f(t,\varphi(t))\, dt \right| \le M\, |x - x_0| \le M\delta.$$

Next, T is a contraction since

$$\begin{aligned} |(T\varphi_1)(x) - (T\varphi_2)(x)| &\le \left| \int_{x_0}^{x} [f(t,\varphi_1(t)) - f(t,\varphi_2(t))]\, dt \right| \\ &\le K\delta \max_{I_\delta} |\varphi_1 - \varphi_2|, \end{aligned}$$

and $K\delta < 1$. Thus, by Theorem 3.8.3, T has a unique fixed point in Y. This completes the proof.

PROBLEMS

3.8.1. If f is a continuous function from a compact metric space X into a metric space Y, then f is uniformly continuous.

3.8.2. If f is a continuous function from a compact metric space X into a metric space Y, then its image $f(X) = \{f(x);\, x \in X\}$ is compact.

3.8.3. Let Ω be a bounded open set in R^n and let $K(x,y)$ be a Lebesgue-measurable function on the product space $\Omega \times \Omega$ such that

$$\int_\Omega \int_\Omega |K(x,y)|^2\, dx\, dy < \infty.$$

Let $f(x)$ be a function in $L^2(\Omega)$. Consider the *integral equation*

$$u(x) = f(x) + \lambda \int_\Omega K(x,y) u(y)\, dy, \tag{3.8.8}$$

where λ is a complex number. Prove that there exists a small positive number λ_0 such that, for any complex number λ, with $|\lambda| < \lambda_0$, there exists a unique solution $u(x)$ of (3.8.8). [*Hint:* Consider the operator $(Tu)(x) = f(x) + \lambda \int K(x,y)u(y)\,dy$ on $L^2(\Omega)$.]

3.8.4. Let Ω be the closed unit ball $\overline{B(0,1)}$ in the Euclidean metric space (R^n,ρ). Let T be a map of Ω into itself such that $\rho(Tx,Ty) \leq \rho(x,y)$ for all x,y in Ω. Then there exists at least one fixed point of T. [*Hint:* Consider $(1 - 1/k)T$.]

3.8.5. Give an example where Ω (in R^n) is not a unit ball (but is a closed set) and the assertion of the last problem is false—that is, T has no fixed points.

CHAPTER 4

ELEMENTS OF FUNCTIONAL ANALYSIS IN BANACH SPACES

4.1 LINEAR NORMED SPACES

We denote by $\mathbb{R}$ the field of the real numbers, and by $\mathbb{C}$ the field of the complex numbers. Let $\mathscr{F}$ denote either $\mathbb{R}$ or $\mathbb{C}$. The elements of $\mathscr{F}$ are often called scalars.

Definition 4.1.1 A *linear vector space* consists of a space X, a mapping $(x,y) \to x + y$ of $X \times X$ into X, and a mapping $(\lambda,x) \to \lambda x$ of $\mathscr{F} \times X$ into X, such that the following conditions are satisfied:

(i) X is an abelian group with the group operation $+$;
(ii) the associative law $\lambda(\mu x) = (\lambda\mu)x$ holds;
(iii) the distributive laws $(\lambda + \mu)x = \lambda x + \mu x, \lambda(x + y) = \lambda x + \lambda y$ hold;
(iv) $1x = x$.

We recall that the condition (i) means that:

(i_1) $x + (y + z) = (x + y) + z$;
(i_2) $x + y = y + x$;
(i_3) there is an element 0 (the *zero* element) such that $x + 0 = x$ for all $x \in X$; and
(i_4) for every $x \in X$ there is an element $(-x)$ (the *inverse* of x) such that $x + (-x) = 0$.

We shall denote by 0 both the scalar 0 and the zero element of the group X; this should cause no confusion.

From the relations $\mu x = \mu(x + 0) = \mu x + \mu 0$ we get $\mu 0 = 0$. Hence if $\lambda x = 0$ and $\lambda \neq 0$, then

$$x = 1x = \left(\frac{1}{\lambda} \cdot \lambda\right)x = \frac{1}{\lambda}(\lambda x) = \frac{1}{\lambda}0 = 0$$

—that is, $x = 0$. The last property is equivalent to any one of the following properties:

$$\begin{array}{ll} \text{if } \lambda x = 0 \text{ and } x \neq 0, & \text{then } \lambda = 0; \\ \text{if } \lambda \neq 0,\ x \neq 0, & \text{then } \lambda x \neq 0; \\ \text{if } \lambda x = \mu x \text{ and } \lambda \neq \mu, & \text{then } x = 0. \end{array}$$

The reader may easily verify that, in a linear vector space,

$$0x = 0, \qquad (-1)x = -x.$$

Definition 4.1.2 A *real* (*complex*) *linear vector space* is a linear vector space with $\mathscr{F} = \mathbb{R}$ ($\mathscr{F} = \mathbb{C}$).

Elements $x_1,\ldots,x_n$ of X are called *linearly independent* if there exist no numbers $\lambda_1,\ldots,\lambda_n$ in $\mathscr{F}$ such that

$$\lambda_1 x_1 + \cdots + \lambda_n x_n = 0 \qquad \text{and} \qquad \sum_{j=1}^{n} |\lambda_j|^2 > 0. \tag{4.1.1}$$

If there exist numbers $\lambda_1,\ldots,\lambda_n$ such that (4.1.1) holds, then we say that $x_1,\ldots,x_n$ are *linearly dependent*. We can then express any one of the x_j, with $\lambda_j \neq 0$, as a linear combination of the others:

$$x_j = -\frac{\lambda_1}{\lambda_j}x_1 - \cdots - \frac{\lambda_{j-1}}{\lambda_j}x_{j-1} - \frac{\lambda_{j+1}}{\lambda_j}x_{j+1} - \cdots - \frac{\lambda_n}{\lambda_j}x_n.$$

If for any positive integer n there exist n linearly independent elements in X, then we say that X is of *infinite dimension*. If X is not infinite-dimensional, then it is easily seen that there is a positive integer n with the following properties:

(a) there is a set of n elements that are linearly independent;
(b) if $m > n$, then any m elements are linearly dependent.

We call n the *dimension* of X. Any set $e_1,\ldots,e_n$ of linearly independent elements is said to form a *basis*. The elements x of X can be represented in a unique way in the form

$$x = \lambda_1 e_1 + \cdots + \lambda_n e_n.$$

The numbers $\lambda_1,\ldots,\lambda_n$ are called the *coordinates* of x with respect to the basis $e_1,\ldots,e_n$.

The *vector sum* $A + B$ of two sets A and B is the set $\{a + b; a \in A, b \in B\}$. We also define $\lambda A = \{\lambda x; x \in A\}$.

Definition 4.1.3 A *normed linear space* is a linear vector space X on which there is defined a function $x \to \|x\|$, called a *norm*, having the following properties:

(a) $\|x\| \geq 0$, and $\|x\| = 0$ if and only if $x = 0$;
(b) $\|\lambda x\| = |\lambda| \, \|x\|$;
(c) $\|x + y\| \leq \|x\| + \|y\|$.

Definition 4.1.4 A *metric linear space* is a linear vector space X that is also a metric space, with metric ρ, having the following additional property: If $\lambda_n \to \lambda$, $\mu_n \to \mu$, $x_n \to x$, $y_n \to y$, then $\lambda_n x_n + \mu_n y_n \to \lambda x + \mu y$.

The last property is equivalent to the statement that the maps $(x,y) \to x + y$ and $(\lambda,x) \to \lambda x$ are continuous maps from the metric spaces $X \times X$ and $\mathscr{F} \times X$, respectively, into X.

Definition 4.1.5 A *Fréchet space* is a metric linear space X having the additional properties:

(α) $\rho(x,y) = \rho(x - y,0)$;
(β) X is complete.

Given a normed linear space we introduce on it a function ρ by

$$\rho(x,y) = \|x - y\|. \tag{4.1.2}$$

We then have:

Theorem 4.1.1 *A normed linear space is a metric linear space* [*with the metric* ρ *defined by* (4.1.2)].

Proof. It is easily seen that ρ satisfies the metric conditions. Thus it remains to show that if $\lambda_n \to \lambda$, $\mu_n \to \mu$, $\|x_n - x\| \to 0$, $\|y_n - y\| \to 0$, then

$$\|(\lambda_n x_n + \mu_n y_n) - (\lambda x + \mu y)\| \to 0. \tag{4.1.3}$$

Noting that $\{\|x_n\|\}$ and $\{\|y_n\|\}$ are bounded sequences, and making use of (b) and (c) in Definition 4.1.3, one can easily verify (4.1.3).

A metric linear space with metric ρ is said to be a normed linear space if there exists a norm $\| \quad \|$ such that $\rho(x,y) = \|x - y\|$. If this is the case, then, of course,

$$\rho(x,y) = \rho(x - y,0), \tag{4.1.4}$$

$$\rho(\lambda x,\lambda y) = |\lambda| \rho(x,y). \tag{4.1.5}$$

Conversely, if ρ satisfies (4.1.4), (4.1.5), then $\|x\| = \rho(x,0)$ is a norm and the space is a normed linear space. The metric space corresponding to it by Theorem 4.1.1 is the originally given metric linear space.

Since (by Theorem 4.1.1) a normed linear space is also a metric space, the theory of Chapter 3 applies to it. We recall that $x_n \to x$ if $\|x_n - x\| \to 0$; $\{x_n\}$ is a Cauchy sequence if $\|x_n - x_m\| \to 0$ as $n,m \to \infty$; if every Cauchy sequence is convergent, then X is called complete. Note that if $x_n \to x$, then $\|x_n\| \to \|x\|$, and if $\{x_n\}$ is a Cauchy sequence, then $\{\|x_n\|\}$ is a Cauchy sequence.

Definition 4.1.6 A real (complex) normed linear space that is complete is called a *real* (*complex*) *Banach space*.

Note that Banach spaces are Fréchet spaces.

Theorem 3.3.1 implies that every normed linear space X can be completed as a metric space. But actually the completed space $\hat{X}$ can be made into a linear vector space by defining

$$\tilde{x} + \tilde{y} = \tilde{z}, \qquad \lambda\tilde{x} = \tilde{u}$$

as follows: if $\{x_n\} \in \tilde{x}$ and $\{y_n\} \in \tilde{y}$, then $\tilde{z}$ is the class containing the sequence $\{x_n + y_n\}$ and $\tilde{u}$ is the class containing the sequence $\{\lambda x_n\}$. $\hat{X}$ is a normed space with the norm given by

$$\|\tilde{x}\| = \lim \|x_n\| \qquad \text{if } \{x_n\} \in \tilde{x}.$$

Thus, *for every normed linear space X there is a Banach space $\hat{X}$ such that X is isomorphic to a dense subset $\sigma(X)$ of $\hat{X}$; the map σ from X into $\hat{X}$ preserves the algebraic operations of addition and of multiplication by scalars.*

Definition 4.1.7 Let $\{x_n\}$ be a sequence in a normed linear space X and let $s_m = \sum_{n=1}^{m} x_n$. If the sequence $\{s_m\}$ has a limit s, then we say that the series $\sum_{n=1}^{\infty} x_n$ is *convergent* and its *sum* is s. We then write

$$s = \sum_{n=1}^{\infty} x_n.$$

If $\sum_{n=1}^{\infty} \|x_n\| < \infty$, then we say that the series $\sum_{n=1}^{\infty} x_n$ is *absolutely convergent*.

Theorem 4.1.2 *Let X be a Banach space and let $\{x_n\} \subset X$. If the series $\sum_{n=1}^{\infty} x_n$ is absolutely convergent, then it is also convergent.*

Proof. Since

$$\|s_m - s_k\| \le \sum_{n=k+1}^{m} \|x_n\| \to 0 \qquad \text{if } m > k \to \infty,$$

the sequence $\{s_m\}$ is a Cauchy sequence in X. Since X is complete, this sequence has a limit.

We note that a set A in a normed linear space X is bounded if and only if there is a positive number R such that $\|x\| \le R$ for all $x \in A$.

Definition 4.1.8 Two norms $\| \quad \|_1$ and $\| \quad \|_2$ on a linear vector space X are said to be *equivalent* if there exist positive numbers α and β such that

$$\alpha\|x\|_1 \le \|x\|_2 \le \beta\|x\|_1 \qquad \text{for all } x \in X.$$

This is clearly the case if and only if the metrics corresponding to them by (4.1.2) are equivalent in the sense defined in Section 3.1.

Example 1. The spaces $L^p(X,\mu)$ $(1 \le p \le \infty)$ are linear vector spaces with $f + g$ and λf defined in the obvious way—that is,

$$\begin{aligned} (f+g)(x) &= f(x) + g(x), \\ (\lambda f)(x) &= \lambda f(x). \end{aligned} \tag{4.1.6}$$

Furthermore, the function $f \to \|f\|$ given by

$$\|f\| = \left\{\int_X |f|^p \, d\mu\right\}^{1/p} \qquad \text{if } 1 \le p < \infty, \tag{4.1.7}$$

and

$$\|f\| = \operatorname*{ess\,sup}_X |f| \qquad \text{if } p = \infty, \tag{4.1.8}$$

is a norm. Unless the contrary is explicitly stated, we shall always consider $L^p(X,\mu)$ to be the normed linear space with the norm given by (4.1.7), (4.1.8). $L^p(X,\mu)$ is a Banach space. If the functions f are taken to be a.e. real-valued, then $L^p(X,\mu)$ is a real Banach space. If the functions f are taken to be complex-valued, then $L^p(X,\mu)$ can be considered to be either a real or a complex Banach space.

Example 2. The spaces l^p $(1 \le p \le \infty)$ are Banach spaces. Here addition and multiplication by scalars are defined in the obvious way, and the norm is given by

$$\|\xi\| = \left\{\sum_{n=1}^{\infty} |\xi_n|^p\right\}^{1/p} \qquad \text{if } 1 \le p < \infty,$$

$$\|\xi\| = \sup_n |\xi_n| \qquad \text{if } p = \infty,$$

where $\xi = (\xi_1, \ldots, \xi_n, \ldots)$.

Example 3. The space $C(X)$, X a compact metric space, is a Banach space with the operations defined in (4.1.6) and with the *uniform norm*

$$\|f\| = \max_{x \in X} |f(x)|.$$

If the functions are real-valued, then it is a real Banach space, and if they are complex-valued, then it can be considered either as a real or as a complex Banach space.

Example 4. The space s (Example 5 of Section 3.1) is not a normed linear space, since its metric fails to satisfy the condition (4.1.5). However, s is a Fréchet space.

Example 5. Let X_i $(1 \le i \le n)$ be linear vector spaces. Their *direct sum* X is a linear vector space whose elements x have the form $x = (x_1, \ldots, x_n)$, with $x_i \in X_i$, and the algebraic operations are defined by

$$\lambda(x_1, \ldots, x_n) + \mu(y_1, \ldots, y_n) = (\lambda x_1 + \mu y_1, \ldots, \lambda x_n + \mu y_n).$$

We write $X = \sum_{i=1}^{n} X_i$. If each X_i has a norm $\| \quad \|_i$, then we can make X a normed linear space with any of the norms

$$\|x\| = \sum_{i=1}^{n} \|x_i\|_i, \qquad \|x\| = \max_{1 \le i \le n} \|x_i\|_i,$$
$$\|x\| = \left\{ \sum_{i=1}^{n} \|x_i\|_i^2 \right\}^{1/2}. \tag{4.1.9}$$

Note that all these norms are equivalent. The linear space with any one of these norms is called the *direct sum of the normed linear spaces* $X_1, \ldots, X_n$ and is denoted by $\sum_{i=1}^{n} X_i$. The direct sum is often called the *direct product* of $X_1, \ldots, X_n$ and it is then denoted by $X_1 \times \cdots \times X_n$.

Definition 4.1.9 A set K in a linear space is called *convex* if for any two points x and y in K, each point $tx + (1 - t)y$, with $0 < t < 1$, also belongs to K.

PROBLEMS

4.1.1. In a normed linear space, every ball is a convex set.

4.1.2. If K and L are convex sets, then their vector sum $K + L$ is also a convex set.

4.1.3. Two norms on a linear vector space are equivalent if and only if every set that is bounded in one of the norms is bounded in the other norm.

4.1.4. If $\{x_n\}$ is a convergent sequence in a normed linear space, with limit x, then also the sequence with elements $(x_1 + \cdots + x_n)/n$ is convergent to x.

4.1.5. The direct sum of a finite number of Banach spaces is a Banach space.

4.1.6. A normed linear space is a Banach space if the following property is satisfied: every absolutely convergent series is convergent.

4.2 SUBSPACES AND BASES

In Section 3.1 we have defined the concepts of a topological subspace and of a metric subspace. We shall now consider the concept of a subspace of a linear vector space.

Definition 4.2.1 A subset Y of a linear vector space X is called a *linear subspace* (or, briefly, a *subspace*) if for any $\lambda, \mu \in \mathscr{F}$ and x,y in Y, the point $\lambda x + \mu y$ is in Y. If X is a normed linear space, and if Y is given the norm induced by the norm of X, then we call Y a *linear subspace* (or, briefly, a *subspace*) *of the normed linear space* X.

The intersection of any number of linear subspaces is again a linear subspace. If K is any subset of X, then the intersection of all the linear subspaces containing K is a subspace that contains K and that is contained in any subspace containing K. Thus it is the minimal subspace containing K. We call it the *linear subspace spanned* (or *generated*) *by* K. It is the set of all points of the form $\lambda_1 x_1 + \cdots + \lambda_m x_m$, where m is any positive integer, the x_i vary in K, and the λ_i vary in $\mathscr{F}$.

A *closed linear subspace* is a subspace that is closed. As in the case of linear subspaces, one can easily show that for any set K there exists a minimal closed subspace containing K. We call it the *closed linear subspace spanned by* K.

Let Y_0 be a linear subspace. Decompose X into cosets Y: two elements x_1 and x_2 belong to the same Y if and only if $x_1 - x_2 \in Y_0$. We make the space of the cosets a linear vector space by defining

$$Y_1 + Y_2 = Y_3, \qquad \lambda Y_1 = Y_4$$

as follows. Take $y_1 \in Y_1$ and $y_2 \in Y_2$. Then Y_3 is the coset containing the element $y_1 + y_2$, and Y_4 is the coset containing the element λy_1. We call this linear space a *factor space* and denote it by X/Y_0. Notice that the set of points (of X) contained in a sum of cosets $Y_1 + Y_2$ coincides with the vector sum of the sets of points contained in Y_1 and Y_2.

If X is normed, then we define

$$\|Y\| = \inf_{x \in Y} \|x\| \qquad \text{for any } Y \in X/Y_0. \tag{4.2.1}$$

Theorem 4.2.1 *Let* X *be a normed linear space and let* Y_0 *be a closed linear subspace. Then* X/Y_0 *is a normed linear space with the norm* (4.2.1).

Proof. We first prove that each coset Y is a closed set. Let $\{x_n\} \subset Y$, $x_n \to x$. Then $x_n - x_m \in Y_0$ and $x_n - x_m \to x_n - x$. Since Y_0 is closed, $x_n - x \in Y_0$. Hence the coset of x is the same as the coset of x_n—that is, it is Y. This means that $x \in Y$.

To prove that $\|Y\|$ is a norm, suppose first that $\|Y\| = 0$. Then there is a sequence $\{x_n\}$ in Y such that $\|x_n\| \to 0$. Since Y is closed and $x_n \to 0$, we get $0 \in Y$—that is, $Y = Y_0$. The homogeneity of the norm—that is, the condition (b) in Definition 4.1.3—follows from

$$\|\lambda Y\| = \inf_{x \in Y} \|\lambda x\| = |\lambda| \inf_{x \in Y} \|x\| = |\lambda| \, \|Y\|.$$

Finally, the condition (c) in Definition 4.1.3 is proved as follows:

$$\|Y_1 + Y_2\| = \inf_{\substack{x_1 \in Y_1 \\ x_2 \in Y_2}} \|x_1 + x_2\| \le \inf_{x_1 \in Y_1} \|x_1\| + \inf_{x_2 \in Y_2} \|x_2\| = \|Y_1\| + \|Y_2\|.$$

Definition A *partially ordered set* is a nonempty set S together with a relation "$\geq$" subject to the following conditions:

(i) $x \le x$;
(ii) if $x \le y$ and $y \le z$, then $x \le z$.

If for every two elements x, y in S at least one of two relations $x \le y$, $y \le x$ holds, then we say that S is a *totally ordered set*.

Definition Let T be a subset of a partially ordered set S. An element x of S is called an *upper bound* of T if $y \le x$ for all $y \in T$.

Definition Let S be a partially ordered set. An element x of S is called *maximal* if for any element $y \in S$, the relation $x \le y$ implies that $y \le x$.

The following result is known as *Zorn's lemma.*

Theorem 4.2.2 *If* S *is a partially ordered set in which every totally ordered subset has an upper bound, then* S *has a maximal element.*

This theorem is equivalent to both the *well-ordering theorem* and the *axiom of choice.* We shall not give here the proof of the theorem.

We shall use Zorn's lemma in the proof of the next theorem.

Theorem 4.2.3 *Every linear vector space* X *contains a set* $\mathfrak{A}$ *of linearly independent elements such that the linear subspace spanned by* $\mathfrak{A}$ *coincides with* X.

A set $\mathfrak{A}$ such as in the assertion of Theorem 4.2.3 is called a *Hamel basis* for X. If we denote a Hamel basis by $\{y_\alpha\}$, then every element $x \in X$ has a unique representation

$$x = \sum \lambda_\alpha y_\alpha \qquad \text{(finite sum)}, \tag{4.2.2}$$

where the λ_α are scalars.

Proof of Theorem 4.2.3. Consider the set S, whose elements are all those subsets of X which consist of linearly independent elements. We denote these subsets by $\mathfrak{A}$, $\mathscr{B}$, and so on. Introduce a partial ordering by inclusion —that is, $\mathfrak{A} \leq \mathscr{B}$ if $\mathfrak{A} \subset \mathscr{B}$. Then every totally ordered set $\{\mathfrak{A}_\alpha\}$ has an upper bound $\mathscr{B}$ given by $\mathscr{B} = \bigcup_\alpha \mathfrak{A}_\alpha$. To prove this we have to show that any finite number of elements of $\mathscr{B}$ are linearly independent. Denote such elements by $x_1,\ldots,x_n$, and let $x_i \in \mathfrak{A}_{\alpha_i}$ for $1 \leq i \leq n$. Since the set $\{\mathfrak{A}_{\alpha_i}\}$ is totally ordered, one of the subsets $\mathfrak{A}_{\alpha_i}$ contains all the others. Denote that set by $\mathfrak{A}_{\alpha_k}$. Since $\{x_1,\ldots,x_n\} \subset \mathfrak{A}_{\alpha_k}$, the elements $x_1,\ldots,x_n$ are linearly independent.

Applying Zorn's lemma, we conclude that S has a maximal element $\mathfrak{A}$. We shall show that every $y \in X$ is a finite linear combination of elements of $\mathfrak{A}$. If this is not the case, then the set $\mathscr{B} = \{\mathfrak{A},y\}$ belongs to S, $\mathfrak{A} \leq \mathscr{B}$, but not $\mathscr{B} \leq \mathfrak{A}$. This contradicts the maximality of $\mathfrak{A}$.

PROBLEMS

4.2.1. Let X be a normed linear space and let Y_0 be a closed linear subspace. Let $\{Y_n\}$ and Y belong to X/Y_0. Prove: $\|Y_n - Y\| \to 0$ as $n \to \infty$ if and only if there exists a sequence $\{y_n\}$ with $y_n \in Y_n$ and a point $x_0 \in Y$ such that $\|y_n - x_0\| \to 0$ as $n \to \infty$. [*Hint:* If $\|Y_n - Y\| = \varepsilon_n$, there exist $x_n \in Y_n$, $x_n' \in Y$ such that $\|x_n - x_n'\| \leq 2\varepsilon_n$. Fix $x_0 \in Y$ and define $y_n = x_0 - x_n' - x_n$.]

4.2.2. Let X be a Banach space and let Y_0 be a closed linear subspace. Prove that X/Y_0 is a Banach space.

4.2.3. Let X be a normed linear space and let Y_0 be a closed linear subspace that is complete. If X/Y_0 is a Banach space, then X is a Banach space.

4.2.4. Show that c is a closed linear subspace of l^∞ and that it is nowhere dense in l^∞.

4.2.5. Is $C[a,b]$ a closed linear subspace of $L^p(a,b)$ $(1 \leq p \leq \infty)$?

4.2.6. If a linear vector space is infinite-dimensional, then there exist on it norms that are not equivalent. [*Hint:* Let $\{y_\alpha\}$ be a Hamel basis and define norms by $\|x\|^2 = \sum_\alpha c_\alpha|\lambda_\alpha|^2$, where x has the form (4.2.2) and c_α are positive numbers.]

4.2.7. If a normed linear space has a countable Hamel basis, then it is separable.

4.3 FINITE-DIMENSIONAL NORMED LINEAR SPACES

We shall characterize in this section the normed linear spaces of finite dimension. We shall need the following lemma.

Lemma 4.3.1 *Let Y be a closed proper linear subspace of a normed linear space X. Then, for any* $\varepsilon > 0$, *there exists a point* $z \in X$, $\|z\| = 1$, *such that* $\|z - y\| > 1 - \varepsilon$ *for all* $y \in Y$.

Proof. Take a point $x_0 \in X$, $x_0 \notin Y$ and define

$$d = \inf_{y \in Y} \|x_0 - y\|.$$

If $d = 0$, then there exists a sequence $\{y_n\}$ in Y such that $y_n \to x_0$. Since Y is closed, we then get $x_0 \in Y$, which is impossible. Thus $d > 0$. For any $\eta > 0$ there is a point $y_0 \in Y$ such that

$$d \leq \|x_0 - y_0\| \leq d + \eta.$$

The point $z = (x_0 - y_0)/\|x_0 - y_0\|$ satisfies $\|z\| = 1$, and, for any $y \in Y$,

$$\|z - y\| = \left\| \frac{x_0 - y_0}{\|x_0 - y_0\|} - y \right\| = \frac{1}{\|x_0 - y_0\|} \|x_0 - y'\|$$

$$\geq \frac{1}{d + \eta} \|x_0 - y'\| \geq \frac{d}{d + \eta} = 1 - \frac{\eta}{d + \eta},$$

where $y' = y_0 + \|x_0 - y_0\| y \in Y$. Taking η such that $\eta/(d + \eta) < \varepsilon$, we get $\|z - y\| > 1 - \varepsilon$. This completes the proof.

Theorem 4.3.2 *If Y is a finite-dimensional linear subspace of a normed linear space X, then Y is closed.*

Proof. Let $\{y_m\} \subset Y$, $y_m \to y$ as $m \to \infty$. We have to show that $y \in Y$. We fix a basis $e_1, \ldots, e_n$ in Y and write every element y_m in the form

$$y_m = \sum_{i=1}^{n} \lambda_i^{(m)} e_i. \tag{4.3.1}$$

We claim that there is a constant C such that

$$\sum_{i=1}^{n} |\lambda_i^{(m)}|^2 \leq C \qquad \text{for all } m. \tag{4.3.2}$$

Indeed, otherwise, the sequence of numbers $\Lambda_m = \sum_i |\lambda_i^{(m)}|^2$ has a subsequence that tends to ∞. For simplicity we denote this sequence again by $\{\Lambda_m\}$. Since $\{y_m\}$ is a convergent sequence, it is also a bounded set. Hence

$$\frac{\|y_m\|}{\Lambda_m} \to 0 \qquad \text{as } m \to \infty. \tag{4.3.3}$$

The sequence $\{Z_m\}$ of points

$$Z_m = \left(\frac{\lambda_1^{(m)}}{\Lambda_m}, \ldots, \frac{\lambda_n^{(m)}}{\Lambda_m}\right)$$

on the unit sphere in n dimensions has a convergent subsequence $\{Z_{m'}\}$. Thus,

$$\lim_{m'} Z_{m'} = (\mu_1, \ldots, \mu_n), \qquad \sum_{i=1}^{n} |\mu_i|^2 = 1. \tag{4.3.4}$$

Dividing both sides of (4.3.1) by Λ_m and taking $m = m' \to \infty$, we get, after using (4.3.3), (4.3.4),

$$\sum_{i=1}^{n} \mu_i e_i = 0, \qquad \sum_{i=1}^{n} |\mu_i|^2 = 1.$$

This is impossible, since the e_i are linearly independent.

Having proved (4.3.2), we can now extract a convergent subsequence from the sequence of vectors $(\lambda_1^{(m)}, \ldots, \lambda_n^{(m)})$. Denoting its limit by $(\nu_1, \ldots, \nu_n)$, we have

$$\lim_{m'} y_{m'} = \sum_{i=1}^{n} \nu_i e_i,$$

where m' is the index of the subsequence. Since, however, $y = \lim_{m} y_m$, we conclude that $y = \sum \nu_i e_i \in Y$.

We can now state the main result on the characterization of finite-dimensional normed linear spaces.

Theorem 4.3.3 *A normed linear space is finite-dimensional if and only if every bounded subset is relatively compact.*

Proof. Suppose first that X is finite-dimensional. We have to show that every bounded sequence has a convergent subsequence. Let $\{y_m\}$ be a bounded sequence. Choose a basis $e_1, \ldots, e_n$ in X and represent the y_m in the form (4.3.1). The arguments used in the proof of Theorem 4.3.2 show that there is a constant C such that (4.3.2) holds. But, again by the arguments used in the proof of Theorem 4.3.2, there is a subsequence $\{m'\}$ of the sequence of natural numbers such that

$$(\lambda_1^{(m')}, \ldots, \lambda_n^{(m')}) \to (\nu_1, \ldots, \nu_n) \qquad \text{as } m' \to \infty.$$

Hence, $y_{m'} \to \sum \nu_i e_i$ as $m' \to \infty$. This completes the proof of one part of the theorem.

Suppose now that X has the property that bounded subsets are relatively compact, and suppose that X is infinite-dimensional. We shall derive a contradiction. First we take a point x_1 such that $\|x_1\| = 1$. We now apply Lemma 4.3.1 with Y being the linear space spanned by x_1. We conclude that there exists a point x_2 linearly independent of x_1 such that $\|x_2\| = 1$, $\|x_2 - x_1\| > \frac{1}{2}$.

Next we apply Lemma 4.3.1 with Y the linear space spanned by x_1 and x_2. We conclude that there exists a point x_3 linearly independent of x_1,x_2 such that

$$\|x_3\| = 1, \qquad \|x_3 - x_1\| > \tfrac{1}{2}, \qquad \|x_3 - x_2\| > \tfrac{1}{2}.$$

Proceeding in this way by induction, we get a sequence $\{x_n\}$ such that

$$\|x_n\| = 1, \qquad \|x_n - x_j\| > \tfrac{1}{2} \qquad \text{if } n > j.$$

Note that at each step, the linear space Y that is used is finite-dimensional and hence (by Theorem 4.3.2) closed. Thus Lemma 4.3.1 can in fact be employed. Since the sequence $\{x_n\}$ is bounded, and since it obviously has no convergent subsequences, we have arrived at a contradiction.

PROBLEMS

4.3.1. Let X be a finite-dimensional linear space. Then any two norms on X are equivalent. (According to Problem 4.2.6, the assertion is false if X is infinite-dimensional.)

4.3.2. Let Y be a finite-dimensional linear subspace of a normed linear space X, and let $x_0 \in X$, $x_0 \notin Y$. Then there exists a point $y_0 \in Y$ such that

$$\inf_{y \in Y} \|x_0 - y\| = \|x_0 - y_0\|.$$

4.3.3. A norm $\| \quad \|$ is said to be *strictly convex* if $\|x\| = 1$, $\|y\| = 1$, $\|x + y\| = 2$ imply that $x = y$. Prove that if the norm of X is strictly convex, then the point y_0 occurring in the assertion of Problem 4.3.2 is unique.

4.3.4. Prove that the norm of $L^p(X,\mu)$ is strictly convex if $1 < p < \infty$, and is not strictly convex if $p = 1$ or if $p = \infty$.

4.3.5. Prove that in $C[a,b]$ the uniform norm is not equivalent to the L^p norm (for $1 \le p < \infty$).

4.3.6. Let Ω be a bounded open set in R^n and let $1 \le q < r \le p < \infty$. Prove that in the linear vector space $L^p(\Omega)$ the norms of $L^q(\Omega)$ and of $L^r(\Omega)$ are not equivalent.

4.3.7. Let n be a positive integer, $1 \le p < \infty$, and let $f(x)$ be a continuous function on $0 \le x \le 1$. Then there exists a unique polynomial Q_n of degree n such that for any other polynomial P_n of degree n

$$\int_0^1 |f(x) - P_n(x)|^p \, dx > \int_0^1 |f(x) - Q_n(x)|^p \, dx.$$

4.4 LINEAR TRANSFORMATIONS

Let X and Y be two linear vector spaces, and let T be a function from a subset D_T of X into Y. We call T also an *operator*, a *mapping*, or a *transformation*. The set D_T where T is defined is called the *domain* of T. If $y = Tx$, then we call y the *image* of x and we call x an *inverse image* of y. For any subset A of D_T we write $T(A) = \{Tx; x \in A\}$. The set $T(D_T)$ is called the *range* of T. We shall often use the notation $T(x) = Tx$. T is called a *linear operator* if D_T is a linear subspace of X and if

$$T(\lambda_1 x_1 + \lambda_2 x_2) = \lambda_1 T x_1 + \lambda_2 T x_2$$

for all x_1, x_2 in D_T and λ_1, λ_2 scalars. It follows that $T(0) = T(2 \cdot 0) = 2T(0)$, so that $T(0) = 0$.

Unless the contrary is explicitly stated, we shall always assume that $D_T = X$—that is, T is defined on the whole space X. If $T(X) = Y$, then we say that T maps X *onto* Y. If $Tx_1 = Tx_2$ implies $x_1 = x_2$, then we say that T is a *one-to-one* map. When T is linear, T is one-to-one if and only if $Tx = 0$ implies $x = 0$.

Let X and Y be metric linear spaces. A *continuous linear transformation* T from X into Y is a linear transformation that is also continuous—that is, if $x_n \to x$, then $Tx_n \to Tx$.

Theorem 4.4.1 *Let* X *and* Y *be normed linear spaces. A linear transformation* T *from* X *into* Y *is continuous if and only if it is continuous at one point.*

Proof. Let T be continuous at a point z. If $x_n \to x$, then $x_n - x + z \to z$. Hence $T(x_n - x + z) \to Tz$, so that $Tx_n - Tx + Tz \to Tz$. Thus $Tx_n \to Tx$.

Let X and Y be normed linear spaces, and let T be a linear map from X into Y. If there is a constant K such that, for all $x \in X$,

$$\|Tx\| \le K\|x\|, \tag{4.4.1}$$

then we say that T is *bounded* and we call T a *bounded linear map*. Note that the norms on the different sides of the last inequality are taken in different spaces.

The g.l.b. of all the constants K for which (4.4.1) is satisfied is denoted by $\|T\|$ and is called the *norm* of T. It is obvious that (4.4.1) holds with $K = \|T\|$—that is,

$$\|Tx\| \le \|T\| \, \|x\|. \tag{4.4.2}$$

We also have

$$\|T\| = \underset{x \neq 0}{\text{l.u.b.}} \frac{\|Tx\|}{\|x\|} = \underset{\|x\|=1}{\text{l.u.b.}} \|Tx\|. \tag{4.4.3}$$

Theorem 4.4.2 *A linear transformation* T *from a normed linear space* X *into a normed linear space* Y *is continuous if and only if it is bounded.*

Proof. If T is bounded and $x_n \to 0$, then, by (4.4.1), $Tx_n \to 0$. Thus T is continuous at $x = 0$. By Theorem 4.4.1, T is then continuous everywhere. Suppose, conversely, that T is continuous. If T is not bounded, then for any positive integer n there is a point x_n' such that $\|Tx_n'\| > n\|x_n'\|$. Let $x_n = x_n'/(n\|x_n'\|)$. Then $x_n \to 0$, $\|Tx_n\| > 1$. Since $T0 = 0$, T cannot be continuous at $x = 0$—a contradiction.

The space of all linear transformations from a linear vector space X into a linear vector space Y can be made into a linear space by defining

$$(T + S)(x) = Tx + Sx,$$

$$(\lambda T)(x) = \lambda Tx.$$

We shall denote this space by $\mathscr{L}(X,Y)$. When X and Y are normed linear spaces, we denote by $\mathscr{B}(X,Y)$ the subspace of $\mathscr{L}(X,Y)$ consisting of all the continuous linear transformations. If $X = Y$, then we write $\mathscr{L}(X) = \mathscr{L}(X,X)$, $\mathscr{B}(X) = \mathscr{B}(X,X)$.

Theorem 4.4.3 *Let* X *and* Y *be normed linear spaces. Then* $\mathscr{B}$(X,Y) *is a normed linear space with the norm* ‖T‖ *defined by* (4.4.3).

Proof. We have to show that T satisfies the conditions (a)–(c) in Definition 4.1.3. As for (a), $\|T\|$ is clearly ≥ 0, and, if $\|T\| = 0$, then by (4.4.2) $Tx = 0$ for all $x \in X$—that is, $T = 0$. The conditions (b), (c) follow from

$$\|\lambda T\| = \sup_{\|x\|=1} \|(\lambda T)x\| = |\lambda| \sup_{\|x\|=1} \|Tx\| = |\lambda|\,\|T\|,$$

$$\|T + S\| = \sup_{\|x\|=1} \|(T + S)x\| \leq \sup_{\|x\|=1} \|Tx\| + \sup_{\|x\|=1} \|Sx\| = \|T\| + \|S\|.$$

Definition 4.4.1 A sequence $\{T_n\}$ of bounded linear operators, from a normed linear space X into a normed linear space Y, is said to be *uniformly convergent* if there exists a bounded linear operator T from X into Y such that $\|T_n - T\| \to 0$ as $n \to \infty$. We then say that $\{T_n\}$ is *uniformly convergent* to T.

Note that uniform convergence means convergence in the norm of $\mathscr{B}(X,Y)$.

Theorem 4.4.4 *If* X *is a normed linear space and* Y *is a Banach space, then* $\mathscr{B}$(X,Y) *is a Banach space.*

Proof. Let $\{T_n\}$ be a Cauchy sequence in $\mathscr{B}(X,Y)$. Then, in particular, the sequence is bounded—that is, there is a constant K such that

$$\|T_n x\| \le K\|x\| \qquad \text{for all } x \in X,\ n \ge 1. \tag{4.4.4}$$

Since $\|T_n x - T_m x\| \le \|T_n - T_m\|\ \|x\| \to 0$ if $n \ge m \to \infty$, the sequence $\{T_n x\}$ is a Cauchy sequence in Y. Since Y is complete, $\lim T_n x$ exists. Define

$$Tx = \lim_n T_n x. \tag{4.4.5}$$

It is clear that T is a linear operator. From (4.4.4), (4.4.5) it also follows that $\|Tx\| \le K\|x\|$ for any $x \in X$. Hence $T \in \mathscr{B}(X,Y)$. It remains to show that $\|T - T_n\| \to 0$ as $n \to \infty$.

Since $\{T_n\}$ is a Cauchy sequence, for any $\varepsilon > 0$ there is a number n_0 such that $\|T_m - T_n\| \le \varepsilon$ if $m \ge n \ge n_0$. Thus, for all $x \in X$,

$$\|T_m x - T_n x\| \le \varepsilon\|x\|.$$

Taking $m \to \infty$, we get

$$\|Tx - T_n x\| \le \varepsilon\|x\|.$$

Hence, $\|T - T_n\| \le \varepsilon$ if $n \ge n_0$. This proves that $\|T - T_n\| \to 0$ as $n \to \infty$.

Definition 4.4.2 Let X be a Banach space, and let there be defined a multiplication $(x,y) \to xy$ from $X \times X$ into X such that (i) X is a ring (not necessarily with identity) with the operations $x + y, xy$, and (ii) $\|xy\| \le \|x\|\ \|y\|$. Then we call X a *Banach algebra*.

Example 1. Let X be a compact metric space. Then $C(X)$ is a Banach algebra with $(fg)(x) = f(x)g(x)$.

Example 2. If X is a Banach space, then $\mathscr{B}(X)$ is a Banach algebra with $(ST)(x) = S(Tx)$. Indeed,

$$\|ST\| = \sup_{\|x\|=1} \|(ST)x\| = \sup_{\|x\|=1} \|S(Tx)\| \le \|S\| \sup_{\|x\|=1} \|Tx\| = \|S\|\ \|T\|.$$

$\mathscr{B}(X)$ has an identity I, given by $Ix = x$.

If $X = R^n$, $\mathscr{B}(X)$ can be identified with the space of $n \times n$ matrices.

PROBLEMS

4.4.1. Let X be a normed linear space and let Y be a Banach space. Let T be a linear operator from X into Y with domain D_T dense in X. If T

is bounded—that is, if (4.4.1) holds for all $x \in D_T$—then there exists a unique bounded linear operator $\hat{T}$ from the whole space X into Y such that $\hat{T}x = Tx$ for all $x \in D_T$. We call $\hat{T}$ the *continuous extension* of T.

4.4.2. Let T be an additive operator [that is, $T(x_1 + x_2) = Tx_1 + Tx_2$] from a real normed linear space X into a normed linear space Y. If T is continuous, then T is homogeneous [that is, $T(\lambda x) = \lambda Tx$]. [*Hint:* Prove that $T[(m/n)x] = (m/n)Tx$, where m,n are integers.]

4.4.3. Let A be a symmetric $n \times n$ matrix. Consider A as an operator in R^n given by $x \to Ax$. Prove that $\|A\| = \max_j |\lambda_j|$, where λ_j are the eigenvalues of A.

4.4.4. Let $f(z) = \sum_{n=0}^{\infty} a_n z^n$ be an entire complex analytic function. Prove that for every $T \in \mathscr{B}(X)$, X a Banach space, the series $\sum_{n=0}^{\infty} a_n T^n$ $(T^0 = I)$ is absolutely convergent in $\mathscr{B}(X)$. One defines $f(T)$ by

$$f(T) = \sum_{n=0}^{\infty} a_n T^n.$$

4.4.5. Let $f(z), g(z)$ be entire complex analytic functions, and let $h(z) = f(z)g(z)$. Prove that for any $T \in \mathscr{B}(X)$, X a Banach space, $h(T) = f(T)g(T)$. In particular we have: $e^{\lambda T}e^{\mu T} = e^{(\lambda+\mu)T}$.

4.4.6. Find the norm of the operator $A \in \mathscr{B}(X)$ given by $(Af)(t) = tf(t)$ $(0 \leq t \leq 1)$, where (a) $X = C[0,1]$, (b) $X = L^p(0,1)$ and $(1 \leq p \leq \infty)$.

4.4.7. A linear operator from a normed linear space X into a normed linear space Y is bounded if and only if it maps bounded sets onto bounded sets.

4.4.8. A linear operator from a normed linear space X into a normed linear space Y is continuous if and only if it maps sequences converging to 0 into bounded sequences.

4.4.9. Let (X,μ) and (Y,ν) be measure spaces and consider the operator A defined by

$$(Af)(x) = \int K(x,y)f(y)\, d\nu(y)$$

from $L^r(Y,\nu)$ into $L^p(X,\mu)$, where $1/p + 1/r = 1$, $1 < p < \infty$. Prove that if $K(x,y) \in L^p(X \times Y, \mu \times \nu)$, then A is a bounded operator.

4.4.10. Prove that $L^1(R^n)$ is a Banach algebra if multiplication is defined by convolution—that is,

$$(f \cdot g)(x) = \int_{R^n} f(x-y)g(y)\, dy.$$

4.5 THE PRINCIPLE OF UNIFORM BOUNDEDNESS

The following important theorem is known as the *Banach-Steinhaus theorem* and also as the *principle of uniform boundedness.*

Theorem 4.5.1 *Let* X *be a Banach space and let* Y *be a normed linear space. Let* $\{T_\alpha\}$ *be a family of bounded linear operators from* X *into* Y. *If for each* $x \in X$ *the set* $\{T_\alpha x\}$ *is bounded, then the set* $\{\|T_\alpha\|\}$ *is bounded.*

Proof. If there exists a ball $B(x_0,\varepsilon)$ on which $\{T_\alpha x\}$ are uniformly bounded—that is, for some constant K,

$$\|T_\alpha x\| \le K \qquad \text{if } \|x - x_0\| < \varepsilon \tag{4.5.1}$$

—then we can easily complete the proof of the theorem. Indeed, for any $y \neq 0$ define

$$z = \frac{\varepsilon}{\|y\|}\, y + x_0.$$

Since $\|z - x_0\| \le \varepsilon$, (4.5.1) implies that $\|T_\alpha z\| \le K$. Hence

$$\frac{\varepsilon}{\|y\|}\|T_\alpha y\| - \|T_\alpha x_0\| \le \left\| \frac{\varepsilon}{\|y\|} T_\alpha y + T_\alpha x_0 \right\| = \|T_\alpha z\| \le K.$$

We thus get

$$\|T_\alpha y\| \le \frac{K + \|T_\alpha x_0\|}{\varepsilon}\|y\| \le \frac{K + K'}{\varepsilon}\|y\|,$$

where $K' = \sup_\alpha \|T_\alpha x_0\| < \infty$. We conclude that $\|T_\alpha\| \le (K + K')/\varepsilon$, and the assertion of the theorem follows.

To complete the proof of the theorem we only have to show that there exists a ball on which the $\{T_\alpha x\}$ are uniformly bounded. We shall suppose that no such ball exists, and derive a contradiction.

Fix a ball B_0. Then there exists a point x_1 in B_0 such that

$$\|T_\alpha x_1\| > 1$$

for some index α_1. By continuity we have $\|T_{\alpha_1} x\| > 1$ in some ball $B(x_1,\varepsilon_1)$ lying in B_0. We take $\varepsilon_1 < 1$. In this new ball the family $\{T_\alpha x\}$ is not uniformly bounded. Hence there exists a point $x_2 \in B(x_1,\varepsilon_1)$ such that

$$\|T_{\alpha_2} x_2\| > 2$$

for some index α_2, $\alpha_2 \neq \alpha_1$. By continuity, $\|T_{\alpha_2} x\| > 2$ in some ball $B(x_2,\varepsilon_2)$ lying in $B(x_1,\varepsilon_1)$. We take $\varepsilon_2 < \frac{1}{2}$. We can proceed in this way to define points $x_3,x_4\dots,x_n\dots$, indices $\alpha_3,\alpha_4,\dots,\alpha_n\dots$, and positive numbers $\varepsilon_3,\varepsilon_4,\dots,\varepsilon_n\dots$, such that $B(x_n,\varepsilon_n) \subset B(x_{n-1},\varepsilon_{n-1})$, $\varepsilon_n < 1/n$, $\alpha_n \neq \alpha_j$ if $1 \leq j < n$, and

$$\|T_{\alpha_n}x\| > n \qquad \text{for all } x \in B(x_n,\varepsilon_n).$$

By Theorem 3.4.1, there exists a point $z \in \bigcap_{n=1}^{\infty} \bar{B}(x_n,\varepsilon_n)$. Since $\|T_{\alpha_n}z\| \geq n$ for all $n \geq 1$, we get a contradiction to the asssumption that $\{T_\alpha z\}$ is a bounded set.

Definition 4.5.1 A sequence $\{T_n\}$ of bounded linear operators, from a normed linear space X into a normed linear space Y, is said to be *strongly convergent* if, for any $x \in X$, $\lim_n T_n x$ exists. If there is a bounded linear operator T such that $\lim_n T_n x = Tx$, then we say that $\{T_n\}$ is *strongly convergent* to T.

Note that uniform convergence implies strong convergence.

Theorem 4.5.2 *Let* X *be a Banach space and let* Y *be a normed linear space. If a sequence* $\{T_n\}$ *of bounded linear operators from* X *into* Y *is strongly convergent, then there exists a bounded linear operator* T *such that* $\{T_n\}$ *is strongly convergent to* T.

Proof. For any $x \in X$ the sequence $\{T_n x\}$ is bounded, since it is convergent. The principle of uniform boundedness implies that $\|T_n\| \leq K$ for all $n \geq 1$, where K is a constant. Hence,

$$\|T_n x\| \leq K\|x\| \qquad \text{for all } x \in X. \tag{4.5.2}$$

Now define $Tx = \lim_n T_n x$. It is clear that T is a linear operator. From (4.5.2) it also follows that $\|Tx\| \leq K\|x\|$. Thus T is a bounded linear operator. Finally, by the definition of T it follows that $\{T_n\}$ is strongly convergent to T.

PROBLEMS

4.5.1. Let X,Y,T_α be as in Theorem 4.5.1. Let

$$X_n = \{x;\ \sup_\alpha \|T_\alpha x\| \leq n\|x\|\}.$$

Then $X = \bigcup_n X_n$. Use the Baire category theorem to give a (somewhat) different proof of Theorem 4.5.1.

4.5.2. Show that

$$\int_0^{2\pi} \left| \frac{\sin (n + \frac{1}{2})x}{\sin \frac{1}{2}x} \right| dx \to \infty \qquad \text{as } n \to \infty.$$

[*Hint:* $|\sin \frac{1}{2}x| \leq \frac{1}{2}x$.]

4.5.3. The *Fourier series* of an integrable function $f(x)$ defined on $(0,2\pi)$ is the series

$$s(x) = \sum_{m=-\infty}^{\infty} a_m e^{imx}, \quad \text{where } a_m = \frac{1}{2\pi}\int_0^{2\pi} f(\xi)e^{-im\xi}\, d\xi.$$

Set $s_n(x) = \sum_{m=-n}^{n} a_m e^{imx}$, $f(x+2\pi) = f(x)$ $(0 < x < 2\pi)$. Prove that

$$s_n(y) = \frac{1}{2\pi}\int_0^{2\pi} f(y+x)D_n(x)\, dx,$$

where

$$D_n(x) = \frac{\sin (n+\frac{1}{2})x}{\sin \frac{1}{2}x}.$$

4.5.4. Denote by X the Banach space of all continuous functions $f(x)$ on $0 \le x \le 2\pi$, with $f(0) = f(2\pi)$, provided with the uniform norm. Denote by Y the Euclidean space R^1. Prove that the linear operator

$$T_n(f) = \frac{1}{2\pi}\int_0^{2\pi} f(x)D_n(x)\, dx$$

from X into Y is bounded, and

$$\|T_n\| = \frac{1}{2\pi}\int_0^{2\pi} |D_n(x)|\, dx.$$

4.5.5. There exists a continuous function $f(x)$ on $0 \le x \le 2\pi$, with $f(0) = f(2\pi)$, such that its Fourier series diverges at $x = 0$. [*Hint:* Use Problems 4.5.2, 4.5.4, and the Banach-Steinhaus theorem.]

4.6 THE OPEN-MAPPING THEOREM AND THE CLOSED-GRAPH THEOREM

The following theorem is called the *open-mapping theorem.*

Theorem 4.6.1 *Let* X *and* Y *be Banach spaces and let* T *be a bounded linear map from* X *onto* Y. *Then* T *maps open sets of* X *onto open sets of* Y.

Proof. We give the proof in three parts.

(a) For any $\varepsilon > 0$, denote by X_ε and Y_ε the balls in X and Y, respectively, with center 0 and radius ε. We shall prove that for any $\varepsilon > 0$ there is a δ such that

$$\overline{TX_{2\varepsilon}} \supset Y_\delta. \tag{4.6.1}$$

Since, obviously, $X = \bigcup_{n=1}^{\infty} nX_\varepsilon$, we have $Y = T(X) = \bigcup_{n=1}^{\infty} nTX_\varepsilon$. By the Baire category theorem (Theorem 3.4.2) it follows that, for some n, $n\overline{TX_\varepsilon}$ contains some ball $B(z,r)$ in Y. $\overline{TX_\varepsilon}$ then contains the ball $B(y_0,\delta)$, where $y_0 = z/n$, $\delta = r/n$. The set

$$P = \{y_1 - y_2;\, y_1 \in B(y_0,\delta),\, y_2 \in B(y_0,\delta)\}$$

is then contained in the closure of the set TQ, where

$$Q = \{x_1 - x_2;\, x_1 \in X_\varepsilon,\, x_2 \in X_\varepsilon\}.$$

Since Q is clearly a subset of $X_{2\varepsilon}$, we get $\overline{TX_{2\varepsilon}} \supset P$. Writing any point $y \in Y_\delta$ in the form $y = (y + y_0) - y_0$, we see that $Y_\delta \subset P$, and (4.6.1) follows.

(b) We shall prove that for any $\varepsilon_0 > 0$ there is a $\delta_0 > 0$ such that

$$TX_{2\varepsilon_0} \supset Y_{\delta_0}. \tag{4.6.2}$$

Choose a sequence $\{\varepsilon_n\}$ of positive numbers such that $\sum_{n=1}^{\infty} \varepsilon_n < \varepsilon_0$. By (a), there exists a sequence $\{\delta_n\}$ of positive numbers such that

$$\overline{TX_{\varepsilon_n}} \supset Y_{\delta_n} \qquad \text{for } n = 0,1,2,\ldots. \tag{4.6.3}$$

We may take the δ_n such that $\delta_n \to 0$ as $n \to \infty$.

Let $y \in Y_{\delta_0}$. By (4.6.3) (with $n = 0$) there is a point $x_0 \in X_{\varepsilon_0}$ such that $\|y - Tx_0\| < \delta_1$. Since $(y - Tx_0) \in Y_{\delta_1}$, (4.6.3) (with $n = 1$) implies that there exists a point $x_1 \in X_{\varepsilon_1}$ such that $\|y - Tx_0 - Tx_1\| < \delta_2$. Proceeding in this way step by step we obtain a sequence $\{x_n\}$ such that $x_n \in X_{\varepsilon_n}$ and

$$\left\| y - T\left(\sum_{k=0}^{n} x_k\right) \right\| < \delta_{n+1}. \tag{4.6.4}$$

The series $\sum_n x_n$ is absolutely convergent, since $\|x_n\| \leq \varepsilon_n$ and $\sum \varepsilon_n < \infty$. By Theorem 4.1.2, the series $\sum_{n=0}^{\infty} x_n$ is then also convergent; denote its sum by x. Then

$$\|x\| \leq \sum_{n=0}^{\infty} \|x_n\| \leq \sum_{n=0}^{\infty} \varepsilon_n < 2\varepsilon_0.$$

Since T is continuous, and since $\delta_n \to 0$ as $n \to \infty$, (4.6.4) gives $y = Tx$. We have thus shown that for any $y \in Y_{\delta_0}$ there exists a point $x \in X_{2\varepsilon_0}$ such that $Tx = y$. This proves (4.6.2).

(c) We shall prove that for any open set G of X and for any point $\bar{y} = T\bar{x}$, $\bar{x} \in G$, there is a ball Y_η $(\eta > 0)$ such that $\bar{y} + Y_\eta \subset T(G)$. This will show that $T(G)$ is open, and thus establish the assertion of the theorem. Since

G is open, there is a ball X_ε such that $\bar{x} + X_\varepsilon \subset G$. By (b), $T(X_\varepsilon) \supset Y_\eta$ for some $\eta > 0$. Hence

$$T(G) \supset T(\bar{x} + X_\varepsilon) = T\bar{x} + T(X_\varepsilon) \supset \bar{y} + Y_\eta.$$

Let T be a one-to-one linear operator from a linear vector space X into a linear vector space Y. We then define the *inverse* T^{-1} of T by

$$T^{-1}y = x \qquad \text{if } Tx = y.$$

It is easily verified that the domain of T^{-1} is a linear subspace of Y and that T^{-1} is a linear operator. T maps X onto Y if and only if the domain of T^{-1} is the whole space Y. We have the relations:

$$T^{-1}Tx = x \qquad \text{for all } x \in X,$$
$$TT^{-1}y = y \qquad \text{for all } y \in D_{T^{-1}}.$$

As an application of the open-mapping theorem we shall prove the following result.

Theorem 4.6.2 *Let* X *and* Y *be Banach spaces and let* T *be a one-to-one bounded linear map from* X *onto* Y. *Then* T^{-1} *is a bounded linear map.*

Proof. Since T^{-1} is a linear operator, we only have to show that it is continuous. By Theorem 4.6.1, $(T^{-1})^{-1} = T$ maps open sets in X onto open sets in Y. Thus T^{-1} has the following property: the inverse images of open sets are open sets. Hence, by the definition of continuity, T^{-1} is continuous.

Corollary 4.6.3 *Let* X *be a Banach space with any one of two norms* $\|\ \|_1$, $\|\ \|_2$. *Assume that there is a constant* K *such that*

$$\|x\|_1 \le K\|x\|_2 \qquad \textit{for all } x \in X. \tag{4.6.5}$$

Then the two norms are equivalent—that is there is a constant K′ *such that*

$$\|x\|_2 \le K'\|x\|_1 \qquad \text{for all } x \in X. \tag{4.6.6}$$

Proof. Denote by X_1 and X_2 the Banach spaces obtained by providing X with the norms $\|\ \|_1$ and $\|\ \|_2$, respectively. Consider the map $Tx = x$ from X_2 onto X_1. By (4.6.5), T is bounded. Hence, by Theorem 4.6.2, T^{-1} is also bounded. This gives (4.6.6).

Definition 4.6.1 Let T be a linear operator from a linear vector space X into a linear vector space Y, having a domain D_T. The *graph* G_T of T is the set of all points (x, Tx) in $X \times Y$, where x varies in D_T. If G_T is a closed set in the Cartesian product $X \times Y$, then we say that T is a *closed operator*.

Note that G_T is a linear subspace in $X \times Y$. Thus T is closed if and only if G_T is a closed linear subspace. Note also that T is closed if and only if $x_n \in D_T$, $x_n \to x$, $Tx_n \to y$ imply that $x \in D_T$ and $Tx = y$.

The following theorem is known as the *closed-graph theorem.*

Theorem 4.6.4 *Let* X *and* Y *be Banach spaces and let* T *be a linear operator from* X *into* Y (*with* $D_T = X$). *If* T *is closed, then it is continuous.*

Proof. The graph G_T of T is a closed linear subspace in the Cartesian product $X \times Y$ (with $\|(x,y)\| = \|x\| + \|y\|$). Hence G_T is itself a Banach space. Consider the bounded linear map J from G_T onto X given by $J(x,Tx) = x$. Since J is one-to-one, Theorem 4.6.2 implies that J^{-1} is a bounded linear operator from X into G_T—that is,

$$\|(x,Tx)\| = \|J^{-1}x\| \le K\|x\| \qquad (\text{for all } x \in X)$$

for some constant K. Consequently, $\|x\| + \|Tx\| \le K\|x\|$, and T is therefore bounded.

PROBLEMS

4.6.1. If T,S,T^{-1},S^{-1} belong to $\mathscr{B}(X)$, then $(TS)^{-1} \in \mathscr{B}(X)$ and $(TS)^{-1} = S^{-1}T^{-1}$.

4.6.2. Let X be a Banach space and let $A \in \mathscr{B}(X)$, $\|A\| < 1$. Prove that $(I + A)^{-1}$ exists and is given by

$$(I + A)^{-1} = \sum_{n=0}^{\infty} (-1)^n A^n,$$

where the series is absolutely convergent [in $\mathscr{B}(X)$]. Show also that

$$\|(I + A)^{-1}\| \le 1/(1 - \|A\|).$$

4.6.3. Let X be a Banach space and let T and T^{-1} belong to $\mathscr{B}(X)$. Prove that if $S \in \mathscr{B}(X)$ and $\|S - T\| < 1/\|T^{-1}\|$, then S^{-1} exists and is a bounded operator, and

$$\|S^{-1} - T^{-1}\| < \frac{\|T^{-1}\|}{1 - \|S - T\|\,\|T^{-1}\|}.$$

[*Hint*: $S = [(S - T)T^{-1} + I]T$.]

4.6.4. Let X and Y be two linear vector spaces. Find necessary and sufficient conditions for a subset G of $X \times Y$ to be the graph of a linear operator from X into Y.

4.6.5. Let X and Y be Banach spaces and let T be a bounded linear map from X into Y. If $T(X)$ is of the second category (in Y), then $T(X) = Y$.

4.6.6. Let X and Y be Banach spaces and let T be a linear map from a linear subspace D_T of X into Y. If D_T (in X) and the graph of T (in $X \times Y$) are closed, then T is bounded—that is, $\|Tx\| \leq K\|x\|$ for all $x \in D_T$ (K constant).

4.6.7. Let X be a normed linear space with any one of two norms $\| \quad \|_1$, $\| \quad \|_2$. If $\|x_n\|_2 \to 0$ implies $\|x_n\|_1 \to 0$, then (4.6.5) holds.

4.7 APPLICATIONS TO PARTIAL DIFFERENTIAL EQUATIONS

Definition 4.7.1 Let T be a linear operator from a linear vector space X into a linear vector space Y, having domain D_T. A linear operator S from X into Y is called an *extension* of T if $D_S \supset D_T$ and $Tx = Sx$ for all $x \in D_T$.

Definition 4.7.2 Let T be a linear operator from a normed linear space X into a normed linear space Y with domain D_T. Suppose there exists an operator S with the following properties: (i) S is a closed linear operator; (ii) S is an extension of T, and (iii) if S' is any operator with the properties (i), (ii), then S' is an extension of S. Then we call S the *closure* of T. When the closure of T exists, we often denote it by $\bar{T}$.

The following theorem gives a necessary and sufficient condition for T to have a closure.

Theorem 4.7.1 *Let* T *be a linear operator from a linear subspace* D_T *of a Banach space* X *into a Banach space* Y. T *has a closure* $\bar{T}$ *if and only if the following condition is satisfied*:

$$x_n \in D_T,\ x_n \to 0,\ Tx_n \to y \qquad \textit{imply} \qquad y = 0. \tag{4.7.1}$$

Proof. If T has a closure $\bar{T}$, and if y and x_n are as in (4.7.1) then $\bar{T}0 = y$. Hence $y = 0$. Suppose conversely that (4.7.1) holds, and denote by A the set of all points x' for which there exists sequences $\{x_n'\}$ such that $x_n' \to x'$, $Tx_n' \to y'$ for some $y' \in Y$. We define $\bar{T}x' = y'$. This definition is unambiguous, for, if $x_n'' \to x'$, $Tx_n'' \to y''$, then $x_n' - x_n'' \to 0$, $T(x_n' - x_n'') \to y' - y''$, and, therefore, by (4.7.1), $y' = y''$. It is easily seen that $\bar{T}$ is a linear operator It is also clear that if S is a closed linear operator that is an extension of T, then S is also an extension of $\bar{T}$. Thus it remains to show that $\bar{T}$ is closed.

Let $z_n \in D_T$, $z_n \to z$, $Tz_n \to u$ for some $u \in Y$. Then, for each n we can find a point x_n in D_T such that

$$\|z_n - x_n\| < \frac{1}{n}, \qquad \|\bar{T}z_n - Tx_n\| < \frac{1}{n}.$$

We then have: $x_n \in D_T$, $x_n \to z$, $Tx_n \to u$. Hence $z \in D_T$ and $\bar{T}z = u$. Thus $\bar{T}$ is indeed a closed operator.

Let $P(z) = P(z_1,\dots,z_n)$ be a polynomial in n variables $z_1,\dots,z_n$ of degree m, say

$$P(z) = \sum_{\Sigma\alpha_j \le m} a_{\alpha_1\cdots\alpha_n} z_1^{\alpha_1}\dots z_n^{\alpha_n}.$$

We shall use the notation: $\alpha = (\alpha_1,\dots,\alpha_n)$, $|\alpha| = \alpha_1 + \cdots + \alpha_n$, $z^\alpha = z_1^{\alpha_1}\cdots z_n^{\alpha_n}$. Then we can write

$$P(z) = \sum_{|\alpha| \le m} a_\alpha z^\alpha.$$

We write $D_j = \partial/\partial x_j$, $D^\alpha = D_1^{\alpha_1}\cdots D_n^{\alpha_n}$, and define

$$P(D) = \sum_{|\alpha| \le m} a_\alpha D^\alpha. \tag{4.7.2}$$

We call $P(D)$ a *partial differential operator with constant coefficients* of *degree* m. $P(D)$ can be considered as an operator from $C^\infty(\Omega)$ into itself, where Ω is any open subset of R^n. It maps a function $u(x)$ into the function

$$[P(D)u](x) = \sum_{|\alpha| \le m} a_\alpha D^\alpha u(x). \tag{4.7.3}$$

We can also consider it as an operator from $C^k(\Omega)$ into $C^{k-m}(\Omega)$ for any $k \ge m$. However, if $k < m$ or if $C^k(\Omega)$ is replaced by the space $L^p(\Omega)$, then $P(D)u$ cannot be defined on the whole space. In some important problems it is necessary to consider $P(D)$ in spaces $L^p(\Omega)$ $(1 \le p < \infty)$. What we then do is define $P(D)$ as in (4.7.3) on the linear subspace $\hat{C}^m(\Omega)$ consisting of all the functions in $C^m(\Omega)$ that have a finite norm

$$\|u\|_{p,m} \equiv \left\{\sum_{|\alpha| \le m} \int_\Omega |D^\alpha u|^p\, dx\right\}^{1/p}. \tag{4.7.4}$$

It is often desirable to extend $P(D)$ to a closed linear operator in $L^p(\Omega)$. We shall prove that this is possible.

Theorem 4.7.2 *The operator* P(D) *from* $\mathrm{L}^\mathrm{p}(\Omega)$ $(1 \le \mathrm{p} < \infty)$ *into itself, with domain* $\hat{\mathrm{C}}^\mathrm{m}(\Omega)$, *has a closure.*

Proof. In view of Theorem 4.7.1 it suffices to show that if $u_k \in \hat{C}^m(\Omega)$, $\|u_k\|_{0,p} \to 0$, $\|P(D)u_k - v\|_{0,p} \to 0$, then $v = 0$. Let $\varphi \in C_0^\infty(\Omega)$. Integration by parts gives

$$\int_\Omega P(D)u_k \cdot \varphi\, dx = \int_\Omega u_k \cdot P(-D)\varphi\, dx.$$

As $k \to \infty$, the integrals on the right converge to zero, whereas the integrals on the left converge to $\int_\Omega v \cdot \varphi \, dx$. Hence

$$\int_\Omega v \cdot \varphi \, dx = 0$$

for all $\varphi \in C_0^\infty(\Omega)$. By the result of Problem 3.3.6 we conclude that $v = 0$.

In the remaining part of this section we shall consider the differential operator $P(iD)$ (where $i = \sqrt{-1}$) rather than $P(D)$.

Definition The polynomial $P(z)$ [or the operator $P(iD)$] is *hypoelliptic* if every function $u \in C^m(\Omega)$ that satisfies $P(iD)u = 0$ in Ω is in $C^\infty(\Omega)$.

There are necessary and sufficient conditions for $P(z)$ to be hypoelliptic. These conditions are independent of Ω. Denote by $N(P)$ the set of complex zeros of the polynomial $P(z)$. We can then state:

Theorem 4.7.3 $P(z)$ *is hypoelliptic if and only if any one of the following conditions holds:*

(i) *For any constant* $A > 0$ *there exists a constant* $B > 0$ *such that all the points* $s = \sigma + i\tau$ *of* $N(P)$ *satisfy the inequality*

$$|\tau| \geq A \log |\sigma| - B.$$

(ii) *If* $|\sigma| \to \infty$, $\sigma + i\tau \in N(P)$, *then* $|\tau| \to \infty$.

(iii) *For any* $\theta \in R^n$,

$$\frac{P(\sigma + \theta)}{P(\sigma)} \to 1 \qquad \text{as } \sigma \in R^n, \ |\sigma| \to \infty.$$

(iv) *For any* α, $|\alpha| \geq 0$,

$$\frac{D^\alpha P(\sigma)}{P(\sigma)} \to 0 \qquad \text{as } \sigma \in R^n, \ |\sigma| \to \infty.$$

The proof of this theorem is beyond the scope of this book. However, one part of it, namely, condition (ii) as a necessary condition for hypoellipticity, will be derived here as an application of the closed-graph theorem. In fact, we shall prove the following stronger theorem.

Theorem 4.7.4 *Assume that there exists a domain* Ω *in* R^n *such that all the solutions of* $P(iD)u = 0$ *in* Ω *that belong to* $C^h(\Omega)$ (*for some* $h \geq m$) *belong also to* $C^{h+1}(\Omega)$. *Then the condition* (ii) *of Theorem* 4.7.3 *holds.*

Proof. For simplicity we may assume that $0 \in \Omega$. Denote by X the space of all the functions of $C^h(\Omega)$ that are solutions in Ω of $P(iD)u = 0$ and that have a finite norm

$$\|u\| = \sup_{x \in \Omega} \left\{ \exp\left[-|x|^2\right] \cdot \sum_{|\alpha| \le h} |D^\alpha u(x)| \right\}. \tag{4.7.5}$$

X is a Banach space. Let Ω_0 be a bounded domain containing 0 and whose closure lies in Ω, and denote by Y the space of functions in $C^{h+1}(\Omega_0)$, having a finite norm

$$\|u\|' = \sup_{x \in \Omega_0} \sum_{|\alpha| \le h+1} |D^\alpha u(x)|.$$

Y is also a Banach space. We shall denote by u^r the restriction of u (in Ω) to Ω_0. By assumption, if $u \in X$, then $u^r \in Y$. Consider now the map $Tu = u^r$ from X into Y. T is closed. Indeed, if

$$\|u_k - u\| \to 0, \qquad \|u_k - v\|' \to 0,$$

then clearly $v = u^r = Tu$. The closed-graph theorem can now be used. It implies that T is a bounded map—that is,

$$\|u\|' = \|u^r\|' \le K\|u\| \qquad (K \text{ constant}). \tag{4.7.6}$$

If $s \in N(P)$, then $u(x) = e^{-ix \cdot s}$ belongs to X. Hence, by (4.7.6),

$$|s| \sup_{x \in \Omega_0} e^{x \cdot \tau} \le K' \sup_{x \in \Omega} \exp\left[-|x|^2 + x \cdot \tau\right],$$

where K' is a constant. Since $0 \in \Omega_0$, the term on the left is $\ge |s|$. Since

$$-|x|^2 + x \cdot \tau \le -|x|^2 + |x|\,|\tau| \le \frac{|\tau|^2}{4},$$

we get

$$|\sigma| \le |s| \le K^* \exp\left[\frac{|\tau|^2}{4}\right] \qquad (K^* \text{ constant}).$$

Hence, $|\tau| \to \infty$ if $|\sigma| \to \infty$.

Definition Let Ω be a bounded domain and let u,v be locally integrable functions in Ω (that is, integrable on compact subsets of Ω). If for any $\varphi \in C_0^\infty(\Omega)$

$$\int_\Omega u D^\alpha \varphi \, dx = (-1)^{|\alpha|} \int v\varphi \, dx, \tag{4.7.7}$$

then we say that v is the αth *weak derivative* of u and write $D^\alpha u = v$ (w.d.).

If u and v are in $L^p(\Omega)$ $(1 \le p < \infty)$ and if for any compact subset K of Ω there exists exists a sequence $\{\varphi_j\}$ in $C^{|\alpha|}(\Omega)$ such that

$$\int_K |\varphi_j - u|^p\,dx \to 0, \qquad \int_K |D^\alpha \varphi_j - v|^p\,dx \to 0 \qquad \text{as } j \to \infty, \tag{4.7.8}$$

then we say that v is the αth L^p *strong derivative* of u and write $D^\alpha u = v$ (s.d.).

For example, the functions $u(x)$ of the space $H^{m,p}(\Omega)$ introduced in 3.3 have L^p strong derivatives of any order $\le m$, and all these derivatives belong to $L^p(\Omega)$.

PROBLEMS

4.7.1. If $D^\alpha u = v$ (s.d.), then $D^\alpha u = v$ (w.d.).

4.7.2. If $D^\alpha u = v$ (w.d.) and $D^\beta v = w$ (w.d.), then $D^{\beta+\alpha} u = w$ (w.d.).

4.7.3. If u and v belong to L^p locally in Ω $(1 \le p < \infty)$, and if $D^\alpha u = v$ (w.d.), then also $D^\alpha u = v$ (s.d.). [*Hint:* Take $\varphi_j = J_{\varepsilon_j}(u)$, $\varepsilon_j \to 0$, and use Problems 3.3.2, 3.3.4 and the relation $D^\alpha J_\varepsilon u = J_\varepsilon v$.]

4.7.4. Denote the closure of $P(D)$ in $L^p(\Omega)$ again by $P(D)$ and denote its domain by D_P. Prove that if the polynomial $P(z)$ has degree m, then $H^{m,p}(\Omega) \subset D_P$.

4.7.5. Prove that $H^{m,p}(\Omega) = \bigcap_{|\alpha| \le m} D_{D^\alpha}$.

4.7.6. Prove that $P(D)$ cannot be extended into a bounded linear operator from $L^p(\Omega)$ into itself.

4.7.7. Introduce in $C^\infty(\Omega)$, Ω a bounded open set in R^n, the norms $|u|_m = \sup_\Omega \sum_{0 \le |\alpha| \le m} |D^\alpha u|$. Denote by $C^*(\Omega)$ the subset of $C^\infty(\Omega)$ consisting of all functions with $|u|_m < \infty$ for all $m \ge 1$. Introduce in $C^*(\Omega)$ a metric $\rho(u,v) = \rho(u - v, 0)$ by

$$\rho(u,0) = \sum_{m=1}^{\infty} \frac{1}{2^m} \frac{|u|_m}{1 + |u|_m}.$$

Then $P(D)$ is a linear operator from the metric linear space $C^*(\Omega)$ into itself. Determine whether it is a continuous operator.

4.7.8. Find which of the following polynomials is hypoelliptic:

(a) $\sum_{j=0}^{k} a_j (z_1^2 + \cdots + z_n^2)^j$ $\quad$ (a_j constants);

(b) $(z_1^2 + z_2^2)(z_2^2 + 1)$;

(c) $z_n + \sum_{j=0}^{k} a_j (z_1^2 + \cdots + z_{n-1}^2)^j$ $\quad$ (a_j constants).

4.8 THE HAHN-BANACH THEOREM

Let X be a linear vector space. An operator (a linear operator) from X into the space $\mathbb{R}$ is called a *real (linear) functional* on X. Next let X be a normed linear space. A bounded linear operator from X into $\mathbb{R}$ is called a *real continuous linear functional* (on X). Similarly we define a *complex (linear) functional* (on X) and a *complex continuous linear functional* (on X) when X is a complex space and $\mathbb{R}$ is replaced by $\mathbb{C}$. We denote by X^* either one of the spaces $\mathscr{B}(X,\mathbb{R})$, $\mathscr{B}(X,\mathbb{C})$ of continuous linear functionals on X, and call it the *conjugate* (or *dual*) *space*. By Theorem 4.4.4, X^* is a Banach space.

The following theorem is one of the most important results in functional analysis. It will enable us to show that the space X^* is "sufficiently rich." It is called the *Hahn-Banach lemma.*

Theorem 4.8.1 *Let* X *be a real linear vector space and let* p *be a real functional on* X*, satisfying*

$$p(x+y) \le p(x) + p(y), \quad p(\lambda x) = \lambda p(x) \qquad \text{for all } \lambda \ge 0,\ x \in X,\ y \in X.$$

Let f *be a real linear functional on a linear subspace* Y *of* X*, satisfying*

$$f(x) \le p(x) \qquad \text{for all } x \in Y.$$

Then there exists a real linear functional F *on X such that*

$$F(x) = f(x) \quad \text{for all } x \in Y, \qquad F(x) \le p(x) \quad \text{for all } x \in X.$$

Proof. Denote by $\mathscr{K}$ the set of all pairs (Y_α, g_α), where Y_α is a linear subspace of X that contains Y, and g_α is a real linear functional on Y_α satisfying

$$g_\alpha(x) = f(x) \quad \text{for all } x \in Y, \qquad g_\alpha(x) \le p(x) \quad \text{for all } x \in Y_\alpha.$$

We define a relation "$\le$" by $(Y_\alpha, g_\alpha) \le (Y_\beta, g_\beta)$ if $Y_\alpha \subset Y_\beta$ and $g_\alpha = g_\beta$ on Y_α. Then $\mathscr{K}$ becomes a partially ordered set. Every totally ordered subset $\{(Y_\beta, g_\beta)\}$ clearly has an upper bound (Y', g') given by: $Y' = \bigcup Y_\beta$, $g' = g_\beta$ on Y_β. Hence, by Zorn's lemma (Theorem 4.2.2), there is a maximal element (Y_0, g_0). If we can show that $Y_0 = X$, then the proof of the theorem is complete (with $F = g_0$).

We shall assume that $Y_0 \ne X$ and derive a contradiction. Let $y_1 \in X$, $y_1 \notin Y_0$, and consider the linear manifold Y_1 spanned by Y_0 and y_1. Each point x in Y_1 has the form $x = y + \lambda y_1$, where $y \in Y_0$ and $\lambda \in \mathbb{R}$ are uniquely determined (for otherwise we get $y_1 \in Y_0$). Define a linear functional g_1 on Y_1 by $g_1(y + \lambda y_1) = g_0(y) + \lambda c$. If we can choose the constant c such that

$$g_0(y) + \lambda c \le p(y + \lambda y_1) \tag{4.8.1}$$

for all $\lambda \in \mathbb{R}$, $y \in Y_0$, then $(Y_1, g_1) \in \mathscr{K}$ and $(Y_0, g_0) \le (Y_1, g_1)$, $Y_0 \ne Y_1$. This contradicts the maximality of (Y_0, g_0).

To choose c that satisfies (4.8.1), we note that for any two points x,y in Y_0,

$$g_0(y) - g_0(x) = g_0(y - x) \le p(y - x) \le p(y + y_1) + p(-y_1 - x).$$

Hence

$$-p(-y_1 - x) - g_0(x) \le p(y + y_1) - g_0(y).$$

This implies that

$$A \equiv \sup_{x \in Y_0} \{-p(-y_1 - x) - g_0(x)\} \le \inf_{y \in Y_0} \{p(y + y_1) - g_0(y)\} \equiv B.$$

Let c be any number satisfying: $A \le c \le B$. Then

$$c \le p(y + y_1) - g_0(y) \qquad \text{for all } y \in Y_0, \tag{4.8.2}$$

$$-p(-y_1 - y) - g_0(y) \le c \qquad \text{for all } y \in Y_0. \tag{4.8.3}$$

Multiplying both sides of (4.8.2) by λ, $\lambda > 0$, and replacing y by y/λ, we obtain

$$\lambda c \le p(y + \lambda y_1) - g_0(y). \tag{4.8.4}$$

Multiplying both sides of (4.8.3) by λ, $\lambda < 0$, and replacing y by y/λ, we again obtain (4.8.4), after using the homogeneity of p. Note that (4.8.4) holds also if $\lambda = 0$. Since (4.8.4) implies (4.8.1), the proof of the theorem is complete.

The following theorem, whose proof is based on Theorem 4.8.1, is called the *Hahn-Banach theorem.*

Theorem 4.8.2 *Let* X *be a normed linear space and let* Y *be a linear subspace. Then to every* $y^* \in Y^*$ *there corresponds an* $x^* \in X^*$ *such that*

$$\|x^*\| = \|y^*\|, \quad x^*(y) = y^*(y) \qquad \textit{for all } y \in Y.$$

Proof. If X is a real space, then the assertion follows from the Hahn-Banach lemma with $p(x) = \|y^*\| \, \|x\|, f(x) = y^*(x)$ and $x^* = F$. To prove that $\|x^*\| \le \|y^*\|$, write for any $x \in X$, $x^*(x) = \theta|x^*(x)|$, where $\theta = \pm 1$. Then

$$|x^*(x)| = \theta x^*(x) = x^*(\theta x) \le p(\theta x) = \|y^*\| \, \|\theta x\| = \|y^*\| \, \|x\|.$$

This shows that $\|x^*\| \le \|y^*\|$. The reverse inequality is obvious.

Suppose now that X is a complex normed linear space and define real functionals f_1, f_2 over Y by

$$y^*(y) = f_1(y) + if_2(y).$$

From the relations

$$y^*(\lambda y + \lambda' y') = f_1(\lambda y + \lambda' y') + if_2(\lambda y + \lambda' y'),$$

$$y^*(\lambda y + \lambda' y') = \lambda y^*(y) + \lambda' y^*(y') = \lambda[f_1(y) + if_2(y)] + \lambda'[f_1(y') + if_2(y')]$$

it follows that if $\lambda, \lambda' \in \mathbb{R}$, then $f_1(\lambda y + \lambda' y') = \lambda f_1(y) + \lambda' f_1(y')$. We also have

$|f_1(y)| \leq |y^*(y)| \leq \|y^*\| \, \|y\|$. Thus, regarding X as a real normed linear space, we can apply the result of Theorem 4.8.2 for X real and conclude that there exists a real linear functional F_1 on X such that

$$\|F_1\| \leq \|y^*\|, \quad F_1(y) = f_1(y) \qquad \text{for all } y \in Y. \tag{4.8.5}$$

Now define x^* on the complex linear space X by

$$x^*(x) = F_1(x) - iF_1(ix).$$

x^* is clearly additive, and $x^*(\lambda x) = \lambda x^*(x)$ if λ is real. Also, $x^*(ix) = F_1(ix) - iF_1(-x) = ix^*(x)$. Hence x^* is linear. To prove that x^* is an extension of y^* it suffices to show that

$$y^*(y) = f_1(y) - if_1(iy) \qquad \text{if } y \in Y$$

—that is, that $f_2(y) = -f_1(iy)$. But this follows from

$$f_1(iy) + if_2(iy) = y^*(iy) = iy^*(y) = if_1(y) - f_2(y).$$

It remains to prove that $\|x^*\| \leq \|y^*\|$. For any $x \in X$, write $x^*(x) = re^{i\theta}$, where $r \geq 0$, θ real. Then, by (4.8.5),

$$|x^*(x)| = |x^*(e^{-i\theta}x)| = F_1(e^{-i\theta}x) \leq \|y^*\| \, \|e^{-i\theta}x\| = \|y^*\| \, \|x\|.$$

Thus $\|x^*\| \leq \|y^*\|$.

We shall use the Hahn-Banach theorem to prove the following useful result.

Theorem 4.8.3 *Let* Y *be a linear subspace of a normed linear space* X, *and let* $x_0 \in X$ *be such that*

$$\inf_{y \in Y} \|y - x_0\| = d > 0. \tag{4.8.6}$$

Then there exists a point $x^* \in X^*$ *such that*

$$x^*(x_0) = 1, \quad \|x^*\| = \frac{1}{d}, \quad x^*(y) = 0 \qquad \textit{for all } y \in Y.$$

Proof. Denote by Y_1 the linear space spanned by Y and x_0. Since $x_0 \notin Y$, every point x in Y_1 has the form $x = y + \lambda x_0$, where $y \in Y$ and the scalar λ are uniquely determined. Consider the linear function z^* defined by $z^*(y + \lambda x_0) = \lambda$. If $\lambda \neq 0$, then

$$\|y + \lambda x_0\| = |\lambda| \left\| \frac{y}{\lambda} + x_0 \right\| \geq |\lambda| \, d.$$

Hence $|z^*(x)| \leq \|x\|/d$ for all $x \in Y_1$—that is, $\|z^*\| \leq 1/d$. Now let $\{y_n\} \subset Y$, $\|x_0 - y_n\| \to d$. Then

$$1 = z^*(x_0 - y_n) \leq \|z^*\| \, \|x_0 - y_n\| \to \|z^*\| \, d.$$

Thus $\|z^*\| = 1/d$. Applying Theorem 4.8.2. with y^* replaced by z^*, we get the assertion of the theorem.

Corollary 4.8.4 *Let* X *be a normed linear space. For any* $x \neq 0$ *there exists an* $x^* \in X^*$ *with* $\|x^*\| = 1$, $x^*(x) = \|x\|$.

Indeed, by Theorem 4.8.3 with $Y = \{0\}$, there exists a $z^* \in X^*$ such that $\|z^*\| = 1/\|x\|$, $z^*(x) = 1$. Take $x^* = \|x\|z^*$.

Corollary 4.8.5 *If* y *and* z *are distinct points in a normed linear space* X, *then there exists an* $x^* \in X^*$ *such that* $x^*(y) \neq x^*(z)$.

This follows from Corollary 4.8.4 with $x = y - z$.

Corollary 4.8.6 *For any* x *in a normed linear space* X,

$$\|x\| = \sup_{x^* \neq 0} \frac{|x^*(x)|}{\|x^*\|} = \sup_{\|x^*\| = 1} |x^*(x)|. \tag{4.8.7}$$

The right-hand side is clearly less than or equal to the left-hand side. On the other hand, by Corollary 4.8.4, there exists an x_0^* such that $x_0^*(x) = \|x\|$, $\|x_0^*\| = 1$. Hence the left-hand side of (4.8.7) is less than or equal to the right-hand side.

Corollary 4.8.7 *Let* Y *be a linear subspace of a normed linear space* X. *If* Y *is not dense in* X, *then there exists a functional* $x^* \neq 0$ *such that* $x^*(y) = 0$ *for all* $y \in Y$.

Indeed, if $\bar{Y} \neq X$, then there is a point x_0 satisfying (4.8.6). Now we can apply Theorem 4.8.3.

Let X be a normed linear space and let $x^* \in X^*$. The *null space* of x^* is the set $N_{x^*} = \{x; x^*(x) = 0\}$. If $x^* \neq 0$, then there exists an element $x_0 \neq 0$ such that $x^*(x_0) = 1$. Each element $x \in X$ can then be written in the form $x = z + \lambda x_0$, where $\lambda = x^*(x)$ and $z = x - \lambda x_0 \in N_{x^*}$. This gives a decomposition of X into a direct sum of N_{x^*} and the one-dimensional space spanned by x_0.

Let X be a normed linear space and let $x^* \in X^*$, $x^* \neq 0$. For any real number c, the set $\{x; \text{Re } x^*(x) = c\}$ is called a *hyperplane*. It divides the space X into two half-spaces defined by the inequalities $\text{Re } \{x^*(x)\} \geq c$ and $\text{Re } \{x^*(x)\} \leq c$. If X is a real space, then this hyperplane coincides with the vector sum $N_{x^*} + cx_0$, where x_0 is any point for which $x^*(x_0) = 1$.

Definition 4.8.1 Let K be a subset of a normed linear space X and let $x_0 \in K$. If an $x^* \in X^*$ satisfies $\text{Re } \{x^*(x)\} \leq \text{Re } \{x^*(x_0)\}$ for all $x \in K$, then we say that x^* *supports* K at x_0 (or that it is *tangent* to K at x_0). The hyperplane Re $\{x^*(x)\} = \|x_0\|$ is called a *supporting* (or a *tangent*) *hyperplane* to K at x_0.

Corollary 4.8.4 can now be rephrased in a geometric language as follows:

Corollary 4.8.8 *Let* X *be a normed linear space and let* x_0 *belong to the boundary of the unit ball (that is,* $\|x_0\| = 1$*). Then there exists a tangent hyperplane to the unit ball at* x_0.

PROBLEMS

4.8.1. Let X be a normed linear space and let $\{x_n\} \subset X$. A point y_0 is the limit of linear combinations $\sum_{j=1}^{n} c_j x_j$ if and only if $x^*(y_0) = 0$ for all x^* for which $x^*(x_j) = 0$ for $1 \le j < \infty$.

4.8.2. Any two Hamel bases in a linear vector space X have the same cardinal number. This number is called the *dimension* of X.

4.8.3. Two linear vector spaces X and Y are called *isomorphic* if there is a one-to-one linear map T from X onto Y. Prove that X and Y are isomorphic if and only if they have the same dimension.

4.8.4. Prove that dim $l^\infty = c$, where c is the cardinal number of the continuum. [*Hint:* The elements $(\lambda, \lambda^2, \ldots, \lambda^n \ldots), 0 < \lambda < 1$, are linearly independent.]

4.8.5. Let X be an infinite-dimensional Banach space. Prove that there exists an infinite, strictly decreasing sequence $\{Y_n\}$ of infinite-dimensional closed linear subspaces of X. [*Hint:* Take Y_1 to be the null space of some $x_1^* \ne 0$ in X^*. Take Y_2 to be the null space of some $x_2^* \ne 0$ in Y_1^*, and so on.]

4.8.6. Let X be an infinite-dimensional Banach space. Prove that the linear space l^∞ is isomorphic to a linear subspace of X. [*Hint:* Take $x_n \in Y_{n-1}$, $x_n \notin Y_n$ with Y_n as in Problem 4.8.5, $Y_0 = X$, such that $\|x_n\| = 1/2^n$, and consider the map $\{\xi_n\} \to \sum \xi_n x_n$.]

4.8.7. Prove that the dimension of an infinite-dimensional Banach space is $\ge c$.

4.8.8. A Banach space X is finite-dimensional if and only if every linear subspace is closed. [*Hint:* Consider the linear space spanned by the x_n in Problem 4.8.6.]

4.8.9. Let $u(t)$ be a function defined on $a < t < b$ with values in a Banach space X. We say that $u(t)$ is *strongly differentiable at* t [*on* (a,b)] if $\lim_{h \to 0} \{[u(t+h) - u(t)]/h\}$ exists [for all $t \in (a,b)$]. The limit is denoted by $du(t)/dt$ and is called the *derivative* of $u(t)$. For functions $A(t)$ with values in $\mathscr{B}(X)$, if $\lim_{h \to 0} \{[A(t+h)x - A(t)x]/h\}$ exists for any $x \in X$, then we say that $A(t)$ has a *strong derivative*. If $\lim_{h \to 0} \{[A(t+h) - A(t)]/h\}$ exists (in the uniform topology), then we say that $A(t)$ is *uniformly differentiable*. Prove that e^{tA} $[A \in \mathscr{B}(X)]$ is uniformly differentiable and $de^{tA}/dt = Ae^{tA}$.

4.8.10. Let X be a real normed linear space, and let $u(t)$ be continuous and strongly differentiable in (a,b). Then for any $a < \alpha < \beta < b$,

$$\|u(\beta) - u(\alpha)\| \le (\beta - \alpha) \sup_{\alpha \le t \le \beta} \left\| \frac{du(t)}{dt} \right\|.$$

[*Hint:* Apply x^* to $u(\beta) - u(\alpha)$.]

4.8.11. Let X be a Banach space and let $A \in \mathscr{B}(X)$. Prove that for any $x_0 \in X$ there exists a unique solution $u(t)$ of

$$\frac{du}{dt} + Au = 0 \quad (0 \le t \le 1), \qquad u(0) = x_0.$$

[*Hint:* To prove uniqueness, show that $e^{tA}u(t)$ is independent of t.]

4.8.12. For every normed linear space X there is a set A such that X is isomorphic to a subspace of the Banach space of functions f on A with norm $\|f\| = \sup_{t \in A} |f(t)|$. If X is separable, A is countable. [*Hint:* Let $\{x_\alpha; \alpha \in A\}$ be dense in X. Let $f(x,\alpha)$ be the bounded linear functional (in $x \in X$) satisfying $\|f(.,\alpha)\| = 1, f(x_\alpha,\alpha) = \|x_\alpha\|$. Define the isomorphism $x \to g_x(\alpha)$, where $g_x(\alpha) = f(x,\alpha)$. Prove: $\big| |f(x,\alpha)| - \|x\| \big| \le 2\|x_\alpha - x\|$.]

4.8.13. A separable normed linear space is isomorphic to a subspace of l^∞.

4.8.14. Let X, Y be Banach spaces and let $f(x,y)$ be a functional on $X \times Y$, continuous and linear in each variable separately. Then $f(x,y)$ is continuous in the variable (x,y). [*Hint:* To prove continuity at (0,0), let $Y_m = \{y; \|y\| \le 1/m\}$, $\varepsilon > 0$, $A_n = \{x \in X; |f(x,y)| \le \varepsilon$ for all $y \in Y_n\}$. A_n is closed [since $f(x,y)$ is continuous in x] and $\bigcup A_n = X$ [since $f(x,y)$ is continuous in y]. Use the Baire category theorem to deduce that $|f(x,y)| \le 2\varepsilon$ if $x \in U$, $y \in Y_m$ for some m and for some neighborhood U of 0 in X.]

4.9 APPLICATION TO THE DIRICHLET PROBLEM

Let Ω be a bounded domain in R^n and let $\Delta = \sum_{j=1}^{n} \partial^2/\partial x_j^2$. The equation $\Delta u = 0$ is called the *Laplace equation*. Any function in $C^2(\Omega)$ that satisfies the Laplace equation in Ω is called a *harmonic function* in Ω. We shall denote by $\partial\Omega$ the boundary of Ω. We say that $\partial\Omega$ is in C^2 if $\partial\Omega$ can be covered by a finite number of open subsets G_i, with each $G_i \bigcap \partial\Omega$ having a parametric representation in terms of functions in C^2.

The *Dirichlet problem* is the following one: given a continuous function f on $\partial\Omega$, find a function u in $C^2(\Omega)$, continuous in $\overline{\Omega}$, and satisfying

$$\Delta u = 0 \qquad \text{in } \Omega, \tag{4.9.1}$$

$$u = f \qquad \text{on } \partial\Omega. \tag{4.9.2}$$

Theorem 4.9.1 *The Dirichlet problem has at most one solution.*

This follows immediately from the following result, known as the *(weak) maximum principle.*

Theorem 4.9.2 *Let* u *be a harmonic function in* Ω, *continuous in* $\overline{\Omega}$. *Then* $\max_{\overline{\Omega}} u = \max_{\partial\Omega} u$.

Proof. If the assertion is not true, then there is a point $y \in \Omega$ such that $u(y) > \max_{\partial\Omega} u$. Consider the function $v(x) = u(x) + \varepsilon\,|x - y|^2$. If $\varepsilon > 0$ is sufficiently small, then $v(y) = u(y) > \max_{\partial\Omega} v$. Hence the maximum of v on $\overline{\Omega}$ is attained at a point z in Ω. At that point,

$$\frac{\partial v}{\partial x_i} = 0, \quad \frac{\partial^2 v}{\partial x_i^2} \le 0 \qquad (i = 1,\ldots,n).$$

Hence $\Delta v \le 0$ at z. However,

$$\Delta v = \Delta u + \varepsilon\Delta(|z - y|^2) = 2n\varepsilon > 0$$

—a contradiction.

As for the existence of a solution to the Dirichlet problem, we quote the following result:

Theorem 4.9.3 *If* $\partial\Omega$ *is in* C^2, *then for any continuous function* f *on* $\partial\Omega$ *there exists a solution to the Dirichlet problem* (4.9.1), (4.9.2).

Theorem 4.9.3 is much deeper than Theorem 4.9.1. There are various methods of proving it, but each of them requires lengthy developments. We shall present one of these methods in Section 5.4.

Consider now the function

$$k(x,y) = \begin{cases} |x - y|^{2-n}, & \text{if } n > 3, \\ -\log|x - y|, & \text{if } n = 2. \end{cases}$$

We call it a *fundamental solution* of the Laplace equation. It satisfies the Laplace equation in x, when x varies in $\Omega - \{y\}$, and it grows to ∞ as $x \to y$.

Definition 4.9.1 A function $G(x,y)$, defined for each $y \in \Omega$ and $x \in \bar{\Omega} - \{y\}$, is called *Green's function* (for the Laplace operator in Ω) if:

(i) $G(x,y) = k(x,y) + h(x,y)$, where $h(x,y)$ is harmonic in x when $x \in \Omega$.
(ii) $G(x,y)$ is continous in x when $x \in \bar{\Omega} - \{y\}$.
(iii) $G(x,y) = 0$ for $x \in \partial\Omega$.

One can use Green's function in order to represent the solution u of the Dirichlet problem (4.9.1), (4.9.2) in terms of an integral involving the boundary values f (compare Problem 4.9.2). What we shall now do is construct Green's function by using, as a tool, the Hahn-Banach theorem. We shall consider only the case $n = 2$, which is somewhat simpler than the case where $n \geq 3$. We shall assume that $\partial\Omega$ consists of a finite number of continuously differentiable closed curves. The following condition then holds (the proof is omitted):

(P) For any z near $\partial\Omega$, denote by T_z the tangent plane to $\partial\Omega$ at the point on $\partial\Omega$ nearest to z. Denote by z' the reflection of z with respect to T_z. Then

$$\max_{x \in \partial\Omega} \frac{|z - x|}{|z' - x|} \to 1 \qquad \text{if } z \to z_0,\ z_0 \in \partial\Omega.$$

Theorem 4.9.4 *If* n = 2 *and* $\partial\Omega$ *is in* C^1, *then Green's function exists.*

Proof. Denote by X the Banach space of all continuous functions on $\partial\Omega$ with the maximum norm, and denote by X' the linear subspace consisting of those functions f for which the Dirichlet problem (4.9.1), (4.9.2) has a solution. For any $y \in \Omega$, consider the linear functional L_y on X' defined by $L_y(f) = u(y)$, where u is the solution of (4.9.1), (4.9.2). From Theorem 4.9.2 it follows that L_y is bounded and that its norm is 1.

By the Hahn-Banach theorem L_y can be extended into a bounded linear functional on X, having norm 1. We denote such an extension again by L_y. For each $z \notin \partial\Omega$ consider the element f_z of X given by

$$f_z(x) = \log|x - z| \qquad (x \in \partial\Omega),$$

and define $k_y(z) = L_y(f_z)$. We claim that $k_y(z)$ is a harmonic function. To prove it let $z' = (z_1 + \delta, z_2)$. Then

$$\frac{k_y(z') - k_y(z)}{\delta} = L_y\left(\frac{f_{z'} - f_z}{\delta}\right).$$

As $\delta \to 0$, $(f_{z'} - f_z)/\delta \to \partial f_z/\partial z_1$, where $\partial f_z/\partial z_1$ is the element $\partial(\log|x - z|)/\partial z_1$ of X. Since L_y is a continuous operator on X, we get

$$\frac{\partial k_y(z)}{\partial z_1} \quad \text{exists and equals} \quad L_y\left(\frac{\partial f_z}{\partial z_1}\right).$$

In the same way one shows that $k_y(z)$ has any number of derivatives and these are equal to the application of L_y to the corresponding derivatives of f_z. In particular, $\Delta k_y(z) = L_y(\Delta f_z)$. Since, however, Δf_z is the function $\sum_{j=1}^{2} \partial^2(\log |x - z|)/\partial z_j^2 = 0$, we conclude that $k_y(z)$ is harmonic outside $\partial\Omega$.

Since $\log |x - z|$ is harmonic in $x \in \Omega$ when $z \notin \overline{\Omega}$, we have

$$k_y(z) = L_y(f_z) = \log |y - z| \qquad \text{if } z \notin \overline{\Omega}. \tag{4.9.3}$$

Now take a point z in Ω and near $\partial\Omega$. Denote by z' its reflection with respect to T_z [see the condition (P) above]. Then

$$\|f_z - f_{z'}\| = \max_{x \in \partial\Omega} \log \frac{|z - x|}{|z' - x|} \to 0 \qquad \text{if } z \to z_0,\ z_0 \in \partial\Omega.$$

It follows that $L_y(f_z - f_{z'}) \to 0$—that is,

$$k_y(z) - k_y(z') \to 0 \qquad \text{if } z \to z_0.$$

But, by (4.9.3), $\lim k_y(z')$ exists and equals $\log |y - z_0|$. Hence $k_y(z)$ $(z \in \Omega)$ can be extended into a continuous function $\hat{k}_y(z)$ in $\overline{\Omega}$, and $\hat{k}_y(z) = \log |y - z|$ if $z \in \partial\Omega$. The function $-\log |x - y| + \hat{k}_y(x)$ satisfies the condition (i)–(iii) in the definition 4.9.1 of Green's function.

PROBLEMS

4.9.1. If $\partial\Omega$ is in C^2 and ν denotes the outward normal to $\partial\Omega$, then

$$\int_\Omega (u\,\Delta v - v\,\Delta u)\,dx = \int_{\partial\Omega} \left(u \frac{\partial v}{\partial \nu} - v \frac{\partial u}{\partial \nu}\right) dS$$

for any two functions u, v that are uniformly continuous in Ω together with their first and second derivatives. [*Hint:* Use the integration-by-parts formula

$$\int_\Omega \frac{\partial w}{\partial x_i}\,dx = \int_{\partial\Omega} w \cos(\nu, x_i)\,dS.]$$

4.9.2. Let u be a solution of (4.9.1), (4.9.2), and assume that its first and second derivatives are uniformly continuous in Ω. Then for any $y \in \Omega$,

$$u(y) = -c \int f(x) \frac{\partial}{\partial \nu} G(x,y)\,dS,$$

where $G(x,y)$ is Green's function (assumed to exist and to be twice differentiable in $x \in \overline{\Omega} - \{y\}$) and c is a positive constant depending only on n.

4.9.3. Green's function, if existing, is unique.

4.9.4. Prove that Green's function is nonnegative.

4.9.5. Green's function is symmetric—that is, $G(x,y) = G(y,x)$. [Assume, in the proof, that $G(x,y)$ has two continuous derivatives in $x \in \overline{\Omega} - \{y\}$.]

4.10 CONJUGATE SPACES AND REFLEXIVE SPACES

Theorem 4.10.1 *Let* X *be a normed linear space. If* X^* *is separable, then* X *is separable.*

The converse is not true in general (see Section 4.14).

Proof. Let $\{x_n^*\}$ be a dense sequence in X^*, and choose $x_n \in X$ such that $\|x_n\| = 1$, $|x_n^*(x_n)| > \|x_n^*\|/2$. Denote by A the set of all linear combinations of the x_n with rational coefficients. A is countable, and it suffices to show that it is dense in X. If $\bar{A} \neq X$, then, by Corollary 4.8.7, there exists an $x^* \neq 0$ such that $x^*(x) = 0$ for all $x \in A$. Let $\{x_{n_i}^*\}$ be a subsequence of $\{x_n^*\}$ that converges to x^*. Then

$$\|x^* - x_{n_i}^*\| \geq |(x^* - x_{n_i}^*)(x_{n_i})| = |x_{n_i}^*(x_{n_i})| \geq \frac{\|x_{n_i}^*\|}{2}.$$

Hence $x_{n_i}^* \to 0$. It follows that $x^* = 0$—a contradiction.

Definition 4.10.1 Let X and Y be normed linear spaces. Suppose there exists a one-to-one linear map T from X onto Y. Then we say that X and Y are *linearly isomorphic* (or, briefly, *isomorphic*) and we call T a *linear isomorphism* (or, briefly, an *isomorphism*). If, further, $\|Tx\| = \|x\|$ for all $x \in X$, then we say that X and Y are *isometrically isomorphic* and we call T an *isometric isomorphism.*

Consider now the conjugate of X^*. We denote it by X^{**}. To each $x \in X$ we can correspond an element $\hat{x}$ in X^{**} by $\hat{x}(x^*) = x^*(x)$ for all $x^* \in X^*$. We write $\hat{x} = \kappa x$ and call κ the *natural imbedding* of X into X^{**}. We also write $\hat{X} = \kappa(X)$.

Theorem 4.10.2 *Let* X *be a normed linear space. Then the natural imbedding from* X *into* X^{**} *is an isometric isomorphism between* X *and* $\hat{X}$.

The linearity of κ is obvious. The relation $\|\kappa x\| = \|x\|$ follows from Corollary 4.8.6.

The natural imbedding of X into X^{**} is useful even for deriving results concerning X and X^* only. We give such a result in the following theorem.

Theorem 4.10.3 *Let* $\{x_\alpha\}$ *be a set of elements in a normed linear space* X. *Suppose that* $\sup_\alpha |x^*(x_\alpha)| < \infty$ *for any* $x^* \in X^*$. *Then* $\sup_\alpha \|x_\alpha\| < \infty$.

Proof. We can apply the principle of uniform boundedness to the family $\{\hat{x}_\alpha\}$. We conclude that $\sup_\alpha \|\hat{x}_\alpha\| < \infty$. From this the assertion follows.

Definition A Banach space X is called *reflexive* if $\kappa(X) = X^{**}$.

We shall prove later on (in Section 4.14) that L^p spaces are reflexive if $1 < p < \infty$, but are not reflexive if $p = 1, p = \infty$.

Lemma 4.10.4 *Let X and Y be isometrically isomorphic normed linear spaces. If Y is reflexive, then X is reflexive.*

Proof. Let σ be the isometric isomorphism from X onto Y. Define a map τ from X^* into Y^* by

$$(\tau x^*)(y) = x^*(\sigma^{-1}y).$$

It is easily seen that τ is an isometric isomorphism from X^* onto Y^*. Similarly define an isometric isomorphism ρ from X^{**} onto Y^{**} by

$$(\rho x^{**})(y^*) = x^{**}(\tau^{-1}y^*).$$

Denote by κ_1 and κ_2 the natural imbeddings of X into X^{**} and of Y onto Y^{**}, respectively. Now take any element x_0^{**} in X^{**} and define $x_0 = \sigma^{-1}\kappa_2^{-1}\rho x_0^{**}$. The point x_0 is in X and, for any $x^* \in X^*$,

$$x^*(x_0) = x^*[\sigma^{-1}(\kappa_2^{-1}\rho x_0^{**})] = (\tau x^*)(\kappa_2^{-1}\rho x_0^{**}) = \rho x_0^{**}(\tau x^*) = x_0^{**}(x^*)$$

Thus $\kappa_1 x_0 = x_0^{**}$. We have thus shown that $\kappa_1(X) = X^{**}$—that is, X is reflexive.

Theorem 4.10.5 *A closed linear subspace of a reflexive Banach space is reflexive.*

Proof. Let Y be a closed linear subspace of a reflexive Banach space X. Consider the map $\sigma: X^* \to Y^*$ defined by $(\sigma x^*)(y) = x^*(y)$ for $y \in Y$. Note that $\|\sigma x^*\| \le \|x^*\|$. Next define for $y^{**} \in Y^{**}$, $(\tau y^{**})(x^*) = y^{**}(\sigma x^*)$ for $x^* \in X^*$. Then $|(\tau y^{**})(x^*)| \le \|y^{**}\| \, \|\sigma x^*\| \le \|y^{**}\| \, \|x^*\|$. Hence $\tau y^{**} \in X^{**}$. Since X is reflexive, the natural imbedding κ maps X onto X^{**}. Hence $\kappa^{-1}(\tau y^{**})$ is a well-defined element of X. We shall prove that it belongs to Y. If $x = \kappa^{-1}(\tau y^{**}) \notin Y$, then there exists an $x^* \in X^*$ with $x^*(y) = 0$ for all $y \in Y$ and $x^*(x) \neq 0$. Since $x^*(y) = 0$ for all $y \in Y$, $\sigma x^* = 0$. Hence

$$0 = y^{**}(\sigma x^*) = (\tau y^{**})(x^*) = (\kappa x)(x^*) = x^*(x)$$

—a contradiction. We have thus proved $\kappa^{-1}[\tau(Y^{**})] \subset Y$.

Now let $y_0^{**} \in Y^{**}$ and define $x_0 = \kappa^{-1}(\tau y_0^{**})$. For any $y^* \in Y^*$, let x^* be any extension of y^* to an element of X^*. Then $y^* = \sigma x^*$, and

$$y_0^{**}(y^*) = (\tau y_0^{**})(x^*) = (\kappa x_0)(x^*) = x^*(x_0) = y^*(x_0),$$

since $x_0 \in Y$. Thus y_0^{**} is the image of x_0 under the natural imbedding of Y into Y^{**}. This proves that Y is reflexive.

Theorem 4.10.6 *A Banach space is reflexive if and only if its conjugate is reflexive.*

Proof. If X is reflexive, then the same is true also of X^{**}, which is then isometrically isomorphic to X (with the map κ). It is obvious that $(X^*)^{**} = [(X^*)^*]^* = (X^{**})^*$. Now let $x_0^{***} \in (X^*)^{**}$. Then $x_0^{***} \in (X^{**})^*$. Thus the functional $x_0^* = x_0^{***}\kappa$ defined by $x_0^*(x) = x_0^{***}(\kappa x)$ for $x \in X$ is in X^*, and

$$x^{**}(x_0^*) = (\kappa x)(x_0^*) = x_0^*(x) = x_0^{***}(\kappa x) = x_0^{***}(x^{**})$$

for any $x^{**} \in X^{**}$. Hence x_0^{***} is the image of x_0^* under the natural imbedding. Since x_0^{***} is an arbitrary element of $(X^*)^{**}$, we conclude that X^* is reflexive. Suppose conversely that X^* is reflexive. Then, by what we have just proved, X^{**} is reflexive. Since $\kappa(X)$ is a closed linear subspace of X^{**}, Theorem 4.10.5 shows that it is reflexive. Hence, by Lemma 4.10.4, also X is reflexive.

We now introduce an important concept of convergence.

Definition 4.10.2 Let X be a normed linear space. A sequence $\{x_n\}$ in X is said to be *weakly convergent* if there exists an element $x \in X$ such that $\lim_n x^*(x_n) = x^*(x)$ for any $x^* \in X^*$. We call x the *weak limit* of $\{x_n\}$ and we say that $\{x_n\}$ *converges weakly* to x. A set $K \subset X$ is called *weakly sequentially compact* if every sequence $\{x_n\}$ in K contains a subsequence that converges weakly to a point in K. A sequence $\{x_n\}$ in X is called a *weak Cauchy sequence* if $\{x^*(x_n)\}$ is a Cauchy sequence for any $x^* \in X^*$. The space X is called *weakly complete* if every weak Cauchy sequence has a weak limit. A set $K \subset X$ is *weakly closed* if the weak limit of any weakly convergent sequence $\{x_n\}$ in K is also in K.

PROBLEMS

4.10.1. A sequence $\{x_n\}$ cannot have two distinct weak limits.

4.10.2. Let X be a normed linear space and let B be a dense subset of X^*. If a sequence $\{x_n\}$ in X is bounded, and if $\lim_n x^*(x_n)$ exists for each $x^* \in B$, then $\lim_n x^*(x_n)$ exists for all x^* in X^*.

Theorem 4.10.7 *Let* $\{x_n\}$ *be a weakly convergent sequence in a normed linear space* X, *then*:

(a) $\{x_n\}$ *is bounded*;

(b) *its weak limit* x *belongs to the closed linear subspace spanned by* $\{x_1,x_2,\ldots,x_n,\ldots\}$;

(c) $\|x\| \leq \underline{\lim}_n \|x_n\|$.

Proof. The assertion (a) follows from Theorem 4.10.3. Next, if (b) is not true, then by Theorem 4.8.3 there is an $x^* \in X^*$ such that $x^*(x_n) = 0$ for all $n \geq 1$, but $x^*(x) \neq 0$. This is impossible, since $\{x_n\}$ converges weakly to x. Finally, from $x^*(x) = \lim_n x^*(x_n)$ we get

$$|x^*(x)| = \lim_n |x^*(x_n)| \leq \underline{\lim}_n \{\|x^*\| \, \|x_n\|\} = \|x^*\| \, \underline{\lim}_n \|x_n\|.$$

Now use Corollary 4.8.6.

The next two results are concerned with important properties of reflexive spaces.

Theorem 4.10.8 *Let* X *be a reflexive Banach space. A set* $K \subset X$ *is weakly sequentially compact if and only if it is both bounded and weakly closed.*

Proof. Suppose K is weakly sequentially compact. Then K is weakly closed. Indeed, let $\{x_n\} \subset K$, $\{x_n\}$ weakly convergent to x. There is a subsequence $\{x_{n_i}\}$ that is weakly convergent to a point y in K. Since $x = y$, $x \in K$. Next, K is bounded. Indeed, otherwise there is a sequence $\{x_n\} \subset K$ such that $\|x_n\| > n$. By our assumption on K, there is a subsequence $\{x_{n_i}\}$ that is weakly convergent. But then, by Theorem 4.10.7 (a), $\{x_{n_i}\}$ is bounded, which is impossible since $\|x_{n_i}\| > n_i$.

Suppose now that K is bounded and weakly closed. Let $\{x_n\}$ be a (bounded) sequence in K. We shall extract a weakly convergent subsequence. Denote by Z the closed linear subspace spanned by $\{x_1,x_2,\ldots,x_n,\ldots\}$. Z is clearly separable and, by Theorem 4.10.5, it is reflexive. Hence Z^{**} is also separable. Theorem 4.10.1 implies that Z^* is separable. Let $\{x_n^*\}$ be a dense sequence in Z^*. Since the sequence $\{x_1^*(x_n)\}$ is bounded, we can extract a convergent subsequence $\{x_1^*(x_{n,1})\}$. Next we extract from the sequence $\{x_2^*(x_{n,1})\}$ a convergent subsequence $\{x_2^*(x_{n,2})\}$. We proceed in this way step by step. In the kth step we extract a convergent subsequence $\{x_k^*(x_{n,k})\}$ from the bounded sequence $\{x_k^*(x_{n,k-1})\}$. Let $y_k = x_{k,k}$. Then $\lim_k x_n^*(y_k)$ exists for any x_n^*. Since $\{x_n^*\}$ is dense in Z^* and since $\{y_k\}$ is a bounded sequence in Z, it follows by Problem 4.10.2 that $\lim_k z^*(y_k)$ exists for each $z^* \in Z^*$. Hence $\lim_k (\kappa y_k)(z^*)$ exists for each $z^* \in Z^*$. By Theorem 4.5.2 it follows that there exists a $y^{**} \in Z^{**}$ such that $\lim_k (\kappa y_k)(z^*) = y^{**}(z^*)$ for all $z^* \in Z^*$. Since Z is reflexive there is then a point $y \in Z$ such that its image in Z^{**} (by the natural imbedding) is y^{**}. Hence,

$$\lim_k z^*(y_k) = z^*(y) \qquad \text{for any } z^* \in Z^*.$$

Now take any $x^* \in X^*$. It determines an element $z^* \in Z^*$ by $z^*(z) = x^*(z)$ for all $z \in Z$. Since $\{y_k\} \subset Z$, $y \in Z$, we conclude that

$$\lim_k x^*(y_k) = x^*(y).$$

Hence $\{y_k\}$ is weakly convergent to y. Since K is weakly closed, $y \in K$. Thus K is weakly sequentially compact.

Theorem 4.10.9 *A reflexive normed linear space* X *is weakly complete.*

Proof. Let $\{x_n\}$ be a weak Cauchy sequence in X. Then $\lim_n x^*(x_n)$ exists for all $x^* \in X$. By Theorem 4.10.7, $\{x_n\}$ is bounded. From the proof of Theorem 4.10.8 it then follows that there is a subsequence $\{x_{n_i}\}$ that converges weakly to some element $x \in X$. But then

$$\lim_n x^*(x_n) = \lim_{n_i} x^*(x_{n_i}) = x^*(x) \qquad \text{for any } x^* \in X$$

Thus $\{x_n\}$ is weakly convergent to x.

PROBLEMS

4.10.3. In a finite-dimensional normed linear space, the concepts of convergence and weak convergence coincide.

4.10.4. A weakly closed set is closed. [The converse is not true; Problem 4.14.6 shows that the closed set $\{x;\ \|x\| = 1\}$ in $L^2(0,1)$ is not weakly closed.]

4.10.5. A sequentially compact set is weakly sequentially compact, but not conversely. [*Hint:* Consider the closed unit ball in an infinite-dimensional reflexive space.]

4.10.6. Let X be a normed linear space and let Y be a dense linear subspace in X. Prove that every $y^* \in Y^*$ can be extended into an $x^* = \sigma y^*$ of X^*, and that σ is an isometric isomorphism of Y^* onto X^*. [*Hint:* See Theorem 3.8.1.]

4.10.7. A normed linear space that is weakly complete is complete.

4.10.8. Prove that either a Banach space X is reflexive or its successive second duals $X^{**}, X^{****}, \ldots$ are all distinct.

4.10.9. Let X and Y be Banach spaces and let $\{T_\alpha\}$ be a family of bounded linear maps from X into Y. If $\sup_\alpha \|y^*(T_\alpha x)\| < \infty$ for any $x \in X$, $y^* \in X^*$, then $\sup_\alpha \|T_\alpha\| < \infty$.

4.10.10. Let X be a normed linear space and let $\{x_n\} \subset X$. If $\{x_n\}$ converges weakly to y, then there exists a sequence $\left\{\sum_{i=1}^{m_n} \lambda_{i,n} x_i\right\}$ (with $\lambda_{i,n}$ scalars) that converges to y.

4.10.11. Let T be a bounded linear operator from a normed linear space X into a normed linear space Y. If $\{x_n\}$ is a sequence in X that is weakly convergent to x_0, then $\{Tx_n\}$ (in Y) is weakly convergent to Tx_0.

4.10.12. A sequence $\{x_n\}$ in a normed linear space X is weakly convergent to x_0 if and only if the following conditions hold: (i) the sequence $\{\|x_n\|\}$ is bounded, and (ii) $x^*(x_n) \to x^*(x_0)$ as $n \to \infty$, for any x^* in a set B dense in X^*.

4.10.13. Let K be a weakly closed set in a reflexive Banach space X. Then there exists an element $\bar{x} \in K$ such that $\inf_{x \in K} \|x\| = \|\bar{x}\|$. Thus the function $f(x) = \|x\|$ takes a minimum on K.

4.10.14. Prove that the unit ball K of a normed linear space X is weakly closed. Prove also that if the norm is strictly convex (for definition, see Problem 4.3.3), then the minimum of $f(x) = \|x\|$ on K is attained at one point only.

4.11 TYCHONOFF'S THEOREM

Let X be a topological space and let $\mathscr{K}$ be the class of its open sets. A subclass $\mathscr{K}_0$ of $\mathscr{K}$ is called a *neighborhood basis* for the topology if every set in $\mathscr{K}$ is a union of sets of $\mathscr{K}_0$. Recall that a *neighborhood* of a point x is any open set containing x. A *neighborhood basis* at a point x is a subclass $\mathscr{K}_0(x)$ of $\mathscr{K}$ having the property that every neighborhood of x contains a set of $\mathscr{K}_0(x)$.

Suppose we are given for every $x \in X$ a neighborhood basis $\mathscr{K}_0(x)$ at x. Then the following properties are clearly satisfied:

(1) $x \in W(x)$ for every $W(x) \in \mathscr{K}_0(x)$.

(2) The intersection $W_1(x) \cap W_2(x)$ of two sets $W_1(x)$, $W_2(x)$ of $\mathscr{K}_0(x)$ contains a set of $\mathscr{K}_0(x)$.

(3) If $y \in W(x)$, $W(x) \in \mathscr{K}_0(x)$, then there is a set $W(y)$ of $\mathscr{K}_0(y)$ such that $W(y) \subset W(x)$.

Conversely, if for any $x \in X$ there is assigned a class $\mathscr{K}_0(x)$ of sets of X such that (1)–(3) hold, then a topology can be defined on X by taking the open sets to be all the possible unions of sets $W(x)$ in $\mathscr{K}_0(x)$ with $x \in X$. The proof of this fact is left to the reader. Note that each set $\mathscr{K}_0(x)$ is then a neighborhood basis at x.

Lemma 4.11.1 *A topological space X is compact if and only if every class* $\{F_\alpha\}$ *of closed sets in X with empty intersection contains a finite subclass* $\{F_{\alpha_1},\ldots,F_{\alpha_n}\}$ *with empty intersection.*

The proof is obtained by noting that the sets $F_\alpha^c = X - F_\alpha$ are open, and using the relations:

$$\left(\bigcap_\alpha F_\alpha\right)^c = \bigcup_\alpha F_\alpha^c, \qquad \left(\bigcap_{i=1}^n F_{\alpha_i}\right)^c = \bigcup_{i=1}^n F_{\alpha_i}^c.$$

Lemma 4.11.2 *A closed subset of a compact topological space is compact.*

Proof. Let B be a closed subset of a compact topological space X. Since (by Problem 3.1.12) a closed subset of B is a closed subset of X, Lemma 4.11.1 yields the assertion.

A class $\{F_\alpha\}$ of sets of X is called *centralized* (or is said to have the *finite intersection property*) if any finite number of sets F_α have nonempty intersection.

Lemma 4.11.1 immediately yields:

Lemma 4.11.3 *A topological space X is compact if and only if every centralized class of closed sets has a nonempty intersection.*

A class $\{M_\alpha\}$ of sets is said to have a *common point of contact* if $\bigcap_\alpha \overline{M}_\alpha \neq \varnothing$. Any point x that belongs to $\bigcap_\alpha \overline{M}_\alpha$ is called a *point of common contact.*

We give a slightly different version of Lemma 4.11.3, the proof of which is left to the reader.

Lemma 4.11.4 *A topological space X is compact if and only if every centralized class of sets in X has a common point of contact.*

A sequence $\{x_n\}$ in a topological space X is said to be *convergent* to x if for any neighborhood U of x there is an n_0 such that $x_n \in U$ for all $n \geq n_0$. If every sequence in X has a subsequence that is convergent to some point of X, then we say that X is *sequentially compact.* X is said to satisfy the *first countability axiom* if there exists a countable neighborhood basis at each point of x.

Theorem 4.11.5 *Let X be a topological space satisfying the first countability axiom. If X is compact, then it is also sequentially compact.*

Proof. Let $\{x_n\}$ be a sequence in X. For every positive integer n let $M_n = \{x_n, x_{n+1}, \ldots\}$. The class $\{M_n\}$ is centralized. Hence, by Lemma 4.11.4, there is a point $\bar{x}$ of common contact—that is, $\bar{x} \in \bigcap_{n=1}^{\infty} \overline{M}_n$. Take a countable basis of neighborhoods U_m at $\bar{x}$. The subsequence $\{x_{n,1}\}$ of $\{x_n\}$ consisting of all points that belong to U_1 must be infinite. Indeed, otherwise $U_1 \cap M_n = \emptyset$ for some n sufficiently large. But this is impossible, since $\bar{x} \in \overline{M}_n$. By the same argument, the subsequence $\{x_{n,2}\}$ of $\{x_n\}$ consisting of all points that belong to $U_1 \cap U_2$ is infinite. Note that $\{x_{n,2}\}$ is a subsequence of $\{x_{n,1}\}$. We proceed in this way step by step. In the kth step we get an infinite subsequence $\{x_{n,k}\}$ that is contained in $U_1 \cap U_2 \cap \cdots \cap U_k$. It is now easily seen that sequence $\{x_{k,k}\}$ converges to $\bar{x}$.

Let X_α ($\alpha \in A$, A an ordered set) be topological spaces. Let

$$X = \prod_{\alpha \in A} X_\alpha$$

be the space whose elements are the ordered sets $x = \{x_\alpha\}$. We define on X a topology by giving a neighborhood basis $\mathscr{K}_0(x^0)$ at each point $x^0 = \{x_\alpha^0\}$. A set $U(x^0)$ is in $\mathscr{K}_0(x^0)$ if there exists a finite number of indices $\alpha_1,\ldots,\alpha_n$ from A and neighborhoods $U(x_{\alpha_i}^0)$ of $x_{\alpha_i}^0$ in X_{α_i} such that $U(x^0)$ is the set of all points $\{x_\alpha\}$ with $x_{\alpha_i} \in U(x_{\alpha_i}^0)$ for $i = 1,\ldots,n$ and x_α arbitrary in X_α if $\alpha \neq \alpha_1,\ldots,\alpha \neq \alpha_n$. It is easy to verify that the conditions (1)–(3) for neighborhood bases are satisfied. Thus X is a topological space with the classes $\mathscr{K}_0(x^0)$ as neighborhood bases. We call X the *Cartesian product* of the topological spaces X_α.

Consider the function $f_\alpha(x) = x_\alpha$ from X onto X_α. The image of a set $M \subset X$ under this map is called the *projection of M on X_α*.

The next result is called the *Tychonoff theorem.*

Theorem 4.11.6 *The Cartesian product of any family of compact topological spaces is a compact topological space.*

Proof. Let $X = \prod X_\alpha$ be the Cartesian product of the compact topological spaces X_α. In view of Lemma 4.11.4 it suffices to show that every centralized class of sets $\{K^\nu\}$ has at least one point of common contact. Consider all the centralized classes of sets $\{N^\mu\}$ that contain $\{K^\nu\}$, and define a partial ordering: $\{N^\mu\} \leq \{M^\lambda\}$ if each set N^μ coincides with some set M^λ. We can apply Zorn's lemma and thus conclude that there exists a maximal centralized class $\{M^\lambda\}$ containing $\{K^\nu\}$. It suffices to show that $\{M^\lambda\}$ has at least one point of common contact.

The maximality of $\{M^\lambda\}$ implies:

(a) The intersection of any finite number of sets of $\{M^\lambda\}$ is again in $\{M^\lambda\}$.

(b) If a set intersects any finite number of sets of $\{M^\lambda\}$, then it belongs to $\{M^\lambda\}$.

Indeed, if B is an intersection of a finite number of sets of $\{M^\lambda\}$, then the class $\{M^\lambda, B\}$ is centralized. By the maximality of $\{M^\lambda\}$ we conclude that $\{M^\lambda, B\} = \{M^\lambda\}$—that is, $B \in \{M^\lambda\}$. The proof of (b) is similar.

Denote the projection of a set M^λ on X_α by M_α^λ. For any α, $\{M_\alpha^\lambda\}$ is a centralized class in X_α. Since X_α is compact, the sets $\{M_\alpha^\lambda\}$ have a point x_α^0 of common contact. Let $x^0 = \{x_\alpha^0\}$. We shall prove that x^0 is a point of common contact of the class $\{M^\lambda\}$—that is, every neighborhood $U(x^0)$ of x^0 in a neighborhood basis at x^0 intersects every set M^λ.

By definition $U(x^0)$ consists of points $\{x_\alpha\}$, where $x_{\alpha_i} \in U(x_{\alpha_i}^0)$ for some indices $\alpha_1, \dots, \alpha_n$ and x_α varies in X_α for all $\alpha \neq \alpha_1, \dots, \alpha \neq \alpha_n$. Here $U(x_{\alpha_i}^0)$ is a neighborhood of $x_{\alpha_i}^0$ in X_{α_i}. Denote by $U_i(x^0)$ the neighborhood of x^0 defined by: $x_{\alpha_i} \in U(x_{\alpha_i}^0)$ and x_α varies in X_α for all $\alpha \neq \alpha_i$. Then

$$U(x^0) = \bigcap_{i=1}^{n} U_i(x^0).$$

Since $x_{\alpha_i}^0$ is a point of contact of the class $\{M_{\alpha_i}^\lambda\}$, $U(x_{\alpha_i}^0)$ intersects each $M_{\alpha_i}^\lambda$. Hence $U_i(x^0)$ intersects each set M^λ. By (a), $U_i(x^0)$ intersects any finite intersection of sets of $\{M^\lambda\}$. But then (b) implies that $U_i(x^0)$ is in $\{M^\lambda\}$. Hence, by (a), $U(x^0)$ is in $\{M^\lambda\}$. It follows that $U(x^0)$ intersects each set M^λ.

PROBLEMS

4.11.1. Let X and Y be topological spaces. A function f from X into Y is called *continuous* if the inverse image of open sets are open sets. Prove that f is continuous if and only if the inverse images of closed sets are closed sets.

4.11.2. A continuous function maps compact sets onto compact sets.

4.11.3. Prove that the mapping $f_\alpha(x) = x_\alpha$ from $X = \prod_{\alpha \in A} X_\alpha$ onto X_α is continuous.

4.11.4. A mapping f from a topological space X into a topological space Y is continuous if for any $y_0 = f(x_0)$ and for any neighborhood N of y_0 there exists a neighborhood M of x_0 such that $f(M) \subset N$.

4.11.5. A compact subset of a Hausdorff topological space is closed.

4.11.6. A compact Hausdorff space is normal.

4.11.7. A real-valued continuous function on a sequentially compact topological space attains its supremum and its infimum.

4.11.8. A function X from a topological space X into a topological space Y is called a *homeomorphism* (of X onto Y) if f is a one-to-one map from X onto Y, and if both f and its inverse are continuous. Prove that a continuous one-to-one map from a compact topological space onto a Hausdorff space is a homeomorphism.

4.11.9. A closed interval on the real line is not homeomorphic to an open interval on the real line.

4.12 WEAK TOPOLOGY IN CONJUGATE SPACES

Definition 4.12.1 A set X is called a *topological linear space* if:

(i) X is a linear vector space.
(ii) X is a Hausdorff topological space.
(iii) The function $(x,y) \to x + y$ from $X \times X$ into X is continuous.
(iv) The function $(\lambda,x) \to \lambda x$ from $\mathscr{F} \times X$ into X is continuous.

Here $\mathscr{F}$ is either $\mathbb{R}$ or $\mathbb{C}$.

Note that metric linear spaces and, in particular, normed linear spaces, are topological linear spaces.

Let X be a normed linear space and let X^* be its conjugate. We shall introduce a new topology on X^*, different from the Banach-space topology considered in previous sections. This new topology will be given by neighborhood bases $\mathscr{K}_0(y^*)$, $y^* \in X^*$. The sets of $\mathscr{K}_0(y^*)$ are given by

$$N(y^*; x_1,\dots,x_n; \varepsilon) = \{x^*; |x^*(x_i) - y^*(x_i)| < \varepsilon \quad \text{for } 1 \le i \le n\},$$

where ε is any positive number, n is any positive integer, and $x_1,\dots,x_n$ are any points of X.

Theorem 4.12.1 *The neighborhood bases $\mathscr{K}_0(\mathrm{y}^*)$ satisfy the conditions* (1)–(3) *of Section* 4.11 *and, therefore, determine a topology on* X*. *With this topology,* X* *is a topological linear space.*

Definition The topology introduced by the neighborhood bases $\mathscr{K}_0(y^*)$ is called the *weak topology*.

Proof. The condition (1) is obvious. The condition (2) follows by noting that the set

$$\{x^*; |x^*(x_i) - y^*(x_i)| < \varepsilon \quad \text{for } 1 \le i \le n\} \cap \{x^*; |x^*(\bar{x}_i) - y^*(\bar{x}_i)| < \varepsilon' \quad \text{for } 1 \le i \le m\}$$

contains the set

$$\{x^*; |x^*(z_i) - y^*(z_i)| < \varepsilon'' \quad \text{for } 1 \le i \le n + m\},$$

where $\varepsilon'' = \min(\varepsilon,\varepsilon')$, $z_i = x_i$ if $1 \le i \le n$, $z_{n+i} = \bar{x}_i$ if $1 \le i \le m$.

To prove (3) let $y_1^* \in N(y^*; x_1,\dots,x_n; \varepsilon)$. Then

$$|y_1^*(x_i) - y^*(x_i)| < \varepsilon' \quad \text{if } 1 \le i \le n,$$

for some $0 < \varepsilon' < \varepsilon$. It follows that

$$N(y_1^*; x_1,\dots,x_n; \varepsilon - \varepsilon') \subset N(y^*; x_1,\dots,x_n; \varepsilon).$$

Having proved (1)–(3), we conclude that X^* is a topological space having the sets $\mathscr{K}_0(y^*)$ for neighborhood bases. To prove that X^* is a Hausdorff space, let $y_1^* \neq y_2^*$. Then there exists an $x_0 \in X$ such that $y_1^*(x_0) \neq y_2^*(x_0)$. Hence, if $\varepsilon = |y_1^*(x_0) - y_2^*(x_0)|/2$, then the two neighborhoods

$$N(y_1^*; x_0; \varepsilon), \qquad N(y_2^*; x_0; \varepsilon)$$

are disjoint. We have thus proved the condition (iv) of Section 3.1 for a topological space to be a Hausdorff space.

To complete the proof of Theorem 4.12.1 we have to establish the properties (iii), (iv) in Definition 4.12.1. To prove (iii), take any neighborhood $N = N(y_1^* + y_2^*; x_1,\ldots,x_n; \varepsilon)$ of $y_1^* + y_2^*$. By Problem 4.11.4, it suffices to find neighborhoods N_1 and N_2 of y_1^* and y_2^*, respectively, such that the vector sum $N_1 + N_2$ is contained in N. If we take

$$N_j = \left\{y_j^*; x_1,\ldots,x_n; \frac{\varepsilon}{2}\right\} \qquad (j = 1,2),$$

then, obviously, $N_1 + N_2 \subset N$.

To prove (iv) take any neighborhood $N' = N(\lambda_0 y^*; x_1,\ldots,x_n; \varepsilon)$ of $\lambda_0 y^*$. It suffices to find a neighborhood M of λ_0 in $\mathscr{F}$ and a neighborhood N'' of y^* such that $\{\lambda x^*; \lambda \in M, x^* \in N''\}$ is contained in N'. Let

$$M = \{\lambda; |\lambda - \lambda_0| < \varepsilon'\}, \qquad N'' = N(y^*; x_1,\ldots,x_n; \varepsilon').$$

Then, if $\lambda \in M$ and $x^* \in N''$,

$$\begin{aligned}|\lambda x^*(x_i) - \lambda_0 y^*(x_i)| &= |\lambda x^*(x_i) - \lambda y^*(x_i)| + |\lambda - \lambda_0|\,|y^*(x_i)| \\ &\leq |\lambda|\varepsilon' + K\varepsilon' \leq (|\lambda_0| + \varepsilon')\varepsilon' + K\varepsilon',\end{aligned}$$

where $K = \max_i |y^*(x_i)|$. Taking $\varepsilon' < 1$, $(|\lambda_0| + 1 + K)\varepsilon' < \varepsilon$, we find that $|\lambda x^*(x_i) - \lambda_0 y^*(x_i)| < \varepsilon$ for $1 \leq i \leq n$. Hence $\lambda x^* \in N'$. This completes the proof of (iv) and of the theorem.

Whereas in the norm topology bounded sets in X^* are not relatively compact in general, they are relatively compact in the weak topology. This will be proved in the next theorem, called the *theorem of Alaoglu*. This result is one of the most important illustrations of the importance of the concept of weak topology.

Theorem 4.12.2 *Let* X *be a normed linear space. Then the closed unit ball in* X* *is compact in the weak topology.*

Proof. To each $x \in X$ we correspond the closed real interval $I_x = [-\|x\|, \|x\|]$ if X is a real space, and the closed disc in the complex plane $I_x = \{z; |z| \leq \|x\|\}$ if X is a complex space. Denote by I the Cartesian product $\prod_{x \in X} I_x$. By Tychonoff's theorem, I is compact.

Consider now the set $\Gamma = \{f\}$ of all the elements f of X^* with $\|f\| \le 1$. We have to prove that Γ is compact. For each $f \in \Gamma$, $|f(x)| \le \|x\|$. Thus we can correspond to f a point $\hat{f}$ in I, having coordinates $f(x)$. We shall denote this correspondence by σ. Thus, $\sigma f = \hat{f}$. It is clear that σ is one-to-one. Let $I' = \sigma(\Gamma)$. From the definition of neighborhoods in X^* and in I it follows that σ and its inverse are both continuous. Hence Γ is compact if and only if I' is compact.

To prove that I' is compact it suffices (by Lemma 4.11.2) to show that I' is closed. Let $\hat{f}_0$ be a point in $\bar{I}'$. We first show that the functional f_0 defined by $f_0(x) = \hat{f}_0(x)$ (= the x-coordinate of $\hat{f}_0$) is linear. Take any points x_1, x_2 in X and any scalars λ_1, λ_2. Let $U_0 = U(\hat{f}_0; x_1 x_2, \lambda_1 x_1 + \lambda_2 x_2; \varepsilon)$ be a neighborhood of $\hat{f}_0$ consisting of the points $\{y_x\}$ with

$$|y_{x_1} - \hat{f}_0(x_1)| < \varepsilon, \qquad |y_{x_2} - \hat{f}_0(x_2)| < \varepsilon,$$

$$|y_{\lambda_1 x_1 + \lambda_2 x_2} - \hat{f}_0(\lambda_1 x_1 + \lambda_2 x_2)| < \varepsilon$$

and y_x arbitrary if $x \ne x_1$, $x \ne x_2$, $x \ne \lambda_1 x_1 + \lambda_2 x_2$. Since $\hat{f}_0 \in \bar{I}'$, there is an element $\hat{f}$ in $I' \cap U_0$. It has the form $\hat{f} = \sigma f$, where f is a continuous linear functional. Thus

$$|\hat{f}(x_1) - \hat{f}_0(x_1)| < \varepsilon, \qquad |\hat{f}(x_2) - \hat{f}_0(x_2)| < \varepsilon,$$

$$|\hat{f}(\lambda_1 x_1 + \lambda_2 x_2) - \hat{f}_0(\lambda_1 x_1 + \lambda_2 x_2)| < \varepsilon,$$

and $\hat{f}(\lambda_1 x_1 + \lambda_2 x_2) = \lambda_1 \hat{f}(x_1) + \lambda_2 \hat{f}(x_2)$. Consequently,

$$\begin{aligned} &|\lambda_1 \hat{f}_0(x_1) + \lambda_2 \hat{f}_0(x_2) - \hat{f}_0(\lambda_1 x_1 + \lambda_2 x_2)| \\ &\quad \le |\lambda_1|\,|\hat{f}_0(x_1) - \hat{f}(x_1)| + |\lambda_2|\,|\hat{f}_0(x_2) - \hat{f}(x_2)| + |\hat{f}_0(\lambda_1 x_1 + \lambda_2 x_2) \\ &\qquad - \hat{f}(\lambda_1 x_1 + \lambda_2 x_2)| \le (|\lambda_1| + |\lambda_2| + 1)\varepsilon. \end{aligned}$$

Since ε is arbitrary, we get $\hat{f}_0(\lambda_1 x_1 + \lambda_2 x_2) = \lambda_1 \hat{f}_0(x_1) + \lambda_2 \hat{f}_0(x_2)$. Thus the functional f_0 defined by $f_0(x) = \hat{f}_0(x)$ is linear. Since $\hat{f}_0(x) \in I_x$, $|f_0(x)| \le \|x\|$. It follows that $f_0 \in X^*$ and $\|f_0\| \le 1$. Hence $f_0 \in \Gamma$. But then $\hat{f}_0 = \sigma f_0 \in I'$. We have thus proved that $\bar{I}' = I'$.

Definition 4.12.2 Let X be a normed linear space. A sequence $\{x_n^*\}$ in X^* is said to be *weakly convergent* if there exists an element $x^* \in X^*$ such that $\lim_n x_n^*(x) = x^*(x)$ for all $x \in X$. We then call x^* the *weak limit* of $\{x_n^*\}$ and we say that $\{x_n^*\}$ is *weakly convergent* to x^*.

Note that a sequence $\{x_n^*\}$ cannot have two distinct weak limits.

Theorem 4.12.3 *Let* X *be a separable normed linear space. Then every bounded sequence of continuous linear functionals in* X^* *has a weakly convergent subsequence.*

Proof. Denote by X_R^* $(R > 0)$ the topological subspace $\{x^* \in X^*; \|x^*\| \le R\}$ (with the weak topology). Its topology can be defined by the neighborhood bases consisting of the sets $X_R^* \cap N$, where N varies over the sets $N(x^*; x_1, \ldots, x_n; \varepsilon)$. Let $\{y_n\}$ be a dense sequence in X. For any $z^* \in X_R^*$, the sets

$$X_R^* \cap N\left(z^*; y_{k_1}, \ldots, y_{k_n}; \frac{1}{m}\right) \qquad (n = 1,2,\ldots; m = 1,2,\ldots)$$

form a countable neighborhood basis (for the weak topology of X_R^*) at z^*. Thus, the topological linear space X_R^* satisfies the first countability axiom. Now let $\{x_n^*\}$ be a sequence of continuous linear functionals with $\|x_n^*\| \le K$, K constant. By the Alaoglu theorem, the set X_K^* is compact in the weak topology. By Theorem 4.11.5, this set is then sequentially compact in the weak topology. It follows that there is a subsequence $\{x_{n_i}^*\}$ that is weakly convergent.

PROBLEMS

4.12.1. Let X be a normed linear space. A sequence $\{f_n\}$ in X^* is weakly convergent to $f \in X^*$ if and only if the following conditions hold: (i) the sequence $\{\|f_n\|\}$ is bounded, and (ii) $\lim_n f_n(x) = f(x)$ for all x in a dense subset of X.

4.12.2. If $\{f_n\}$ is weakly convergent to f, then $\|f\| \le \underline{\lim}_n \|f_n\|$.

4.12.3. Give a direct proof of Theorem 4.12.3. [*Hint:* Choose a subsequence $\{x_{n,1}^*\}$ such that $\lim x_{n,1}^*(x_1)$ exists. Choose a subsequence $\{x_{n,2}^*\}$ of $\{x_{n,1}^*\}$ such that $\lim x_{n,2}^*(x_2)$ exists, and so on. Then $\{x_{n,n}^*\}$ is weakly convergent.]

4.12.4. Weak topology can be introduced in the normed linear space X by taking as a neighborhood basis of any point y the class of sets

$$N(y; x_1^*, \ldots, x_n^*; \varepsilon) = \{x; |x_j^*(x) - x_j^*(y)| < \varepsilon \quad \text{for } 1 \le j \le n\}.$$

Prove that this turns X into a topological linear sequence. Prove also that a sequence $\{x_n\}$ is convergent to x in the weak topology if and only if $\{x_n\}$ is weakly convergent to x according to Definition 4.10.2.

4.12.5. If a Banach space X is reflexive, then its closed unit ball is compact in the weak topology. (The converse is also true.)

4.12.6. Any Banach space X is isometrically isomorphic to a closed linear subspace of the space $C(\Gamma)$ of continuous functions on a compact Hausdorff space with the uniform topology. [*Hint:* Let Γ be the closed unit ball in X^* provided with the weak topology. To each $x \in X$ corresponds a function $\hat{x} \in C(\Gamma)$ by $\hat{x}(x^*) = x^*(x)$.]

4.13 ADJOINT OPERATORS

Definition 4.13.1 Let X and Y be normed linear spaces and let $T \in \mathscr{B}(X,Y)$. The *adjoint* T^* of T is an operator from Y^* into X^* defined by $(T^*y^*)(x) = y^*(Tx)$.

Theorem 4.13.1 *The mapping* $T \to T^*$ *is an isometric isomorphism of* $\mathscr{B}(X,Y)$ *into* $\mathscr{B}(Y^*,X^*)$.

Proof. Take any $y^* \in Y^*$. T^*y^* is clearly a linear functional on X. Furthermore,

$$|(T^*y^*)(x)| \leq \|y^*\| \, \|Tx\| \leq \|y^*\| \, \|T\| \, \|x\|.$$

Hence $T^*y^* \in X^*$. The map $T \to T^*$ is obviously linear. Thus it remains to show that $\|T^*\| = \|T\|$. By Corollary 4.8.6 we have

$$\|Tx\| = \sup_{\|y^*\|=1} \|y^*(Tx)\|.$$

Hence

$$\|T^*\| = \sup_{\|y^*\|=1} \|T^*y^*\| = \sup_{\|y^*\|=1} \sup_{\|x\|=1} |(T^*y^*)(x)|$$

$$= \sup_{\|y^*\|=1} \sup_{\|x\|=1} |y^*(Tx)| = \sup_{\|x\|=1} \sup_{\|y^*\|=1} |y^*(Tx)| = \sup_{\|x\|=1} \|Tx\| = \|T\|.$$

Theorem 4.13.2 *Let* X,Y,Z *be normed linear spaces and let* $T \in \mathscr{B}(X,Y)$, $S \in \mathscr{B}(Y,Z)$. *Then* $(ST)^* = T^*S^*$. *The adjoint of the identity* I *in* $\mathscr{B}(X)$ *is the identity in* $\mathscr{B}(X^*)$.

Proof. Let $z^* \in Z^*$. Then, for any $x \in X$,

$$[(ST)^*z^*](x) = z^*[(ST)x] = z^*[S(Tx)] = (S^*z^*)(Tx)$$

$$= [T^*(S^*z^*)](x) = [(T^*S^*)z^*](x).$$

Thus $(ST)^*z^* = (T^*S^*)z^*$ for any $z^* \in Z^*$—that is, $(ST)^* = T^*S^*$. Next $(I^*x^*)(x) = x^*(Ix) = x^*(x)$—that is, $I^*x^* = x^*$. Thus I^* is the identity in $\mathscr{B}(X^*)$.

Denote by $\hat{X},\hat{Y}$ the images under the natural imbedding of X,Y into X^{**},Y^{**}, respectively. If $T \in \mathscr{B}(X,Y)$, then define $\hat{T} \in \mathscr{B}(\hat{X},\hat{Y})$ by $\hat{T}\hat{x} = \hat{y}$, where $y = Tx$. If S is a linear operator from a set D_S in X^{**} into Y^{**} such that $D_S \supset \hat{X}$ and $S\hat{x} = \hat{T}\hat{x}$ for all $\hat{x} \in \hat{X}$, then we call S (which is an extension of $\hat{T}$) *an extension of* T. If $D_S = \hat{X}$, then we write $S = T$.

Theorem 4.13.3 *Let* X,Y *be normed linear spaces and let* $T \in \mathscr{B}(X,Y)$. *Then the second adjoint* $T^{**}: X^{**} \to Y^{**}$ *is an extension of* T. *If* X *is reflexive, then* $T^{**} = T$.

Proof. Let $x \in X$, $y^* \in Y^*$. Then

$$(T^{**}\hat{x})(y^*) = \hat{x}(T^*y^*) = (T^*y^*)(x) = y^*(Tx)$$
$$= \widehat{Tx}(y^*) = (\hat{T}\hat{x})(y^*),$$

—that is, $T^{**}\hat{x} = \hat{T}\hat{x}$. This completes the proof.

Theorem 4.13.4 *Let* X *be a Banach space and let* Y *be a normed linear space. A linear operator* $T \in \mathscr{B}(X,Y)$ *has a bounded inverse* T^{-1} *(with domain* Y*) if and only if* T^* *has a bounded inverse* $(T^*)^{-1}$ *(with domain* X^**). In that case,* $(T^{-1})^* = (T^*)^{-1}$.

Proof. Suppose T^{-1} exists. By Theorem 4.13.2, $I^* = (TT^{-1})^* = (T^{-1})^*T^*$, where I^* is the identity operator in $\mathscr{B}(Y^*,Y^*)$. Similarly $T^*(T^{-1})^*$ is the identity operator in $\mathscr{B}(X^*,X^*)$. It follows that $(T^*)^{-1}$ exists and is equal to $(T^{-1})^*$.

Suppose conversely that $(T^*)^{-1}$ is a bounded operator in $\mathscr{B}(X^*,Y^*)$. Then, by what we have just proved, $(T^{**})^{-1}$ exists and is in $\mathscr{B}(Y^{**},X^{**})$. By Theorem 4.13.3, T^{**} is an extension of T. Hence T is one-to-one. If we can prove that $T(X) = Y$, then the proof is complete, since, on $T(X)$, $T^{-1} = (T^{**})^{-1}$ is a bounded operator.

Since T^{**} has a bounded inverse and since $\hat{X}$ is a closed subset of X^{**}, the set $T^{**}(\hat{X})$ is closed [compare Problem 4.13.3(a)]. Hence $T(X)$ is closed. Suppose now that $T(X) \neq Y$. Then $\overline{T(X)} \neq Y$. By Theorem 4.8.7 there exists a $y^* \in Y^*$ with $y^* \neq 0$, $y^*(Tx) = 0$ for all $x \in X$. Hence $T^*y^* = 0$. This is impossible, since T^* is one-to-one.

Definition 4.13.2 Let X be a normed linear space and let A be a subset of X. We shall denote by $A^\perp$ the set of all $x^* \in X^*$ such that $x^*(x) = 0$ for all $x \in A$. We call $A^\perp$ the *orthogonal complement* of A. Similarly, for a set $\Gamma \subset X^*$ we define the *orthogonal complement* $\Gamma^\perp$ of Γ to be the set of all $x \in X$ such that $x^*(x) = 0$ for all $x^* \in \Gamma$.

Notice that $A^\perp$ and $\Gamma^\perp$ are closed linear subspaces in X^* and X, respectively.

Let $T \in \mathscr{B}(X,Y)$. We shall denote by N_T the null space of T—that is, $N_T = \{x; Tx = 0\}$. Similarly we denote by N_{T^*} the null space of T^*. Finally, we denote by R_T the range of an operator T. Thus, if $T \in \mathscr{B}(X,Y)$, then $R_T = T(X)$.

Theorem 4.13.5 *Let* X,Y *be normed linear spaces and let* $T \in \mathscr{B}(X,Y)$. *Then*

$$\bar{R}_T = N_{T^*}^\perp. \tag{4.13.1}$$

Proof. We first prove that $\bar{R}_T \supset N_{T^*}^{\perp}$. What we have to show is that if $y_0 \notin \bar{R}_T$, then $y_0 \notin N_{T^*}^{\perp}$. By Theorem 4.8.3 there exists a $y^* \in Y^*$ such that $y^*(y_0) \neq 0$ and $y^*(Tx) = 0$ for all $x \in X$. Hence $T^*y^* = 0$. Since

$$N_{T^*}^{\perp} = \{y \in Y;\ y^*(y) = 0 \qquad \text{for any } y^* \text{ such that } T^*y^* = 0\},$$

it follows that $y_0 \notin N_{T^*}^{\perp}$.

We next prove that $\bar{R}_T \subset N_{T^*}^{\perp}$. Let $y \in \bar{R}_T$. Then there exists a sequence $\{y_n\} \in R_T$ such that $\lim y_n = y$. If $T^*y^* = 0$, then $y^*(y_n) = y^*(Tx_n) = (T^*y^*)(x_n) = 0$, where $Tx_n = y_n$. Hence also $y^*(y) = 0$. We conclude that $y \in N_{T^*}^{\perp}$.

Remark. Theorem 4.13.5 can be interpreted as an existence theorem for the equation $Tx = y$. It then implies the following result: *if T^* is one-to-one and R_T is closed, then $R_T = Y$—that is, for any given $y \in Y$ there exists a solution x in X of the equation $Tx = y$.* In Section 5.2 we shall solve such equations more systematically.

The dual of (4.13.1) is $\bar{R}_{T^*} = N_T^{\perp}$. But this relation is not true in general. We shall prove, however, the following weaker form of it.

Theorem 4.13.6 *Let* X *and* Y *be Banach spaces and let* $T \in \mathscr{B}(X,Y)$. *If* R_T *is closed, then* R_{T^*} *is closed and*

$$R_{T^*} = N_T^{\perp}. \tag{4.13.2}$$

The proof depends upon the following lemma.

Lemma 4.13.7 *Let* X, Y *be Banach spaces and let* $T \in \mathscr{B}(X,Y)$. *If* R_T *is closed, then there exists a constant* K *such that for any* $y \in R_T$ *there is a point* $x \in X$ *with* $Tx = y$ *and* $\|x\| \le K\|y\|$.

Proof. By the open-mapping theorem (Theorem 4.6.1) T maps the unit ball B_1 of X onto a set containing a ball in R_T with center 0—that is,

$$TB_1 \supset \{y;\ y \in R_T,\ \|y\| < \delta\} \qquad \text{for some } \delta > 0.$$

Now let $0 \neq y \in R_T$. Then $\delta y/2\|y\|$ is in TB_1—that is, $Tz = \delta y/2\|y\|$ for some z, $\|z\| < 1$. Let $x = 2\|y\|z/\delta$. Then $Tx = y$ and $\|x\| < (2/\delta)\|y\|$.

Proof of Theorem 4.13.6. We first prove that $N_T^{\perp} \subset R_{T^*}$. Let $x^* \in N_T^{\perp}$. Define a linear functional g on R_T by $g(Tx) = x^*(x)$. g is defined unambiguously. Indeed, if $Tx_1 = Tx_2$, then $(x_1 - x_2) \in N_T$ and therefore $x^*(x_1 - x_2) = 0$. Hence $g(Tx_1) - g(Tx_2) = x^*(x_1) - x^*(x_2) = 0$. By Lemma 4.13.7 there exists a constant K such that for any $y \in R_T$ there is an x with $Tx = y$, $\|x\| \leq K\|y\|$. Hence

$$|g(y)| = |x^*(x)| \leq K\|x^*\| \, \|y\|.$$

Thus g is a bounded linear functional on R_T. By the Hahn-Banach theorem we can extend g into a continuous linear functional $y^* \in Y^*$. Since, for any $x \in X$,

$$(T^*y^*)(x) = y^*(Tx) = g(Tx) = x^*(x),$$

we have $T^*y^* = x^*$. Thus $x^* \in R_{T^*}$.

Suppose conversely that $x^* \in R_{T^*}$. We shall prove that $x^* \in N_T^\perp$. Let $y^* \in Y^*$ be such that $T^*y^* = x^*$. If $x \in N_T$, then

$$x^*(x) = (T^*y^*)(x) = y^*(Tx) = y^*(0) = 0.$$

This shows that $x^* \in N_T^\perp$.

The assertion that R_{T^*} is closed follows from (4.13.2) and the fact that $N_T^\perp$ is closed.

PROBLEMS

4.13.1. Choose a basis $\{e_1,\ldots,e_n\}$ in R^n and choose a basis $\{e_1^*,\ldots,e_n^*\}$ in R^{n*}. To every linear operator T in $\mathscr{B}(R^n,R^n)$ corresponds a matrix (a_{ij}), where $Te_j = \sum_{i=1}^{n} a_{ij}e_i$. Similarly, to every operator $S \in \mathscr{B}(R^{n*},R^{n*})$ corresponds a metrix (a_{ij}^*), where $Se_j^* = \sum_{i=1}^{n} a_{ij}^* e_i^*$. Find the relation between (a_{ij}) and (a_{ij}^*) if $S = T^*$.

4.13.2. Let $f(z)$ be an entire complex analytic function and let $T \in \mathscr{B}(X)$. Prove that $[f(T)]^* = f(T^*)$.

4.13.3. Let X and Y be Banach spaces and let $T \in \mathscr{B}(X,Y)$. Prove:

(a) If T has a continuous inverse, then R_T is closed.

(b) T^* is one-to-one if and only if R_T is dense in Y.

(c) If T maps X onto Y, then T^* has a bounded inverse with domain R_{T^*}.

4.13.4. Let X,Y be normed linear spaces. Denote by $d(x,L)$ the distance (in X) from x to a set L. Let $T \in \mathscr{B}(X,Y)$. Prove that

$$d(x,N_T) = \max\,\{x^*(x);\ \|x^*\| \leq 1 \text{ and } x^* \in N_T^\perp\}. \tag{4.13.3}$$

[*Hint:* Use Theorem 4.8.3.]

4.13.5. Prove the converse of Theorem 4.13.6—that is, if (4.13.2) holds, then R_T is closed. [*Hint:* It suffices to show that $d(x,N_T) \leq K\|Tx\|$. Now for any $x^* \in R_{T^*}$ there exists a y^* such that $T^*y^* = x^*$, $\|y^*\| \leq K\|x^*\|$. Deduce that $|x^*(x)| \leq K\|Tx\|$. Now use (4.13.2), (4.13.3).]

4.14 THE CONJUGATES OF L^p AND $C[0,1]$

Let (X,μ) be a measure space. We shall denote by $\|f\|_p$ the norm of f in the space $L^p(X,\mu)$—that is,

$$\|f\|_p = \left\{\int |f|^p \, d\mu\right\}^{1/p} \qquad (1 \le p < \infty).$$

We also write

$$\|f\|_\infty = \operatorname*{ess\,sup}_X |f|.$$

Theorem 4.14.1 *Assume that* (X,μ) *is a* σ*-finite measure space and let* $1 < p < \infty$, $1/p + 1/q = 1$. *Then to every continuous linear functional* x^* *on* $L^p(X,\mu)$ *there corresponds a unique element* g *in* $L^q(X,\mu)$ *such that*

$$x^*(f) = \int fg \, d\mu \qquad \textit{for all } f \in L^p(X,\mu). \tag{4.14.1}$$

Furthermore,

$$\|x^*\| = \|g\|_q. \tag{4.14.2}$$

Proof. We shall give the proof only for the complex $L^p(X,\mu)$ space. The proof for the real space is similar. The uniqueness of g is obvious. We shall need the following lemma:

Lemma 4.14.2 *Let* f *be a measurable function.*

(a) *If* $f \in L^p(X,\mu)$, *then*

$$\|f\|_p = \max_{\|g\|_q=1} \left|\int fg \, d\mu\right| = \max_{\|g\|_q=1} \int |fg| \, d\mu. \tag{4.14.3}$$

(b) *If* $f \notin L^p(X,\mu)$, *then*

$$\sup_{\|g\|_q=1} \int |fg| \, d\mu = \infty.$$

Proof. To prove (a) we may clearly assume that $\|f\|_p \neq 0$. By Hölder's inequality,

$$\int |fg| \, d\mu \le \|f\|_p \|g\|_q = \|f\|_p \qquad \text{if } \|g\|_q = 1. \tag{4.14.4}$$

Next let $g = \overline{\theta(f)} |f|^{p-1} \|f\|_p^{-p/q}$, where $\theta(f)$ is defined by $f = \theta(f) \cdot |f|$ if $f \neq 0$ and $\theta(f) = 0$ if $f = 0$. It is easily seen that $\theta(f)$ is measurable. We have

$$\int |g|^q \, d\mu = \int |f|^{(p-1)q} \|f\|_p^{-p} \, d\mu = 1.$$

Also,

$$\int fg\,d\mu = \int |f|^p \|f\|_p^{-p/q}\,d\mu = \|f\|_p^{p-p/q} = \|f\|_p.$$

This, together with (4.14.4), completes the proof of (4.14.3).

To prove (b), write $X = \bigcup_{n=1}^{\infty} X_n$, where $X_n \subset X_{n+1}$, $\mu(X_n) < \infty$, and introduce, for any positive integer n, the function

$$f_n(x) = \begin{cases} f(x), & \text{if } |f(x)| \le n,\ x \in X_n, \\ n, & \text{if } |f(x)| > n,\ x \in X_n, \\ 0, & \text{if } x \notin X_n. \end{cases}$$

Then $f_n \in L^p(X,\mu)$ $|f_n| \le |f|$ a.e., and $\|f_n\|_p \nearrow \infty$ as $n \nearrow \infty$. Hence,

$$\sup_{\|g\|_q=1} \int |fg|\,d\mu \ge \max_{\|g\|_q=1} \int |f_n g|\,d\mu = \|f_n\|_p \to \infty$$

as $n \to \infty$.

Corollary 4.14.3 *Let* $g \in L^q(X,\mu)$. *Then the functional*

$$G(f) = \int fg\,d\mu \qquad [f \in L^p(X,\mu)]$$

is a continuous linear functional on $L^p(X,\mu)$ *and* $\|G\| = \|g\|_q$.

It is obvious that G is linear. The equality $\|G\| = \|g\|_q$ follows from Lemma 4.14.2(a) with p and q interchanged.

We proceed with the proof of Theorem 4.14.1. We first make the assumption that $\mu(X) < \infty$. For any measurable set E, let

$$\nu(E) = x^*(\chi_E).$$

ν obviously is finitely additive. We shall prove that it is completely additive. Let $\{E_k\}$ be a sequence of mutually disjoint measurable sets and let $S_n = \bigcup_{k=1}^{n} E_k$, $S = \bigcup_{k=1}^{\infty} E_k$. By the Lebesgue bounded convergence theorem [here we use the assumption that $\mu(X) < \infty$] we get

$$\|\chi_{S_n} - \chi_S\|_p \to 0 \quad \text{as } n \to \infty.$$

Hence $x^*(\chi_{S_n}) \to x^*(\chi_S)$. It follows that

$$\nu(S) = x^*(\chi_S) = \lim_n x^*(\chi_{S_n}) = \lim_n \nu(S_n)$$

$$= \lim_n \sum_{m=1}^{n} \nu(E_m) = \sum_{m=1}^{\infty} \nu(E_m).$$

We have thus proved that ν is a complex measure. If $\mu(E) = 0$, then χ_E is zero as an element in $L^p(X,\mu)$. Hence $\nu(E) = x^*(\chi_E) = x^*(0) = 0$. Thus the complex measure ν is absolutely continuous with respect to the measure μ. Finally, ν is a finite complex measure, since any function χ_E is in $L^p(X,\mu)$, and thus $|\nu(E)| = |x^*(\chi_E)| < \infty$.

By the Radon-Nikodym theorem there exists an integrable function g such that

$$\nu(E) = \int_E g \, d\mu$$

for any measurable set E. Hence

$$x^*(\chi_E) = \int \chi_E \, g \, d\mu.$$

By linearity of both sides with respect to χ_E we get

$$x^*(f) = \int fg \, d\mu \tag{4.14.5}$$

for any simple function f. We shall extend this relation to any bounded measurable function f. Writing f in the form $\hat{f}_1 - \hat{f}_2 + i\hat{f}_3 - i\hat{f}_4$, where the $\hat{f}_i$ are bounded, nonnegative measurable functions, and applying Theorem 2.2.5 to each $\hat{f}_i$, we conclude that there exists a sequence $\{f_n\}$ of uniformly bounded simple functions that converges to f everywhere on X. By the Lebesgue bounded convergence theorem it follows that

$$\lim_n \int f_n g \, d\mu = \int fg \, d\mu.$$

Since (by the Lebesgue bounded convergence theorem) $\|f_n - f\|_p \to 0$ as $n \to \infty$, we also have

$$\lim_n x^*(f_n) = x^*(f).$$

Combining the last two relations with the fact that (4.14.5) holds with f replaced by each f_n, we obtain the equality (4.14.5) for the bounded measurable function f.

We next show that $g \in L^q(X,\mu)$. Let $f \in L^p(X,\mu)$ and define

$$f_n(x) = \begin{cases} |f(x)| \overline{\theta(g)}, & \text{if } |f(x)| \le n, \\ 0, & \text{if } |f(x)| > n. \end{cases}$$

Then $\|f_n\|_p \le \|f\|_p$, and, therefore,

$$|x^*(f_n)| \le \|x^*\| \, \|f_n\|_p \le \|x^*\| \, \|f\|_p. \tag{4.14.6}$$

Since $f_n g \geq 0$ and $\lim f_n g = |fg|$ a.e., Fatou's lemma gives

$$\int |fg| \, d\mu \leq \underline{\lim}_n \int f_n g \, d\mu = \underline{\lim}_n \, x^*(f_n) \leq \|x^*\| \, \|f\|_p.$$

Here we have used (4.14.5) with f replaced by f_n (f_n is bounded) and (4.14.6). We can now apply Lemma 4.14.2(b) (with the roles of p and q interchanged) and conclude that $g \in L^q(X,\mu)$.

Now let $f \in L^p(X,\mu)$ and define $f_n(x) = f(x)$ if $|f(x)| \leq n, f_n(x) = 0$ if $|f(x)| > n$. Then $\{f_n\}$ converges to f in $L^p(X,\mu)$. (This follows using, for instance, Corollary 2.8.5 with f replaced by $|f|^p$.) It follows that

$$x^*(f) = \lim_n x^*(f_n).$$

By Hölder's inequality we also find that $\|f_n g - fg\|_1 \to 0$ as $n \to \infty$. Since $x^*(f_n) = \int f_n g \, d\mu$, we conclude that (4.14.5) holds. Having proved (4.14.1), with $g \in L^q(X,\mu)$, the assertion (4.14.2) follows from Lemma 4.14.2(a). We have thus completed the proof in case $\mu(X) < \infty$.

In the general case where X is σ-finite, write X as a countable union of monotone-increasing measurable sets X_j, where $\mu(X_j) < \infty$. Denote by x_n^* the restriction of x^* to X_n—that is, for any f in $L^p(X_n,\mu)$, $x_n^*(f) = x^*(\hat{f}_n)$, where $\hat{f}_n$ is the extension of f to X by 0. By what we have already proved, there exists a functional g_n in $L^q(X_n,\mu)$ such that, for any f in $L^p(X_n,\mu)$,

$$x^*(\hat{f}_n) = x_n^*(f) = \int_{X_n} fg_n \, d\mu \qquad \text{and} \qquad \|g_n\|_q = \|x_n^*\|. \tag{4.14.7}$$

For every $f \in L^p(X_m,\mu)$ and for every $n > m$, denote by F the extension of f to X_n by 0. Then $\hat{f}_m = \hat{F}_n$. It follows that

$$\int_{X_m} fg_m \, d\mu = x_m^*(f) = x^*(\hat{f}_m) = x^*(\hat{F}_n) = x_n^*(F) = \int_{X_m} fg_n \, d\mu.$$

Therefore, $g_n = g_m$ on X_m if $n > m$. Hence there is a function g on X defined by $g = g_n$ on each set X_n. The Lebesgue monotone convergence theorem implies that

$$\int_X |g|^q \, d\mu = \lim_n \int_{X_n} |g_n|^q \, d\mu.$$

Since

$$\int_{X_n} |g_n|^q \, d\mu = \|x_n^*\| \leq \|x^*\|,$$

we conclude that $g \in L^q(X,\mu)$ and $\|g\|_q \leq \|x^*\|$.

Let f be any function in $L^p(X,\mu)$ and let $f_n = \chi_n f$, where χ_n is the characteristic function of X_n. By the Lebesgue monotone convergence theorem, $\lim \int_{X_n} |f|^p \, d\mu = \int |f|^p \, d\mu$. Hence

$$\int |f - f_n|^p \, d\mu = \int_{X - X_n} |f|^p \, d\mu \to 0 \qquad \text{as } n \to \infty.$$

We conclude that

$$\lim_n x^*(f_n) = x^*(f).$$

Also, by Hölder's inequality, $\|f_n g - fg\|_1 \to 0$ as $n \to \infty$. Thus, in particular,

$$\lim_n \int f_n g \, d\mu = \int fg \, d\mu.$$

The last two equations together with (4.14.7) give $x^*(f) = \int fg \, d\mu$ for any $f \in L^p(X,\mu)$. We have thus completed the proof of (4.14.1), with $g \in L^q(X,\mu)$. The assertion (4.14.2) follows from Lemma 4.14.2(a).

Denote by τ the map $x^* \to g$ given by Theorem 4.14.1. Then τ is an isometry from $[L^p(X,\mu)]^*$ into $L^q(X,\mu)$. By Corollary 4.14.3 the range of τ is the whole space $L^q(X,\mu)$. It is also clear that τ is a linear map. Hence:

Corollary 4.14.4 *Let* (X,μ) *be a* σ*-finite measure space and let* $1 < p < \infty$, $1/p + 1/q = 1$. *Then there is an isometric isomorphism* τ *from* $[L^p(X,\mu)]^*$ *onto* $L^q(X,\mu)$.

Corollary 4.14.5 *Let* (X,μ) *be a* σ*-finite measure space and let* $1 < p < \infty$. *Then* $L^p(X,\mu)$ *is a reflexive space.*

Proof. Take any x^{**} in $[L^p(X,\mu)]^{**}$. For any x^* in $[L^p(X,\mu)]^*$, let $\tau x^* = g$ and define a functional $\tilde{x}^{**}$ on $L^q(X,\mu)$ by $\tilde{x}^{**}(g) = x^{**}(x^*)$. This is a bounded linear functional. Hence by Theorem 4.14.1, there exists a function f in $L^p(X,\mu)$ such that $\tilde{x}^{**}(g) = \int fg \, d\mu$. Since the last integral is equal to $x^*(f)$ (by the definition of τ), we get $x^{**}(x^*) = x^*(f)$. This shows that x^{**} is the image of f under the natural imbedding.

Theorem 4.14.6 *Let* (X,μ) *be a* σ*-finite measure space. Then for every* $x^* \in [L^1(X,\mu)]^*$ *there corresponds a unique function* g *in* $L^\infty(X,\mu)$ *such that*

$$x^*(f) = \int fg \, d\mu \qquad \textit{for all } f \in L^1(X,\mu).$$

Furthermore,

$$\|x^*\| = \|g\|_\infty,$$

and the map $\tau: x^* \to g$ *is an isometric isomorphism from* $[L^1(X,\mu)]^*$ *onto* $L^\infty(X,\mu)$.

Proof. The proof is the same as that of Theorem 4.14.1, except for one difference, which occurs in part (a) of Lemma 4.14.2, when $p = \infty$ (but not when $p = 1$). The assertion now is that

$$\|f\|_\infty = \sup_{\|g\|_1 = 1} \left| \int fg \, d\mu \right| = \sup_{\|g\|_1 = 1} \int |fg| \, d\mu. \tag{4.14.8}$$

In proving (4.14.8) we may clearly assume that $\|f\|_\infty \neq 0$. It is clear that the term on the right is not larger than the term on the left. Next, from the definition of the essential supremum it follows that for any $\varepsilon > 0$ there is a set E of positive measure such that $|f(x)| > \|f\|_\infty - \varepsilon$ for all $x \in E$. Take

$$g(x) = \begin{cases} \dfrac{1}{\mu(E)} \theta(f), & \text{if } x \in E, \\ 0, & \text{if } x \notin E. \end{cases}$$

Then

$$\int fg \, d\mu = \int_E fg \, d\mu > \|f\|_\infty - \varepsilon.$$

Since ε is arbitrary, the term on the left in (4.14.8) is not larger than the term on the right. We have thus completed the proof of (4.14.8). As already stated before, all the other steps in the proof of Theorem 4.14.1 immediately extend to the present case.

By Problem 4.14.2, $L^\infty(\Omega)$ is not separable. Hence, by Theorem 4.10.1, also its dual is not separable. Since $L^1(\Omega)$ is separable, we conclude:

Corollary 4.14.7 *The space* $\mathrm{L}^1(\Omega)$ *is not reflexive.*

We shall now find the dual of the real space $C[0,1]$. We denote by BV the space of all functions $g(x)$ on $[0,1]$ of bounded variations, satisfying the condition $g(0) = 0$. Denote by $V(g)$ the total variation of g.

Let x^* be a continuous linear functional over $C[0,1]$. Since $C[0,1]$ is a closed linear subspace of $L^\infty(0,1)$, there exists a continuous linear functional Φ over $L^\infty(0,1)$ that satisfies:

$$\Phi(f) = x^*(f) \qquad \text{if } f \in C[0,1] \quad \text{and} \quad \|\Phi\| = \|x^*\|.$$

For each $x \in [0,1]$ we define a function ψ_x as follows:

$$\psi_x(t) = \begin{cases} 1, & \text{if } 0 \leq t \leq x, \\ 0, & \text{if } x < t \leq 1. \end{cases}$$

These functions belong to $L^\infty(0,1)$. Define

$$g(x) = \Phi(\psi_x). \tag{4.14.9}$$

Since $\psi_0 = 0$, $g(0) = \Phi(0) = 0$. We shall prove that $g \in BV$ and that

$$V(g) \leq \|x^*\|. \tag{4.14.10}$$

Let $0 = x_0 < x_1 < \cdots < x_{n-1} = x_n = 1$, and write $\varepsilon_i = \operatorname{sgn}[g(x_i) - g(x_{i-1})]$. Then

$$\sum_{i=1}^{n} |g(x_i) - g(x_{i-1})| = \sum_{i=1}^{n} \varepsilon_i[g(x_i) - g(x_{i-1})]$$
$$= \sum_{i=1}^{n} \varepsilon_i[\Phi(\psi_{x_i}) - \Phi(\psi_{x_{i-1}})] = \Phi(\psi),$$

where $\psi = \sum_{i=1}^{n} \varepsilon_i[\psi_{x_i} - \psi_{x_{i-1}}]$. Since $\|\psi\|_\infty = 1$, we get

$$\sum_{i=1}^{n} |g(x_i) - (g_{i-1})| \leq \|\Phi\| \, \|\psi\|_\infty = \|\Phi\| = \|x^*\|.$$

This proves (4.14.10).

Now let $f(t) \in C[0,1]$, and define

$$f_n(t) = \sum_{k=1}^{n} f\left(\frac{k}{n}\right)[\psi_{k/n}(t) - \psi_{(k-1)/n}(t)].$$

It is easily seen that $\|f_n - f\|_\infty \to 0$ as $n \to \infty$. Hence

$$\lim_n \Phi(f_n) = \Phi(f) = x^*(f).$$

On the other hand (compare Problem 2.11.8)

$$\Phi(f_n) = \sum_{k=1}^{n} f\left(\frac{k}{n}\right)\left[g\left(\frac{k}{n}\right) - g\left(\frac{k-1}{n}\right)\right] \to \int_0^1 f(t)\, dg(t)$$

as $n \to \infty$. We conclude that

$$x^*(f) = \int_0^1 f(t)\, dg(t).$$

It follows (by Problem 2.11.10) that

$$|x^*(f)| = \left|\int_0^1 f(t)\, dg(t)\right| \leq \max_{0 \leq t \leq 1} |f(t)| \cdot V(g).$$

Recalling (4.14.10), we get $\|x^*\| = V(g)$. We sum up:

Theorem 4.14.8 *For every continuous linear functional* x* *over* C[0,1], *there corresponds a function* g *in* BV *such that*

$$x^*(f) = \int_0^1 f\, dg \qquad \textit{for every } f \in C[0,1]. \tag{4.14.11}$$

Further,

$$\|x^*\| = V(g). \tag{4.14.12}$$

By Problem 4.14.3, if $\hat{g} \in BV$, then

$$\int_0^1 f\,dg = \int_0^1 f\,d\hat{g} \qquad \text{for all } f \in C[0,1],$$

if and only if $g = \hat{g}$ at all points where one of the functions is continuous. Let us define equivalence " $\sim$ " in the space BV by saying that $g \sim \hat{g}$ if and only if g and $\hat{g}$ differ only at the points of discontinuity. Note that this set of points is countable. We shall denote by BV_0 the space of equivalent classes and by $[g]$ the class containing g.

Given any class $[\hat{g}]$, consider the bounded linear functional x^* on $C[0,1]$ defined by

$$x^*(f) = \int_0^1 f\,d\hat{g}.$$

By the proof of Theorem 4.14.8, there exists a function g in BV such that (4.14.11) holds and $\|x^*\| = V(g)$. Since

$$\|x^*\| = \sup_{\|f\|=1} \left|\int_0^1 f\,d\hat{g}\right| \le V(\hat{g})$$

and since $g \sim \hat{g}$, we see that

$$V(g) = \min_{\bar{g} \in [\hat{g}]} V(\bar{g}) = \underset{\bar{g} \in [\hat{g}]}{\text{g.l.b.}}\, V(\bar{g}). \tag{4.14.13}$$

The right-hand side of (4.14.13) defines a norm on BV_0. We shall designate the norm of $[\hat{g}]$ by $\|[\hat{g}]\|$. Combining (4.14.13) with (4.14.12), we can now state:

Corollary 4.14.9 *The map* $\tau: x^* \to [g]$ *defined in Theorem* 4.14.8 *is an isometric isomorphism of the dual of* $C[0,1]$ *onto the space* BV_0 *with norm* $\|[g]\| = \underset{\hat{g} \in [g]}{\text{g.l.b.}}\, V(\hat{g})$.

In every class $[\bar{g}]$ there is a function g as in (4.14.13). There is also a unique g_0 having the following properties: (i) $g_0(t)$ is right continuous for all $0 \le t < 1$; (ii) $g_0(0) = 0$, and (iii) $g_0(t)$ is left continuous at $t = 1$. In fact, $g_0(t) = g(t+0) - g(0+)$ if $0 \le t < 1$. We shall call g_0 the *normalization* of $[\bar{g}]$, or of g. We shall denote by NBV the space of all these functions g_0. This space is clearly a normed linear space with norm $\|g_0\| = V(g_0)$.

We claim that if g_0 is a normalization of g, then

$$V(g_0) = V(g). \tag{4.14.14}$$

In view of (4.14.13), we only have to show that

$$V(g_0) \le V(g). \tag{4.14.15}$$

To prove (4.14.15), take any $\varepsilon > 0$ and any partition

$$0 = t_0 < t_1 < \cdots < t_{n-1} < t_n = 1$$

of [0,1]. For each t_i $(0 \le i \le n-1)$ choose $\tau_i > t_i$ and $\tau_i - t_i$ so small that

$$|g(t_i + 0) - g(\tau_i)| < \frac{\varepsilon}{2n}.$$

Choose also τ_n such that $\tau_{n-1} < \tau_n < 1$ and $1 - \tau_n$ so small that

$$|g(1-0) - g(\tau_n)| < \frac{\varepsilon}{2n}.$$

Then, if $1 \le i \le n-1$,

$$\begin{aligned} g_0(t_i) - g_0(t_{i-1}) & \\ &= [g(t_i + 0) - g(0+)] - [g(t_{i-1} + 0) - g(0+)] \\ &= g(t_i + 0) - g(t_{i-1} + 0) \\ &= [g(t_i + 0) - g(\tau_i)] - [g(t_{i-1} + 0) - g(\tau_{i-1})] + [g(\tau_i) - g(\tau_{i-1})]. \end{aligned}$$

Similarly,

$$\begin{aligned} g_0(1) - g_0(t_{n-1}) = {} & [g(1-0) - g(\tau_n)] \\ & - [g(t_{n-1} + 0) - g(\tau_{n-1})] + [g(\tau_n) - g(\tau_{n-1})]. \end{aligned}$$

It follows that

$$\sum_{i=1}^{n} |g_0(t_i) - g_0(t_{i-1})| \le \sum_{i=1}^{n} |g(\tau_i) - g(\tau_{i-1})| + \varepsilon \le V(g) + \varepsilon.$$

Hence $V(g_0) \le V(g) + \varepsilon$. Since ε is arbitrary, (4.14.15) follows.

Combining Corollary 4.14.9 with (4.14.13) and (4.14.14), we get:

Corollary 4.14.10 *Consider the map* $\sigma : x^* \to g_0$ *where* g_0 *is the normalization of* g *and* g *is defined by Theorem* 4.14.8. *This map* σ *is an isometric isomorphism from the conjugate of* C[0,1] *onto the space* NBV.

Recall that g_0 is related to x^* by

$$x^*(f) = \int_0^1 f(t)\,dg_0(t).$$

PROBLEMS

4.14.1. If $f \notin L^p(X,\mu)$ $(1 < p < \infty)$, then there is a function g in $L^q(X,\mu)$ $(1/p + 1/q = 1)$ such that $fg \ge 0$ and $\int fg\, d\mu = \infty$.

4.14.2. Let Ω be an open set in R^n. Prove that $L^\infty(\Omega)$ is not separable. [*Hint:* Compare with Problem 3.1.7.]

4.14.3. Let g and $\hat{g}$ belong to BV. Prove that

$$\int_0^1 f\,dg = \int_0^1 f\,d\hat{g} \qquad \text{for all } f \in C[0,1]$$

if and only if $g(t) = \hat{g}(t)$ at each point t, where either g or $\hat{g}$ is continuous.

4.14.4. Every continuous linear functional x^* over $l^p\,(1 < p < \infty)$ can be represented in the form

$$x^*(x) = \sum_{i=1}^{\infty} \eta_i x_i \qquad [x = (x_1, x_2, \ldots) \in l^p]$$

where

$$\eta = (\eta_1, \eta_2, \ldots) \in l^q \left(\frac{1}{p} + \frac{1}{q} = 1\right) \quad \text{and} \quad \|x^*\| = \left\{\sum_{i=1}^{\infty} |\eta_i|^q\right\}^{1/q}.$$

4.14.5. Every linear functional f on R^n has the form

$$f(x) = \sum_{i=1}^{n} \eta_i x_i \qquad [x = (x_1, \ldots, x_n) \in R^n],$$

where $(\eta_1, \ldots, \eta_n) \in R^n$ and $\|f\| = (\sum \eta_i^2)^{1/2}$.

4.14.6. The functions $x_n(t) = \sin n\pi t$ belong to $L^2(0,1)$. Prove that $\{x_n\}$ does not converge to 0 but that it converges weakly to 0. [*Hint:* Use the Riemann-Lebesgue theorem, which states that $\int_0^1 g(t) \sin n\pi t\,dt \to 0$ as $n \to \infty$, for any integrable function g.]

4.14.7. A sequence $\{x_n\}$ $(x_n = \{\xi_i^{(n)}\})$ in l^p converges weakly to $x_0 = \{\xi_i^{(0)}\}$ in l^p if and only if: (i) $\{\|x_n\|\}$ is a bounded sequence, and (ii) for any i, $\lim_n \xi_i^{(n)} = \xi_i^{(0)}$.

4.14.8. A sequence $\{f_n(t)\}$ in $L^p(0,1)$ $(1 < p < \infty)$ is weakly convergent to $f_0(t) \in L^p(0,1)$ if and only if (i) $\{\|f_n\|_p\}$ is a bounded sequence, and (ii) $\lim_n \int_0^\tau f_n(t)\,dt = \int_0^\tau f_0(t)\,dt$ for any $\tau \in (0,1]$.

4.14.9. Prove *Helly's principle*: if $\{g_n\}$ is a sequence of monotone and uniformly bounded functions, then there exists a subsequence $\{g_{n_i}\}$ that converges everywhere to a bounded monotone function. [*Hint:* By Alaoglu's theorem and Theorem 4.14.8 we may assume that $\int_0^1 f(\tau)\,dg_n(\tau) \to \int_0^1 f(\tau)\,dg_0(\tau)$ for any continuous function f, where g_0 is monotone. Let t be a point of continuity of g_0. Take f to run over a sequence φ_m, $\varphi_m(\tau) = 1$ if $0 \le \tau \le t$, $\varphi_m(\tau) = 0$ if $t + (1/m) \le \tau < \infty$ and $\varphi_m(\tau)$ linear in $[t, t + (1/m)]$. Deduce $\overline{\lim}\, [g_n(t) - g_n(0)] \le g_0(t) - g_0(0)$. Next take $f(\tau) = \varphi_m[\tau + (1/m)]$, and deduce $\underline{\lim}\, [g_n(t) - g_n(0)] \ge g_0(t) - g_0(0)$.]

4.14.10. Find the adjoint of the operator given in Problem 4.4.9.

CHAPTER 5

COMPLETELY CONTINUOUS OPERATORS

5.1 BASIC PROPERTIES

Definition Let X, Y be normed linear spaces. An operator $T \in \mathscr{B}(X,Y)$ is called *completely continuous* or *compact* if it maps bounded sets of X into compact sets of Y. Note that T is compact if and only if it maps bounded sets of X onto relatively compact sets of Y.

Theorem 5.1.1 *A completely continuous linear operator maps weakly convergent sequences into convergent sequences.*

Proof. Let T be a completely continuous operator in $\mathscr{B}(X,Y)$ and let $\{x_n\}$ be a sequence in X weakly convergent to a point $x_0 \in X$. Since T is bounded, $\{Tx_n\}$ is weakly convergent to Tx_0. If the sequence $\{y_n\}$, where $y_n = Tx_n$, is not convergent to $y_0 = Tx_0$, then there exists a subsequence $\{y_{n_k}\}$ such that $\|y_{n_k} - y_0\| \geq \varepsilon_0$ for some $\varepsilon_0 > 0$ and all k. Now, the sequence $\{x_{n_k}\}$ is weakly convergent and it is therefore also bounded. Since T is compact, there exists a subsequence $\{y_{n'_k}\}$ of $\{y_{n_k}\}$ that is convergent to some point $\bar{y} \in Y$. But then $\{y_{n'_k}\}$ is also weakly convergent to $\bar{y}$. It follows that $\bar{y} = y_0$. Hence $\|y_{n'_k} - y_0\| \to 0$ as $k \to \infty$—a contradiction.

Theorem 5.1.2 *Let* X *be a normed linear space and let* Y *be a Banach space. If* $\{T_n\}$ *is a sequence of completely continuous operators in* $\mathscr{B}$(X,Y) *uniformly convergent to some* T ∈ $\mathscr{B}$(X,Y), *then* T *is also completely continuous.*

Proof. Let A be a bounded set in X. For any $\varepsilon > 0$ there exists an operator T_n such that $\|T_n x - Tx\| < \varepsilon/3$ for every $x \in A$. Since $\overline{T_n(A)}$ is a compact set, it is also totally bounded. It follows that there is a finite number of points $x_1,\ldots,x_m$ such that

$$\inf_{1 \le i \le m} \|T_n x - T_n x_i\| < \frac{\varepsilon}{3}$$

for every $x \in$ A. But then

$$\inf_{1 \le i \le m} \|Tx - Tx_i\| < \varepsilon.$$

Thus $T(A)$ is totally bounded. Since Y is a Banach space, it follows that $\overline{T(A)}$ is a compact set. Hence T is compact.

Theorem 5.1.3 *Let* X *be a normed linear space. If* S $\in \mathscr{B}$(X), T $\in \mathscr{B}$(X) *and* T *is completely continuous, then* ST *and* TS *are completely continuous.*

The proof is left to the reader.

Theorem 5.1.4 *Let* X *be a normed linear space and let* Y *be a Banach space. An operator* T $\in \mathscr{B}$(X,Y) *is completely continuous if and only if its adjoint* T* *is completely continuous.*

Proof. Let T be a compact operator. Each of the sets $\{Tx;\ \|x\| < n\}$ $(n = 1,2,\ldots)$ is relatively compact and, hence, separable. It follows that $T(X)$ is separable. Let A be a dense sequence in $T(X)$. Take any bounded sequence $\{y_n^*\}$ in Y^*. By the diagonal method, there exists a subsequence $\{y_{n_i}^*\}$ such that $\{y_{n_i}^*(y)\}$ is convergent for any $y \in A$. It follows (compare Problem 4.12.1) that $\{y_{n_i}^*(y)\}$ is convergent for any $y \in \overline{T(X)}$.

Let $x_i^* = T^* y_{n_i}^*$. Then $x_i^*(x) = y_{n_i}^*(Tx)$. It follows that $\{x_i^*(x)\}$ is convergent for any $x \in X$. By Theorem 4.5.2 we conclude that there exists an element x^* in X^* such that $\lim x_i^*(x) = x^*(x)$ for any $x \in X$. If we prove that

$$\|x_i^* - x^*\| \to 0 \qquad \text{as } i \to \infty, \tag{5.1.1}$$

then it follows that T^* is compact.

Choose $x_i \in X$ such that

$$\|x_i\| = 1, \qquad |x^*(x_i) - x_i^*(x_i)| \ge \tfrac{1}{2}\|x^* - x_i^*\|. \tag{5.1.2}$$

Suppose now that (5.1.1) is not true. Then there exists an $\eta > 0$ and a subsequence $\{x_{m_j}^*\}$ of $\{x_i^*\}$ such that

$$\|x_{m_j}^* - x^*\| \ge \eta \qquad \text{for all } j.$$

Using (5.1.2), we then get

$$|y_{m_j}^*(Tx_{m_j}) - \lim_k y_{m_k}^*(Tx_{m_j})| \geq \tfrac{1}{2}\eta. \tag{5.1.3}$$

Since $\|x_{m_j}\| = 1$ and T is compact, there exists a subsequence $\{m_j'\}$ of $\{m_j\}$ such that $\lim Tx_{m_j'} = y_0$ exists. Hence, for any $\varepsilon > 0$ there is an n_0 such that, if $m_j' \geq n_0$,

$$\|y_0 - Tx_{m_j'}\| < \varepsilon, \qquad |y_{m_j'}^*(y_0) - \lim_k y_{m_k}^*(y_0)| < \varepsilon.$$

[Here we have used the facts that $y_0 \in \overline{T(X)}$ and $\{y_{m_i}^*(y)\}$ is convergent if $y \in \overline{T(X)}$.] Therefore,

$$\begin{aligned} &|y_{m_j'}^*(Tx_{m_j'}) - \lim_k y_{m_k'}^*(Tx_{m_j'})| \\ &\leq |y_{m_j'}^*(Tx_{m_j'} - y_0)| + |y_{m_j'}^*(y_0) - \lim_k y_{m_k'}^*(y_0)| + |\lim_k y_{m_k'}^*(y_0 - Tx_{m_j'})| \\ &\leq M\varepsilon + \varepsilon + M\varepsilon, \end{aligned}$$

where $\|y_n^*\| \leq M$. Since $\varepsilon > 0$ is arbitrary, we get a contradiction to (5.1.3).

Suppose conversely that T^* is compact. By what we have just proved it follows that T^{**} is compact.

Now let $\{x_n\}$ be any bounded sequence in X. Then the sequence $\{\hat{x}_n\}$, where $\hat{x}_n = \kappa x_n$ (κ the natural imbedding of X into X^{**}), is also bounded. It follows that there exists a subsequence $\{\hat{x}_{n_i}\}$ such that $\{T^{**}\hat{x}_{n_i}\}$ is a convergent sequence in X**. Since, by Theorem 4.13.3, $\kappa(Tx_{n_i}) = T^{**}\hat{x}_{n_i}$ (here κ is the natural imbedding of Y into Y^{**}), and since κ is an isometry, it follows that $\{Tx_{n_i}\}$ is a Cauchy sequence in Y. Since Y is complete, this sequence is convergent. Hence T is compact.

PROBLEMS

5.1.1. A continuous linear transformation with finite-dimensional range is compact.

5.1.2. A linear combination of completely continuous linear transformations is completely continuous.

5.1.3. Let A be the operator defined in Problem 4.4.9 and assume that X, Y are compact metric spaces, that open sets are measurable in X and Y, that μ,ν are finite measures and that $K(x,y)$ is continuous on $X \times Y$. Prove that A is a compact linear operator from $L^r(Y,\nu)$ into $L^p(X,\mu)$.

5.1.4. Let Ω_1,Ω_2, be bounded closed sets in R^n and let $X = \Omega_1$, $Y = \Omega_2$. Denote by μ the Lebesgue measure. If $K(x,y) \in L^p(X \times Y, \mu \times \mu)$, then A, defined in Problem 5.1.3, is compact. [*Hint*: Approximate K by continuous functions and use Theorem 5.1.2.]

5.1.5. Is the operator T defined by $(Tx)(t) = tx(t)$ $(0 < t < 1)$ completely continuous in $L^2(0,1)$?

5.1.6. Let $f(z)$ be an entire complex analytic function with $f(0) = 0$, and let X be a Banach space. If $T \in \mathscr{B}(X)$ is completely continuous, then $f(T)$ is also completely continuous.

5.1.7. Let T be a compact linear operator from a Banach space X onto itself. If T^{-1} is a bounded operator, then X is finite-dimensional.

5.1.8. Let T be a bounded linear operator in a reflexive Banach space. If T maps weakly convergent sequences onto convergent sequences, then T is completely continuous.

5.2 THE FREDHOLM-RIESZ-SCHAUDER THEORY

Throughout this section we shall denote by T a completely continuous linear operator in a Banach space X, and by λ a nonzero scalar. From the proof of Theorem 5.1.4 we conclude that T^* is completely continuous (X need not be complete). We shall consider the bounded linear operator $\lambda I - T$ and write

$$N_\lambda = \{x;\ \lambda x - Tx = 0\}, \qquad N_\lambda^* = \{y^*;\ \lambda y^* - T^*y^* = 0\},$$
$$R_\lambda = \{y;\ y = \lambda x - Tx,\ x \in X\}, \qquad R_\lambda^* = \{x^*;\ x^* = \lambda y^* - T^*y^*, y^* \in X^*\},$$

where T^* is the adjoint of T.

We recall the following fact proved in Section 4.3.

Lemma 5.2.1 *A closed linear subspace* X_0 *of* X *is finite-dimensional if and only if every bounded sequence in* X_0 *has a convergent subsequence.*

We shall now prove:

Lemma 5.2.2 N_λ *and* N_λ^* *are finite-dimensional subspaces of* X *and* X^*, *respectively.*

Proof. It is clear that N_λ is a closed linear subspace of X. Let $\{x_i\}$ be any bounded sequence in N_λ. Then $\lambda x_i = Tx_i$. Since T is completely continuous, there is a subsequence $\{Tx_{i'}\}$ of $\{Tx_i\}$ that is convergent. Hence $\{x_{i'}\}$ is convergent. In view of Lemma 5.2.1 it follows that N_λ is finite-dimensional. The assertion for N_λ^* is proved in a similar way.

Lemma 5.2.3 R_λ *and* R_λ^* *are closed linear subspaces of* X *and* X^*, *respectively.*

Proof. It is clear that R_λ and R_λ^* are linear subspaces. If we prove that R_λ is closed, then it also follows, by Theorem 4.13.6, that R_λ^* is closed. Suppose then that $\{y_n\} \subset R_\lambda$, $y_n \to y_0$. We shall prove that $y_0 \in R_\lambda$. The proof will be given in two steps.

(a) Let $y_n = \lambda x_n - Tx_n$. There exists a bounded sequence $\{z_n\}$ such that $\lambda x_n - Tx_n = \lambda z_n - Tz_n$. To prove this, denote by $d(x)$ the distance from x to N_λ. If the sequence $\{d(x_n)\}$ is not bounded, then $d(x_{n_i}) \to \infty$ for some sequence of n_i's. Let $\bar{x}_n = x_n/d(x_n)$. Then $d(\bar{x}_{n_i}) = 1$. Thus there exist points v_{n_i} in N_λ such that the points $w_{n_i} = \bar{x}_{n_i} - v_{n_i}$ satisfy: $\|w_{n_i}\| \leq 2$. But then

$$(\lambda I - T)w_{n_i} = (\lambda I - T)\bar{x}_{n_i} = \frac{(\lambda I - T)x_{n_i}}{d(x_{n_i})} = \frac{y_{n_i}}{d(x_{n_i})} \to 0.$$

Since T is compact, it follows that a subsequence $\{Tw_{n'_i}\}$ of $\{Tw_{n_i}\}$ is convergent. Hence $w_{n'_i} \to w$ as $n'_i \to \infty$. But then $(\lambda I - T)w = 0$ and $d(w) = \lim d(w_{n'_i}) = \lim d(\bar{x}_{n'_i}) = 1$. This is impossible. We have thus proved that $\{d(x_n)\}$ is a bounded sequence, say $d(x_n) \leq C$ for all n. We can therefore choose $\hat{z}_n$ in N_λ such that $\|x_n - \hat{z}_n\| \leq C + 1$. Now take $z_n = x_n - \hat{z}_n$.

(b) From the compactness of T it follows that there exists a subsequence $\{z_{n_i}\}$ of $\{z_n\}$ such that $\{Tz_{n_i}\}$ is convergent. Hence also $\{z_{n_i}\}$ is convergent. Denoting its limit by x_0, we get $y_0 = \lambda x_0 - Tx_0$. This shows that $y_0 \in R_\lambda$.

Combining Lemma 5.2.3 with Theorems 4.13.5, 4.13.6, we get:

Theorem 5.2.4 (a) $R_\lambda = N_\lambda^{*\perp}$—*that is, the equation* $\lambda x - Tx = y$ *has a solution if and only if* y *lies in the orthogonal complement of* N_λ^*.

(b) $R_\lambda^* = N_\lambda^\perp$—*that is, the equation* $\lambda y^* - T^*y^* = x^*$ *has a solution if and only if* x^* *lies in the orthogonal complement of* N_λ.

Let $L = \lambda I - T$. Then

$$L^n = (\lambda I - T)^n = \lambda^n I + \sum_{\nu=1}^{n} \binom{n}{\nu} \lambda^{n-\nu}(-1)^\nu T^\nu.$$

It follows that L^n has the form $\mu I - T_0$, where $\mu = \lambda^n$ is a scalar and T_0 is a compact operator. Hence the null space N_λ^n of L^n is finite-dimensional. Notice that $N_\lambda^n \subset N_\lambda^{n+1}$.

Lemma 5.2.5 *There exists a positive integer* k *such that* $N_\lambda^n \neq N_\lambda^{n+1}$ *for all* $n < k$ *and* $N_\lambda^n = N_\lambda^k$ *for all* $n > k$.

Proof. We first show that if $N_\lambda^n = N_\lambda^{n+1}$, then $N_\lambda^{n+2} = N_\lambda^n$. Let $x \in N_\lambda^{n+2}$. Then $0 = L^{n+2}x = L^{n+1}(Lx)$. Hence $Lx \in N_\lambda^{n+1} = N_\lambda^n$. This means that $L^n(Lx) = 0$—that is, $x \in N_\lambda^{n+1} = N_\lambda^n$. Continuing in this way step by step, we deduce that $N_\lambda^{n+h} = N_\lambda^n$ for any positive integer h. Thus, it remains to show that $N_\lambda^n \neq N_\lambda^{n+1}$ cannot occur for all n. If this does occur, then, by Lemma 4.3.1, there exists a sequence $\{x_n\}$ in X such that $x_n \in N_\lambda^{n+1}$, $\|x_n\| = 1$, and $\|x_n - x\| > \frac{1}{2}$ for all $x \in N_\lambda^n$.

If $n > m$, then

$$L^n(Lx_n + Tx_m) = L^{n+1}x_n + TL^n x_m = 0.$$

Hence $Lx_n + Tx_m \in N_\lambda^n$. Using the relation $Tx_n - Tx_m = \lambda x_n - (Lx_n + Tx_m)$, it then follows that

$$\frac{1}{|\lambda|}\|Tx_n - Tx_m\| = \left\| x_n - \frac{1}{\lambda}(Lx_n + Tx_m) \right\| > \tfrac{1}{2}.$$

Thus the sequence $\{Tx_n\}$ has no convergent subsequences. Since, however, $\{x_n\}$ is a bounded sequence, we have derived a contradiction to the assumption that T is compact.

Lemma 5.2.6 *If* $\mathrm{R}_\lambda = \mathrm{X}$, *then* $\mathrm{N}_\lambda = 0$.

Proof. If $N_\lambda \neq 0$, then there exists a point $x_0 \neq 0$ in N_λ. Since $R_\lambda = X$, we can construct a sequence $\{x_n\}$ in X such that $Lx_1 = x_0$, $Lx_2 = x_1$, $Lx_3 = x_2, \ldots$. But then $L^n x_n = x_0 \neq 0$, $L^{n+1}x_n = Lx_0 = 0$. Thus $N_\lambda^{n+1} \neq N_\lambda^n$ for all n. This contradicts Lemma 5.2.5.

Theorem 5.2.7 $\mathrm{R}_\lambda = \mathrm{X}$ *if and only if* $\mathrm{N}_\lambda = 0$.

Proof. Suppose $N_\lambda = 0$. By Theorem 5.2.4, $R_\lambda^* = X^*$. Since T^* is compact, we can apply Lemma 5.2.6 to T^* (instead of T) and conclude that $N_\lambda^* = 0$. Theorem 5.2.4 then gives $R_\lambda = X$.

Lemma 5.2.8 *Let* $\mathrm{x}_1^*, \ldots, \mathrm{x}_n^*$ *be linearly independent elements of* X^*. *Then there exist points* $\mathrm{x}_1, \ldots, \mathrm{x}_n$ *in* X *such that* $\mathrm{x}_i^*(\mathrm{x}_j) = \delta_{ij}$ ($\delta_{ij} = 0$ *if* $\mathrm{i} \neq \mathrm{j}$, $\delta_{ij} = 1$ *if* $\mathrm{i} = \mathrm{j}$) *for all* $1 \leq \mathrm{i}, \mathrm{j} \leq \mathrm{n}$.

Proof. Denote by N_j the null space of x_j^* and let $M_j = N_1 \cap \cdots \cap N_{j-1} \cap N_{j+1} \cap \cdots \cap N_n$. We shall prove that

$$M_j \text{ is not contained in } N_j, \qquad 1 \leq j \leq n. \tag{5.2.1}$$

It suffices to prove (5.2.1) for $j = 1$. If $M_1 \subset N_1$, then $x_1^*(x) = 0$ whenever $x_2^*(x) = \cdots = x_n^*(x) = 0$. Consider the linear map $A\colon X \to \mathscr{F}^{n-1}$ ($\mathscr{F}$ the field of scalars of X) given by

$$Ax = [x_2^*(x), \ldots, x_n^*(x)],$$

and define on $A(X)$ a linear functional ψ by

$$\psi(Ax) = \psi[x_2^*(x), \ldots, x_n^*(x)] = x_1^*(x).$$

ψ is defined unambiguously. Indeed, if $Ax = A\bar{x}$, then $x_2^*(x - \bar{x}) = \cdots = x_n^*(x - \bar{x}) = 0$. Hence $x_1^*(x - \bar{x}) = 0$—that is, $\psi(Ax) = \psi(A\bar{x})$.

By the Hahn-Banach theorem, we can extend the linear functional ψ into a linear functional ψ_1 in $\mathscr{F}^{n-1}$. But then ψ_1 has the form

$$\psi_1(y_2,\dots,y_n) = \sum_{i=2}^{n} \alpha_i y_i,$$

where α_i are constants. It follows that

$$x_1^*(x) = \sum_{i=2}^{n} \alpha_i x_i^*(x),$$

thus contradicting the linear independence of the x_j^*.

From (5.2.1) it follows that there exist points $\bar{x}_j \in M_j$ such that $\bar{x}_j \notin N_j$. This means that $x_i^*(\bar{x}_j) = 0$ if $i \neq j$, and $x_j^*(\bar{x}_j) \neq 0$. Now take $x_j = \bar{x}_j / x_j^*(\bar{x}_j)$.

Theorem 5.2.9 N_λ *and* N_λ^* *have the same finite dimension.*

Proof. In view of Lemma 5.2.2, $n = \dim N_\lambda$ and $\nu = \dim N_\lambda^*$ are finite numbers. Let $x_1,\dots,x_n$ be a basis in N_λ and let $y_1^*,\dots,y_\nu^*$ be a basis in N_λ^*. By the Hahn-Banach theorem there exist elements $x_1^*,\dots,x_n^*$ in X^* such that $x_i^*(x_j) = \delta_{ij}$ for all $1 \leq i, j \leq n$. In view of Lemma 5.2.8 there exist elements $y_1,\dots,y_\nu$ in X such that $y_i^*(y_j) = \delta_{ij}$ for all $1 \leq i, j \leq \nu$.

Suppose $n < \nu$. Let

$$Sx = Tx + \sum_{i=1}^{n} x_i^*(x) y_i.$$

S is a finite sum of compact operators and therefore it is also a compact operator. We shall prove that

$$N_{\lambda I - S} = 0. \tag{5.2.2}$$

Suppose $x \in N_{\lambda I - S}$. Then $\lambda x = Sx$—that is,

$$\lambda x - Tx = \sum_{i=1}^{n} x_i^*(x) y_i.$$

It follows that

$$0 = (\lambda y_j^* - T^* y_j^*)(x) = y_j^*(\lambda x - Tx) = x_j^*(x).$$

Thus $x_j^*(x) = 0$ for $1 \leq j \leq n$. Hence $\lambda x - Tx = 0$—that is, $x \in N_\lambda$. Consequently, $x = \sum_{i=1}^{n} \lambda_i x_i$. But then $0 = x_j^*(x) = \lambda_j$. Thus $x = 0$. This proves (5.2.2).

From (5.2.2) and Theorem 5.2.7 it follows that the range of $\lambda I - S$ is X. Thus, in particular, there is an $x \in X$ such that $\lambda x - Sx = y_{n+1}$. But then

$$1 = y_{n+1}^*(y_{n+1}) = y_{n+1}^*(\lambda x - Sx) = y_{n+1}^*\left((\lambda x - Tx) - \sum_{i=1}^{n} x_i^*(x)y_i\right) = 0.$$

since $y_{n+1}^*(y_i) = 0$ $(1 \leq i \leq n)$ and since $y_{n+1}^* \in N_\lambda^*$. The contradiction obtained proves that $n \geq \nu$.

Applying this result to T^*, we conclude that the dimension m of the null space of $\lambda I - T^{**}$ is not larger than the dimension ν of the null space of $\lambda I - T^*$—that is, $\nu \geq m$. Since obviously $m \geq n$, we get $\nu \geq n$. This completes the proof.

Theorems 5.2.4, 5.2.7, and 5.2.9 constitute the *Fredholm-Riesz-Schauder theory*. We summarize these theorems, using the notation N_S = null set of S, R_S = range of S:

Theorem 5.2.10 *Let* X *be a Banach space and let* T *be a completely continuous linear operator in* X. *For any* $\lambda \neq 0$,

$$\dim N_{\lambda I - T} = \dim N_{\lambda I - T^*} < \infty, \tag{5.2.3}$$

$$R_{\lambda I - T} = N_{\lambda I - T^*}^{\perp}, \qquad R_{\lambda I - T^*} = N_{\lambda I - T}^{\perp}. \tag{5.2.4}$$

Theorem 5.2.10 can also be stated in the following form (called the *Fredholm alternative*).

Either (a) for any $y \in X$ there exists a unique solution x of $(\lambda I - T)x = y$, or (b) there is an $x \neq 0$ such that $(\lambda I - T)x = 0$. If (a) holds, then for any $y^* \in X^*$ there exists a unique solution x^* of $(\lambda I - T^*)x^* = y^*$. If (b) holds, then $\dim N_{\lambda I - T} = \dim N_{\lambda I - T^*}$ is finite and, furthermore, the equation $(\lambda I - T)x = y$ $[(\lambda I - T^*)x^* = y^*]$ has a solution if and only if y (y^*) is in the orthogonal complement of $N_{\lambda I - T^*}$ $(N_{\lambda I - T})$.

PROBLEMS

5.2.1. Let X be a compact set in R^n and let μ denote the Lebesgue measure. Let $K(x,y) \in L^2(X \times X, \mu \times \mu)$. For any given $g \in L^2(X,\mu)$, consider the equation, in $L^2(X,\mu)$,

$$f(x) = g(x) + \lambda \int K(x,y)f(y)\, d\mu(y). \tag{5.2.5}$$

We call this equation an *integral equation of Fredholm's type*. Prove that if $g = 0$ implies $f = 0$, then there exists a unique solution of (5.2.5) for any $g \in L^2(X,\mu)$.

5.2.2. Let X be a compact metric space and let μ be a finite measure on X. Let $K(x,y)$ be continuous on $X \times X$. Assume that the only continuous solution of

$$f(x) = \lambda \int K(x,y)f(y)\, d\mu(y)$$

is $f = 0$. Prove that for every continuous function $g(x)$ on X there exists a unique continuous solution $f(x)$ of (5.2.5).

5.2.3. Let X,μ,K be as in Problem 5.2.2. Denote by $\varphi_1,\ldots,\varphi_m$ a basis for the space of solutions φ of the equation

$$\varphi(x) = \lambda \int K(y,x)\varphi(y)\, d\mu(y).$$

Then the equation (5.2.5) has a solution if and only if $\int g(x)\varphi_i(x)\, d\mu(x) = 0$ for $1 \le i \le n$.

5.2.4. Consider the *Volterra integral equation*

$$f(s) = g(s) + \int_0^t K(s,t)f(t)\, dt \qquad (0 \le t \le 1), \tag{5.2.6}$$

where $K(s,t)$ is continuous for $0 \le s,\ t \le 1$. Prove that for any continuous function g there exists a unique continuous solution f of (5.2.6).

5.2.5. The integral equation

$$\varphi(t) = \lambda \int_0^\infty (\sin st)\varphi(y)\, dy$$

for $\lambda = \pm\sqrt{2/\pi}$ has an infinite number of solutions, namely,

$$\varphi(t) = \sqrt{\pi/2}\, e^{-ax} \pm x/(a^2 + x^2), \qquad a > 0.$$

5.3 ELEMENTS OF SPECTRAL THEORY

Let T be a bounded linear operator in a normed linear space X.

Definition The *resolvent set* $\rho(T)$ of T consists of all complex numbers λ for which $(\lambda I - T)^{-1}$ is a bounded operator (with domain X). The *spectrum* $\sigma(T)$ of T consists of the complement of $\rho(T)$ in $\mathbb{C}$.

The operator $R(\lambda;T) = (\lambda I - T)^{-1}$ if bounded (and with domain X) is called the *resolvent* of T.

Let $\lambda \in \sigma(T)$. Then there are three possibilities:

(a) The range of $\lambda I - T$ is dense in X and $(\lambda I - T)^{-1}$ exists but is unbounded. We then say that λ belongs to the *continuous spectrum* of T.

(b) $(\lambda I - T)^{-1}$ exists, but its domain is not dense in X. We then say that λ belongs to the *residual spectrum* of T.

(c) $\lambda I - T$ does not have an inverse—that is, there is an $x \neq 0$ satisfying $\lambda x - Tx = 0$. We then say that λ is an *eigenvalue* of T.

If λ is an eigenvalue, then any $x \neq 0$ satisfying $\lambda x - Tx = 0$ is called an *eigenvector* (or *eigenelement*) of T (corresponding to λ).

Theorem 5.3.1 *Let* $T \in \mathscr{B}(X)$, X *a Banach space. Then* $\rho(T)$ *is an open set and*

$$R(\mu;T) - R(\lambda;T) = (\lambda - \mu)R(\lambda;T)R(\mu;T) \tag{5.3.1}$$

if $\lambda,\mu \in \rho(T)$. *Furthermore,* $R(\lambda;T)$ *is analytic in* $\lambda \in \rho(T)$.

Proof. If $\lambda_0 \in \rho(T)$ and $|\lambda - \lambda_0| < \|R(\lambda_0;T)\|^{-1}$, then $\lambda \in \rho(T)$. Indeed,

$$\lambda I - T = (\lambda_0 I - T) + (\lambda - \lambda_0)I = (\lambda_0 I - T)[I - (\lambda_0 - \lambda)R(\lambda_0;T)].$$

It follows that $(\lambda I - T)^{-1}$ exists and is equal to

$$\left[\sum_{n=0}^{\infty} (\lambda_0 - \lambda)^n (R(\lambda_0;T))^n\right](\lambda_0 I - T)^{-1}.$$

Next,

$$(\lambda I - T)^{-1} - (\mu I - T)^{-1} = (\lambda I - T)^{-1}[(\mu I - T) - (\lambda I - T)](\mu I - T)^{-1}$$
$$= (\mu - \lambda)R(\lambda;T)R(\mu;T).$$

It remains to prove that $R(\lambda;T)$ is analytic in $\lambda \in \rho(T)$. This follows from (5.3.1). Indeed, if $\mu \to \lambda$,

$$\frac{R(\mu;T) - R(\lambda;T)}{\mu - \lambda} = -R(\mu;T)R(\lambda;T) \to -[R(\lambda;T)]^2$$

in the uniform topology.

Theorem 5.3.2 *Let* X *be a Banach space and let* T *be a completely continuous linear operator in* X. *Then the equation* $x - \lambda Tx = 0$ *has nontrivial solutions only for a countable set of complex numbers having no finite points of accumulation.*

Proof. Suppose there exists a bounded sequence $\{\lambda_n\}$ of mutually distinct numbers such that $x - \lambda_n Tx = 0$ has a solution $x = x_n \neq 0$ for each n. We first claim that, for each n, $x_1,\cdots,x_n$ are linearly independent. Indeed, if this is false, then there exists an integer k such that $x_1,\cdots,x_{k-1}$ are linearly independent and $x_1\cdots,x_k$ are linearly dependent, say

$$\sum_{j=1}^{k} c_j x_j = 0 \quad \left(\sum_{j=1}^{k} |c_j|^2 > 0\right). \tag{5.3.2}$$

Then

$$0 = \lambda_k \sum_{j=1}^{k} c_j T x_j = \sum_{j=1}^{k} \frac{\lambda_k c_j}{\lambda_j} x_j.$$

Comparing this with (5.3.2), we get

$$\sum_{j=1}^{k-1} \left(1 - \frac{\lambda_k}{\lambda_j}\right) c_j x_j = 0.$$

Since $1 - \lambda_k/\lambda_j \neq 0$ if $j \neq k$, we get $c_j = 0$ if $j \neq k$. But then also $c_k = 0$. This contradicts the assumption that $\sum_{j=1}^{k} |c_j|^2 > 0$.

Denote by Y_n the linear space spanned by $x_1,\ldots,x_n$. By Lemma 4.3.1 there exists a sequence $\{y_n\}$ in X such that $\|y_n\| = 1$, $y_n \in Y_n$ and $\|y_n - y\| > \frac{1}{2}$ for any $y \in Y_{n-1}$.

Let $w \in Y_n$. Then $w = \sum_{i=1}^{n} \beta_i x_i$, so that

$$w - \lambda_n Tw = \sum_{i=1}^{n} \beta_i x_i - \lambda_n \sum_{i=1}^{n} \frac{\beta_i}{\lambda_i} x_i = \sum_{i=1}^{n-1} \left(1 - \frac{\lambda_n}{\lambda_i}\right) \beta_i x_i.$$

Thus $w - \lambda_n Tw \in Y_{n-1}$. If $n > m$, then

$$\|T(\lambda_n y_n) - T(\lambda_m y_m)\| = \|y_n - (y_n - \lambda_n T y_n + \lambda_m T y_m)\| > \tfrac{1}{2},$$

since the last element in parentheses belongs to Y_{n-1}. Thus $\{T(\lambda_n y_n)\}$ has no convergent subsequence. Since, however, $\{\lambda_n y_n\}$ is a bounded sequence, we get a contradiction to the compactness of T.

Denote by S the set of all complex numbers λ for which the equation $x - \lambda Tx = 0$ has a nontrivial solution. By what we have proved above, $S \cap \{\lambda, |\lambda| < m\}$ is a finite set for any $m = 1,2,3,\ldots$. It follows that S is countable and that it has no finite points of accumulation.

If T is compact, then, by the Fredholm-Reisz-Schauder theory, if $\lambda \neq 0$, then either $\lambda \in \rho(T)$ or λ is an eigenvalue. If X is infinite dimensional, then $0 \in \sigma(T)$, since if T^{-1} were bounded, then $I = TT^{-1}$ would have been a compact operator. Combining these remarks with Theorem 5.3.2, we get:

***Corollary* 5.3.3** *Let* T *be a compact linear operator in an infinite-dimensional Banach space. Then* σ(T) *consists of* 0 *and of either a finite number of eigenvalues or an infinite sequence of eigenvalues that converges to* 0.

In Problems 5.3.1–5.3.10, T is a bounded linear operator in a Banach space X.

PROBLEMS

5.3.1. $\sigma(T)$ is nonempty.

5.3.2. $\rho(T) = \rho(T^*)$ and $R(\lambda;T^*) = R(\lambda;T)^*$.

5.3.3. If $\lambda_0 \in \sigma(T)$ and $p(z)$ is a polynomial, then $p(\lambda_0) \in \sigma[p(T)]$. [*Hint:* $p(\lambda_0)I - p(T) = (\lambda_0 I - T)q(T)$, $q(z)$ a polynomial. If $p(\lambda_0)I - p(T)$ has a bounded inverse, then also $\lambda_0 I - T$ has a bounded inverse.]

5.3.4. If $\lambda' \in \sigma[p(T)]$, p a polynomial, then $\lambda' = p(\lambda_1)$ for some $\lambda_1 \in \sigma(T)$. [*Hint:* $p(T) - \lambda' I = \prod_{i=1}^{n} (T - \lambda_i I)$, the λ_i being the zeros of p.]

5.3.5. The series $\sum_{n=0}^{\infty} T^n/\lambda^{n+1}$ is uniformly convergent if $|\lambda| > \overline{\lim_n} \|T^n\|^{1/n}$. Its sum is equal to the resolvent $R(\lambda; T)$.

5.3.6. If $\lambda_0 \in \sigma(T)$, then $|\lambda_0| \le \|T\|$.

5.3.7. If $\lambda_0 \in \sigma(T)$, then $|\lambda_0^n| \le \|T^n\|$. Hence $|\lambda_0| \le \underline{\lim} \|T^n\|^{1/n}$.

5.3.8. Let $s(T) = \max_{\lambda \in \sigma(T)} |\lambda|$. Then $R(\lambda;T)$ is analytic for $|\lambda| > s(T)$. It can be shown that the series $\sum T^n/\lambda^{n+1}$ for $R(\lambda;T)$ is then convergent if $|\lambda| > s(T)$. Prove that $\lim_n \|T^n\|^{1/n}$ exists and equals $s(T)$.

5.3.9. If T^n is compact, then $\sigma(T)$ is either a finite set or a countable set with 0 as the only point of accumulation.

5.3.10. Denote by $d(\lambda)$ the distance from λ to the spectrum of T. Then $\|R(\lambda;T)\| \ge 1/d(\lambda)$ for $\lambda \in \rho(T)$.

5.3.11. Prove the identity $R(\lambda;S) - R(\lambda;T) = R(\lambda;S)(S - T)R(\lambda;T)$, where S,T are bounded linear operators in a normed linear space X.

5.4 APPLICATION TO THE DIRICHLET PROBLEM

Let G be a bounded closed set in R^n and let $K(x,y)$ be a continuous function of (x,y) for $x \in G$, $y \in G$, $x \neq y$. Assume that

$$|K(x,y)| \le \frac{C}{|x - y|^\alpha} \qquad (C \text{ constant}) \tag{5.4.1}$$

for some $0 < \alpha < n$. Then we call K an *integrable kernel*. Consider the continuous kernels

$$K_\varepsilon(x,y) = \begin{cases} K(x,y), & \text{if } |x - y| > \varepsilon, \\ K(x,y)\,|x - y|^\beta \varepsilon^{-\beta}, & \text{if } |x - y| \le \varepsilon, \end{cases}$$

for some $\alpha < \beta < n$. It is easily seen that

$$\sup_{x \in G} \int_G |K(x,y) - K_\varepsilon(x,y)|\, dy \to 0 \qquad \text{as } \varepsilon \to 0. \tag{5.4.2}$$

Theorem 5.4.1 *The following alternative holds: either* (a) *for every* g ∈ C(G) *there exists a unique solution* f *in* C(G) *of*

$$f(x) = g(x) + \lambda \int_G K(x,y) f(y)\, dy, \tag{5.4.3}$$

or, (b) *there is a nontrivial solution in* C(G) *of* (5.4.3) *when* g = 0. *If* (b) *holds, then the spaces of continuous solutions of the equations*

$$\varphi(x) = \lambda \int_G K(x,y)\varphi(y)\, dy, \tag{5.4.4}$$

$$\psi(x) = \lambda \int_G K(y,x)\psi(y)\, dy, \tag{5.4.5}$$

have the same finite dimension. Furthermore, there exists a solution of (5.4.3) *if and only if* $\int_G f(x)\psi(x)\,dx = 0$ *for any solution* ψ *of* (5.4.5).

A proof of Theorem 5.4.1 can be given on the basis of the results of Problems 5.2.2 and 5.2.3 and (5.4.2). The details will not be given here. Instead, we shall show how Theorem 5.4.1 (or rather a result similar to it) can be used to solve the Dirichlet problem. For simplicity we take $n = 3$.

Let Ω be a bounded domain in R^3 with the boundary S in C^2. Consider the functions, in Ω,

$$V(x) = \frac{1}{2\pi} \int_S \frac{\sigma(y)}{|x-y|} dS_y,$$

$$W(x) = \frac{1}{2\pi} \int_S \mu(y) \frac{\partial}{\partial \nu_y} \frac{1}{|x-y|}\, dS_y,$$

where ν_y is the outward normal to S at y. We call $V(x)$ a *simple layer potential* and we call $W(x)$ a *double layer potential.* If μ is continuous, then it can be proved that, for any $x_0 \in S$,

$$\lim_{x \to x_0} W(x) = -\mu(x_0) + \frac{1}{2\pi} \int_S \mu(y) \frac{\partial}{\partial \nu_y} \frac{1}{|x_0-y|}\, dS_y,$$

The function $W(x)$ is harmonic in Ω. Hence $W(x)$ is the solution of the Dirichlet problem with boundary data $-f$ if

$$\mu(x_0) = f(x_0) + \int_S K(x_0,y)\mu(y)\, dS_y, \tag{5.4.6}$$

where

$$K(x,y) = \frac{1}{2\pi} \frac{\partial}{\partial \nu_y} \frac{1}{|x-y|}.$$

Using Theorem 5.4.1 (or rather a variant of it, with G, dx replaced by S, dS), we conclude:

Theorem 5.4.2 *If* $\lambda = 1$ *is not an eigenvalue of the operator* $\mu(x) \to \int K(x,y)\mu(y)\, dS_y$, *then there exists a unique solution of the Dirichlet problem.*

It can be shown (cf. Problem 5.4.5) that $\lambda = 1$ is indeed not an eigenvalue.

Consider now the *Neumann problem*—that is, the problem of finding a harmonic function u in Ω satisfying the boundary condition

$$\frac{\partial u}{\partial \nu} = f \qquad \text{on } S.$$

We now employ the simple layer potential V. It satisfies

$$\lim_{x \to x_0} \frac{\partial V(x)}{\partial \nu_{x_0}} = \sigma(x_0) + \frac{1}{2\pi} \int_S \sigma(y) \frac{\partial}{\partial \nu_{x_0}} \frac{1}{|x_0 - y|}\, dS_y. \tag{5.4.7}$$

This leads to an integral equation for σ with the kernel $K(y,x)$:

$$\sigma(x_0) = f(x_0) - \int_S K(y,x)\sigma(y)\, dS_y.$$

The number $\lambda = -1$ is an eigenvalue. But it can be shown that the only nontrivial solutions of the homogeneous equation $\psi(x) = -\int K(x,y)\psi(y)\, dS_y$ are the nonzero constants. Hence the Neumann problem has a solution if and only if $\int_S f\, dS = 0$.

PROBLEMS

5.4.1. Let K be an integrable kernel. Prove that the operator T defined by

$$(Tf)(x) = \int_G K(x,y)f(y)\, dy$$

is a compact linear operator in $C(G)$.

5.4.2. The simple layer potentials $V(x)$ (with continuous σ) are continuous functions in R^3.

5.4.3. Let Ω_0 denote the complement of $\overline{\Omega}$ in R^3. Show that if $V(x) \equiv$ const. in Ω_0, then $V(x) \equiv$ const. in Ω.

5.4.4. Denote by μ_y the outward normal to S at y, S considered as the boundary of Ω_0. (Thus $\mu_y = -\nu_y$.) Let $G_m = \Omega_0 \cap \{x; |x| < m\}$, m large. One can show that the simple layer potentials V (with continuous σ) satisfy

$$\int_{G_m} \nabla V \cdot \nabla V\, dx = \int_S V \frac{\partial V}{\partial \mu_y}\, dS_y + \int_{|x|=m} V \frac{\partial V}{\partial \nu}\, dS,$$

where ν is the outward normal. Use this relation to prove that if $\partial V/\partial \mu_y = 0$ for all $y \in S$, then $V \equiv$ const.

5.4.5. It is known that

$$\lim_{x \to y} \frac{\partial V(x)}{\partial \nu_y} - \lim_{z \to y} \frac{\partial V(z)}{\partial \mu_y} = 2\sigma(y) \qquad (y \in S). \tag{5.4.8}$$

Prove that $\lambda = 1$ is not an eigenvalue of the map $\sigma(x) \to \int K(y,x)\sigma(y)\, dS_y$. It follows that $\lambda = 1$ is not an eigenvalue of the map $\mu(x) \to \int K(x,y)\mu(y)\, dS_y$. [*Hint:* Use (5.4.7), (5.4.8).]

CHAPTER 6

HILBERT SPACES AND SPECTRAL THEORY

6.1 HILBERT SPACES

Definition 6.1.1 A nonempty set H is called a *Hilbert space* if H is a complex linear vector space, together with a complex-valued function $(\cdot,\cdot)$ from $H \times H$ into $\mathbb{C}$ having the following properties:

(i) $(x,x) \geq 0$, and $(x,x) = 0$ if and only if $x = 0$;
(ii) $(x + y,z) = (x,z) + (y,z)$ for all x,y,z in H;
(iii) $(\lambda x,y) = \lambda(x,y)$ for all x,y in H and $\lambda \in \mathbb{C}$;
(iv) $(x,y) = \overline{(y,x)}$ for all x,y in H;
(v) If $\{x_n\} \subset H$, $\lim (x_n - x_m, x_n - x_m) \to 0$ as $n,m \to \infty$, then there is an element $x \in H$ such that $\lim_n (x_n - x, x_n - x) = 0$.

The function $(\cdot,\cdot)$ is called the *scalar product* or the *inner product* of the space. The number (x,y) is called the *scalar* (or *inner*) *product of x and y*. If the condition (v) is not required, then we call H a *scalar-product space* (or an *inner-product space*). The number $\|x\| = (x,x)^{1/2}$ is called the *norm* of x.

If the scalar product is a real-valued function and if H is a real linear vector space, then we call H a *real Hilbert space*. We shall consider in the sequel only Hilbert spaces. But many of the results—in particular, all the results of Sections 6.1 through 6.4—extend, with minor modifications in the proofs, to real Hilbert spaces.

Note that (iii) and (iv) imply that $(x,\lambda y) = \bar{\lambda}(x,y)$. We also have: $(x,0) = (0,x) = 0$.

Theorem 6.1.1 *In a Hilbert space* H,

$$|(x,y)| \le \|x\| \, \|y\|. \tag{6.1.1}$$

This inequality is called the *Schwarz inequality*.

Proof. If $x = 0$ or $y = 0$, then (6.1.1) is obvious. Suppose then that $x \ne 0$, $y \ne 0$. For any complex number λ,

$$\begin{aligned} 0 &\le (x + \lambda y, x + \lambda y) \\ &= \|x\|^2 + |\lambda|^2 \|y\|^2 + \lambda(y,x) + \bar{\lambda}(x,y) \\ &= \|x\|^2 + |\lambda|^2 \|y\|^2 + 2 \operatorname{Re} \{\lambda(x,y)\}. \end{aligned}$$

Let $\lambda = -re^{i\theta}$, $r > 0$, and choose θ such that $\theta = -\arg(x,y)$ if $(x,y) \ne 0$. Then we get $\|x\|^2 + r^2\|y\|^2 \ge 2r|(x,y)|$. Taking $r = \|x\|/\|y\|$, the assertion (6.1.1) follows.

Theorem 6.1.2 *A Hilbert space* H *is a Banach space with the norm* $\| \quad \| = (\, , \,)^{1/2}$.

Proof. The only nontrivial property of the norm that has to be verified is the inequality $\|x + y\| \le \|x\| + \|y\|$. Since

$$(x,y) + (y,x) = 2 \operatorname{Re} \{(x,y)\} \le 2\|x\| \, \|y\|,$$

we have

$$\begin{aligned} \|x + y\|^2 &= \|x\|^2 + \|y\|^2 + (x,y) + (y,x) \\ &\le \|x\|^2 + \|y\|^2 + 2\|x\| \, \|y\| = (\|x\| + \|y\|)^2. \end{aligned}$$

Hence $\|x + y\| \le \|x\| + \|y\|$. We have proved that H is a normed linear space. The condition (v) in Definition 6.1.1 means that H is complete.

From the proof of (6.1.1) we find that if equality holds and $y \ne 0$, then $x = -\lambda y$. From the proof of Theorem 6.1.2 it then follows that if $\|x\| + \|y\| = \|x + y\|$ and $y \ne 0$, then $x = -\lambda y$. Hence if $\|x\| = \|y\| = 1$, $\|x + y\| = 2$, then $|\lambda| = 1$, $|1 - \lambda| = 2$. Therefore $\lambda = -1$—that is, $x = y$. We conclude (see Problem 4.3.3):

Corollary 6.1.3 *The norm of a Hilbert space is strictly convex.*

Theorem 6.1.4 *For any two elements* x,y *in a Hilbert space* H,

$$\|x + y\|^2 + \|x - y\|^2 = 2\|x\|^2 + 2\|y\|^2. \tag{6.1.2}$$

Indeed, the left-hand side is equal to $(x+y,x+y)+(x-y,x-y)=2(x,x)+2(y,y)=2\|x\|^2+2\|y\|^2$.

The equation (6.1.2) is called the *parallelogram law*. Its interest is that it characterizes the norms that are derived from scalar products. In fact, we have:

Theorem 6.1.5 *Let* H *be a complex Banach space with norm* $\| \ \|$ *satisfying* (6.1.2). *Then* H *is a Hilbert space with scalar product* (,) *such that* $(\ , \ )^{1/2}$ *coincides with the given norm on* H.

Proof. Define

$$(x,y)=\tfrac{1}{4}\{[\|x+y\|^2-\|x-y\|^2]+i[\|x+iy\|^2-\|x-iy\|^2]\}. \quad (6.1.3)$$

(If H is a real Banach space, then the expression with the second square brackets is to be omitted.) Since

$$(x,x)=\|x\|^2+\frac{i}{4}\|x\|^2|1+i|^2-\frac{i}{4}\|x\|^2|1-i|^2=\|x\|^2,$$

$(x,x)^{1/2}$ coincides with the norm $\|x\|$. To prove that (,) satisfies the condition (ii) in Definition 6.1.1 of the scalar product, we use the parallelogram law to obtain:

$$\|u+v+w\|^2+\|u+v-w\|^2=2\|u+v\|^2+2\|w\|^2,$$
$$\|u-v+w\|^2+\|u-v-w\|^2=2\|u-v\|^2+2\|w\|^2.$$

Hence

$$(\|u+v+w\|^2-\|u-v+w\|^2)+(\|u+v-w\|^2-\|u-v-w\|^2)$$
$$=2\|u+v\|^2-2\|u-v\|^2.$$

Thus,

$$\text{Re}\,(u+w,v)+\text{Re}\,(u-w,v)=2\,\text{Re}\,(u,v).$$

The relation with "Re" replaced by "Im" is proved similarly. Hence

$$(u+w,v)+(u-w,v)=2(u,v).$$

Taking $w=u$, we get $(2u,v)=2(u,v)$. Taking $u+w=x$, $u-w=y, v=z$, we then obtain

$$(x,z)+(y,z)=2\left(\frac{x+y}{2},z\right)=(x+y,z).$$

To prove the condition (iii) in Definition 6.1.1 we first note that, by (ii),

$$(mx,y)=((m-1)x+x,y)=((m-1)x,y)+(x,y)$$
$$=((m-2)x,y)+2(x,y)=\cdots=m(x,y).$$

Hence also $n(x/n,y) = (n(x/n),y) = (x,y)$. It follows that $(x/n,y) = (1/n)(x,y)$. If $r = m/n$, where m,n are any positive integers, then

$$r(x,y) = \frac{m}{n}(x,y) = m\left(\frac{x}{n}, y\right) = \left(\frac{m}{n}x,y\right) = (rx,y).$$

Since (x,y) is a continuous functional in x, we obtain $\lambda(x,y) = (\lambda x,y)$ for any positive real number. If $\lambda < 0$, then

$$\begin{aligned}\lambda(x,y) - (\lambda x,y) &= \lambda(x,y) - (|\lambda|(-x),y) = \lambda(x,y) - |\lambda|(-x,y)\\ &= \lambda(x,y) + \lambda(-x,y) = \lambda(0,y) = 0.\end{aligned}$$

Hence (iii) holds for any real λ. It is easily seen that (iii) also holds for $\lambda = i$. Finally, for any complex number $\lambda = \mu + i\nu$,

$$\begin{aligned}\lambda(x,y) &= \mu(x,y) + i\nu(x,y) = (\mu x,y) + i(\nu x,y)\\ &= (\mu x,y) + (i\nu x,y) = ((\mu + i\nu)x,y) = (\lambda x,y).\end{aligned}$$

The condition (iv) is easily verified, and (v) follows from the completeness assumption on the space H. Thus the proof is complete.

PROBLEMS

6.1.1. Let $f \in L^2(0,1)$. Show that for any positive integer n there exists a unique polynomial p_n of degree $\leq n$ such that $\|f - p\| \geq \|f - p_n\|$ for all polynomials p of degree $\leq n$.

6.1.2. Prove that the norm of $L^p(0,1)$ does not satisfy the parallelogram law if $p \neq 2$.

6.1.3. If $\{x_n\}$ is weakly convergent to x, in a Hilbert space X, and if $\lim_n \|x_n\| = \|x\|$, then $\lim_n x_n = x$.

6.1.4. Is the assertion of Problem 6.1.3 true in $L^p(0,1), p \neq 2$?

6.1.5. If $\{x_n\}$ is weakly convergent to x, in a Hilbert space H, then there is a subsequence $\{y_m\}$ whose arithmetic means z_n [that is, $z_n = (y_1 + \cdots + y_n)/n$] converge to x in the norm of H. [*Hint:* Suppose $x = 0$. Take $m_1 = 1$. Let m_2 $(> m_1)$ be such that $|(x_{m_1},x_m)| < \frac{1}{2}$ if $m \geq m_2$. Let $m_3 (> m_2)$ be such that $|(x_{m_1},x_m)| < \frac{1}{3}$, $|(x_{m_2},x_m)| < \frac{1}{3}$ if $m \geq m_3$, and so on. Take $y_n = x_{m_n}$.]

6.2 THE PROJECTION THEOREM

Definition Let x,y be two points in a Hilbert space H. If $(x,y) = 0$, then we say that x is *orthogonal* to y and write $x \perp y$. Let M be any subset of H. If x is orthogonal to all the elements of M, then we say that x is *orthogonal* to M, and write $x \perp M$. Let N be a subset of H. If $x \perp M$ for all $x \in N$, then we say that N is *orthogonal* to M, and write $N \perp M$. We then also have

$M \perp N$. We denote that $M^\perp$ the set of all elements that are orthogonal to M. We call $M^\perp$ the *orthogonal complement* of M. It is easily seen that $M^\perp$ is a closed linear space. If $N \perp M$, then, clearly, $N \subset M^\perp$.

Lemma 6.2.1 *Let* M *be a closed convex set in a Hilbert space* H. *For every point* $x_0 \in H$ *there exists a unique point* y_0 *in* M *such that*

$$\|x_0 - y_0\| = \inf_{y \in M} \|x_0 - y\|. \tag{6.2.1}$$

Proof. Denote the right-hand side of (6.2.1) by d. Then there exists a sequence $\{y_n\}$ in M such that $\|x_0 - y_n\| \to d$ as $n \to \infty$. By the parallelogram law,

$$4\|x_0 - \tfrac{1}{2}(y_m + y_n)\|^2 + \|y_m - y_n\|^2 = 2\|x_0 - y_m\|^2 + 2\|x_0 - y_n\|^2$$
$$\to 4d^2 \qquad \text{if } m,n \to \infty.$$

Since $\frac{1}{2}(y_m + y_n) \in M$, we also have

$$4\|x_0 - \tfrac{1}{2}(y_m + y_n)\|^2 \geq 4d^2.$$

Hence $\|y_m - y_n\| \to 0$ if $m,n \to \infty$. Since H is complete and M is closed, $\lim y_n = y_0$ exists and belongs to M. Since $\|x_0 - y_0\| = \lim_n \|x_0 - y_n\| = d$, (6.2.1) follows.

Suppose now that there exists another point y_1 satisfying

$$\|x_0 - y_1\| = \inf_{y \in M} \|x_0 - y\|.$$

Then

$$2\left\| x_0 - \frac{y_0 + y_1}{2} \right\| \leq \|x_0 - y_0\| + \|x_0 - y_1\| .$$

$$\leq 2 \inf_{y \in M} \|x_0 - y\| \leq 2\left\| x_0 - \frac{y_0 + y_1}{2} \right\|,$$

since $(y_0 + y_1)/2$ belongs to M. Hence,

$$2\left\| x_0 - \frac{y_0 + y_1}{2} \right\| = \|x_0 - y_0\| + \|x_0 - y_1\|.$$

Since the Hilbert norm is strictly convex (by Corollary 6.1.3), we get: $x_0 - y_0 = x_0 - y_1$. Thus, $y_1 = y_0$.

We shall now prove the *projection theorem*.

Theorem 6.2.2 *Let* M *be a closed linear subspace of a Hilbert space* H. *Any* $x_0 \in H$ *can be written in the form* $x_0 = y_0 + z_0$, *where* $y_0 \in M$, $z_0 \in M^\perp$. *The elements* y_0, z_0 *are uniquely determined by* x_0.

Proof. If $x_0 \in M$, then $y_0 = x_0, z_0 = 0$ give the asserted decomposition of x_0. If $x_0 \notin M$, let y_0 be the point in M satisfying

$$\|x_0 - y_0\| = \inf_{y \in M} \|x_0 - y\|.$$

The existence of y_0 is asserted in Lemma 6.2.1. Take now any point $y \in M$ and any scalar λ. Then $y_0 + \lambda y \in M$. It follows that

$$\|x_0 - y_0\|^2 \le \|x_0 - y_0 - \lambda y\|^2 = \|x_0 - y_0\|^2 - 2 \operatorname{Re} \lambda(y, x_0 - y_0) + |\lambda|^2 \|y\|^2.$$

Hence

$$-2 \operatorname{Re} \lambda(y, x_0 - y_0) + |\lambda|^2 \|y\|^2 \ge 0.$$

Taking $\lambda = \varepsilon > 0$, dividing by ε, and letting $\varepsilon \to 0$, we obtain the inequality

$$\operatorname{Re}(y, x_0 - y_0) \le 0.$$

If we take $\lambda = -i\varepsilon$, $\varepsilon > 0$, then we obtain, after dividing by ε and letting $\varepsilon \to 0$,

$$\operatorname{Im}(y, x_0 - y_0) \le 0.$$

Since the two inequalities obtained for y hold also for the point $-y$ (since $-y \in M$), we conclude that $(y, x_0 - y_0) = 0$ for any $y \in M$. Thus, the point $z_0 = x_0 - y_0$ is in $M^\perp$.

It remains to prove uniqueness. Suppose $x_0 = y_1 + z_1$,where $y_1 \in M$, $z_1 \in M^\perp$. Then the point $y_0 - y_1 = z_1 - z_0$ lies in both M and $M^\perp$. Hence it must be equal to 0—that is, $y_0 = y_1$ and $z_0 = z_1$.

Corollary 6.2.3 *If* M *is a closed linear substance and* $\mathrm{M} \neq \mathrm{H}$, *then there exists an element* $\mathrm{z}_0 \neq 0$ *such that* $\mathrm{z}_0 \perp \mathrm{M}$.

Indeed, there exists a point $x_0 \in H$, $x_0 \notin M$. Let $x_0 = y_0 + z_0$, where $y_0 \in M$, $z_0 \in M^\perp$. Then $z_0 \perp M$ and $z_0 \neq 0$.

Theorem 6.2.4 *For every bounded linear functional* x^* *on a Hilbert space* H *there exists a unique element* z *of* H *such that* $\mathrm{x}^*(\mathrm{x}) = (\mathrm{x},\mathrm{z})$ *for all* $\mathrm{x} \in \mathrm{H}$, *and* $\|\mathrm{x}^*\| = \|\mathrm{z}\|$.

This theorem is called the *Riesz theorem.*

Proof. Denote by N the null space of x^*. N is clearly a closed linear subspace of H. If $N = H$, then $x^* = 0$ and $x^*(x) = (x,0)$. If $N \neq H$, then by Corollary 6.2.3 there exists a point $z_0 \in N^\perp$, $z_0 \neq 0$. Clearly $\alpha = x^*(z_0) \neq 0$. For any $x \in H$, the point $x - x^*(x)z_0/\alpha$ is in N. Hence

$$\left(x - \frac{x^*(x)z_0}{\alpha}, z_0\right) = 0.$$

Consequently,

$$x^*(x)\left(\frac{z_0}{\alpha}, z_0\right) = (x,z_0).$$

Taking $z = [\bar{\alpha}/(z_0,z_0)]z_0$, we get $x^*(x) = (x,z)$ for all $x \in H$.

Suppose that there is another point z' such that $x^*(x) = (x,z')$ for all $x \in H$. Then $(x,z - z') = 0$ for all $x \in H$. Taking $x = z - z'$, we find that $\|z - z'\| = 0$—that is, $z = z'$.

Next,

$$\|x^*\| = \sup_{\|x\|=1} |x^*(x)| = \sup_{\|x\|=1} |(x,z)| \leq \sup_{\|x\|=1} (\|x\|\,\|z\|) = \|z\|.$$

On the other hand,

$$\|z\|^2 = (z,z) = |x^*(z)| \leq \|x^*\|\,\|z\|;$$

hence $\|z\| \leq \|x^*\|$. We have thus proved that $\|x^*\| = \|z\|$. This completes the proof of the theorem.

Corollary 6.2.5 *The map* $\sigma: H \to H^*$ *given by* $(\sigma x)(y) = (y,x)$ *is an isometric imbedding of* H *onto* H^*. *Furthermore,* $\sigma(\lambda x + \mu z) = \bar{\lambda}\sigma x + \bar{\mu}\sigma z$.

From this corollary we shall deduce:

Corollary 6.2.6 *A Hilbert space* H *is reflexive. Thus, in particular,* H *is weakly complete, and a set in* H *is weakly compact if and only if it is bounded and weakly closed.*

Proof. It is easily seen that the Banach space H^* is a Hilbert space with the scalar product $(\sigma x,\sigma y) = (y,x)$. [In verifying the condition (iii) of Definition 6.1.1 we use the fact that $\sigma(\lambda x) = \bar{\lambda}\sigma x$.] Denote by τ the isometric imbedding from the Hilbert space H^* onto its conjugate H^{**}. Thus $(\tau\sigma y)(\sigma x) = (\sigma x,\sigma y)$ for all x,y in H. Now let f be any element of H^{**}. Then $f = \tau\sigma y$ for some $y \in H$ and, for all $x \in H$,

$$f(\sigma x) = (\tau\sigma y)(\sigma x) = (\sigma x,\sigma y) = (y,x) = (\sigma x)(y).$$

It follows that f is the image of y under the natural inbedding κ of H into H^{**}. Thus κ maps H onto H^{**}; H is therefore reflexive.

The following extension of the Riesz theorem is called the *Lax–Milgram lemma.*

Theorem 6.2.7 *Let* B(x,y) *be a bilinear functional* [*that is,* B(x,y) *is linear in* x *and* B(x,y) *is linear in* y] *in* H × H, *where* H *is a Hilbert space.*

Assume that

$$|B(x,y)| \leq C\|x\| \, \|y\| \qquad (C > 0), \tag{6.2.2}$$

$$|B(x,x)| \geq c\|x\|^2 \qquad (c > 0), \tag{6.2.3}$$

for all x,y *in* H. *Then for any bounded linear functional* x* *in* H *there exists a unique point* z *in* H *such that* x*(x) = B(x,z) *for all* x ∈ H.

Proof. For fixed $y \in H$, $|B(x,y)| \leq C'\|x\|$ $(C' = C\|y\|)$. Hence, by Theorem 6.2.4 there exists a unique element Ty such that

$$B(x,y) = (x,Ty).$$

T is a linear map of H into itself. Since

$$\|Ty\|^2 = (Ty,Ty) = B(Ty,y) \leq C\|Ty\| \, \|y\|,$$

$\|Ty\| \leq C\|y\|$—that is, T is bounded. Next,

$$c\|x\|^2 \leq |B(x,x)| = |(x,Tx)| \leq \|x\| \, \|Tx\|,$$

so that $\|Tx\| \geq c\|x\|$. It follows that T maps H in a one-to-one way onto a closed linear subspace H_0 of H. We claim that $H_0 = H$. Indeed, otherwise there exists (by Corollary 6.2.3) an element $z \neq 0$ orthogonal to all the points Tx, $x \in X$. In particular

$$0 = |(z,Tz)| = |B(z,z)| \geq c\|z\|^2.$$

Thus $z = 0$, which is impossible.

Now let $x^* \in H^*$. By Theorem 6.2.4 there exists a point $z_0 \in H$ such that $x^*(x) = (x,z_0)$. Since $T(H) = H$, there exists a point z such that $Tz = z_0$. Therefore

$$x^*(x) = (x,Tz) = B(x,z).$$

To prove uniqueness, suppose $B(x,z) = B(x,z')$ for all $x \in H$. Then also $B(x,z - z') = 0$ for all $x \in H$. Taking $x = z - z'$, we obtain

$$0 = |B(z - z', z - z')| \geq c\|z - z'\|^2.$$

Hence $z' = z$.

Definition Let M and N be linear subspaces of a Hilbert space H and assume that every element in the vector sum $M + N$ has a unique representation of the form $x + y$, where $x \in M$, $y \in N$. Then we call $M + N$ the *direct sum* of M and N. If $M \perp N$, then we write $M + N$ also in the form $M \oplus N$. Note that $M \oplus N$ is the direct sum of M and N. If $Y = M \oplus N$, then we write $N = Y \ominus M$ and call N the *orthogonal complement* of M in Y. Similarly we write $M = Y \ominus N$ and call M the orthogonal complement of N in Y.

Notice that the projection theorem states in effect that if M is a closed linear subspace, then $H = M \oplus M^\perp$. Thus $M^\perp$ (M) is the orthogonal complement of M $(M^\perp)$ in H.

PROBLEMS

6.2.1. If M and N are closed linear spaces and $M \perp N$, then $M \oplus N$ is a closed linear space.

6.2.2. Let M be any subset of a Hilbert space H. Then $(M^\perp)^\perp$ is the closed linear space spanned by M.

6.3 PROJECTION OPERATORS

Definition Let M be a closed linear subspace. By the projection theorem, every $x \in H$ can be written uniquely in the form $x = y + z$, where $y \in M$, $z \in M^\perp$. The point y is called the *projection* of x in M, and the operator P given by $Px = y$ is called the *projection on* M. We also denote this operator by P_M. We call M the *subspace of the projection* P.

Definition Let T be a bounded linear operator in a Hilbert space H. The *adjoint* T^* of T is defined by the relation $(Tx,y) = (x,T^*y)$ for all x,y in H. If $T = T^*$, then we say that T is *self-adjoint*. When we speak of self-adjoint operators we always mean bounded, linear self-adjoint operators.

The present definition of adjoint operators is not the same as the one given by Definition 4.13.1, with $X = Y$, in the case of a normed linear space X. In view of Corollary 6.2.5, the two definitions are related by an isometric map σ; σ is additive and $\sigma(\lambda T) = \bar{\lambda}\sigma T$.

Note that if T is a self-adjoint operator in a Hilbert space H, then (Tx,x) is a real number for any $x \in H$.

PROBLEMS

6.3.1. Let T and S be bounded linear operators in a Hilbert space H, and let λ be a scalar. Prove: $(T + S)^* = T^* + S^*$, $(TS)^* = S^*T^*$, $(\lambda T)^* = \bar{\lambda}T^*$, $I^* = I$, $T^{**} = T$, $\|T^*\| = \|T\|$. If, further, T^{-1} is a bounded linear map (with domain H), then also $(T^*)^{-1}$ is a bounded linear map (with domain H) and $(T^{-1})^* = (T^*)^{-1}$.

6.3.2. State Theorem 5.2.10 in a Hilbert space, using the definition of a self-adjoint operator as given in the present section.

Theorem 6.3.1 *A projection* P *is a self-adjoint operator satisfying* $P^2 = P$, *and* $\|P\| = 1$ *if* $P \neq 0$.

Proof. Let $P = P_M$ and let $x_i = y_i + z_i$ (for $i = 1,2$), where $y_i \in M$, $z_i \in M^\perp$. Then $\lambda_1 x_1 + \lambda_2 x_2 = (\lambda_1 y_1 + \lambda_2 y_2) + (\lambda_1 z_1 + \lambda_2 z_2)$ and

$$(\lambda_1 y_1 + \lambda_2 y_2) \in M, (\lambda_1 z_1 + \lambda_2 z_2) \in M^\perp.$$

It follows that

$$P(\lambda_1 x_1 + \lambda_2 x_2) = \lambda_1 y_1 + \lambda_2 y_2 = \lambda_1 P x_1 + \lambda_2 P x_2.$$

Thus P is a linear operator. Next, $P^2 x_1 = P(Px_1) = Py_1 = y_1 = Px_1$, so that $P^2 = P$. Since $\|x_1\|^2 = \|y_1\|^2 + \|z_1\|^2 \geq \|y_1\|^2 = \|Px_1\|^2$, it also follows that $\|P\| \leq 1$. If $P \neq 0$, there is a point $x \neq 0$ in the subspace of the projection P. Thus, $\|Px\| = \|x\|$. This gives $\|P\| = 1$. It remains to show that P is self-adjoint. This is proved as follows:

$$(Px_1,x_2) = (y_1,x_2) = (y_1,y_2) = (x_1,y_2) = (x_1,Px_2).$$

The converse of Theorem 6.3.1 is also true. In fact, we have:

Theorem 6.3.2 *A self-adjoint operator* P *with* $P^2 = P$ *is a projection.*

Proof. Let $M = P(H)$. M is a linear subspace of H. It is also closed. Indeed, if $y_n = Px_n, y_n \to z$, then $Py_n = P^2 x_n = Px_n = y_n$. Hence $z = \lim y_n = \lim Py_n = Pz \in M$. Next, since P is self-adjoint and $P^2 = P$,

$$(x - Px,Py) = (Px - P^2x,y) = 0 \qquad \text{for all } y \in H,$$

so that $x - Px$ lies in $M^\perp$. Therefore $x = Px + (x - Px)$ is the unique decomposition of x as a sum $y + z$ with $y \in M$ and $z \in M^\perp$. This shows that P is the projection on M.

Definition Two projections P_1 and P_2 are *orthogonal* if $P_1 P_2 = 0$. Since $(P_1 P_2)^* = P_2^* P_1^* = P_2 P_1, P_1 P_2 = 0$ if and only if $P_2 P_1 = 0$.

Theorem 6.3.3 *Two projections* P_M *and* P_N *are orthogonal if and only if* $M \perp N$.

Proof. Let $P_M P_N = 0$ and let $x \in M$, $y \in N$. Then

$$(x,y) = (P_M x,P_N y) = (P_N P_M x,y) = 0.$$

Thus $M \perp N$. Suppose conversely that $M \perp N$. For any $x \in H$, $P_N x \in N$ and therefore $(P_N x) \perp M$. It follows that $P_M(P_N x) = 0$. Thus $P_M P_N = 0$.

Theorem 6.3.4 *The sum of two projections* P_M *and* P_N *is a projection if and only if* $P_M P_N = 0$. *In that case,* $P_M + P_N = P_{M \oplus N}$.

Proof. If $P = P_M + P_N$ is a projection, then $P^2 = P$. Hence, $P_M P_N + P_N P_M = 0$. Multiplying on the left by P_M, we get

$$P_M P_N + P_M P_N P_M = 0.$$

Multiplying on the right by P_M, we get $2P_M P_N P_M = 0$. Hence $P_M P_N = 0$. Conversely, if $P_M P_N = 0$, then also $P_N P_M = 0$. It follows that $P^2 = P$. Since P is also self-adjoint, it is a projection. Finally, $Px = P_M x + P_N x$ varies over $M \oplus N$ as x varies in H. Hence $P = P_{M \oplus N}$.

Theorem 6.3.5 *The product of two projections* $\mathrm{P_M}$ *and* $\mathrm{P_N}$ *is a projection if and only if they commute—that is,* $\mathrm{P_M P_N = P_N P_M}$. *In that case,* $\mathrm{P_M P_N = P_{M \cap N}}$.

Proof. If $P = P_M P_N$ is a projection, then $P^* = P$. Hence $P_M P_N = P_N^* P_M^* = P_N P_M$. Conversely, let $P_M P_N = P_N P_M = P$. Then $P^* = P$, so that P is self-adjoint. Also, $P^2 = P_M P_N P_M P_N = P_M^2 P_N^2 = P_M P_N = P$. Thus P is a projection. Further, $Px = P_M(P_N x) = P_N(P_M x)$. Therefore, $Px \in M \cap N$. On the other hand, if $x \in M \cap N$, then $Px = P_M(P_N x) = P_M x = x$. Thus $P = P_{M \cap N}$.

Definition A projection P_L is a *part* of a projection P_M if $L \subset M$.

Theorem 6.3.6 *A projection* $\mathrm{P_L}$ *is a part of a projection* $\mathrm{P_M}$ *if and only if one of the following conditions holds*:

(i) $\mathrm{P_M P_L = P_L}$;
(ii) $\mathrm{P_L P_M = P_L}$;
(iii) $\|\mathrm{P_L x}\| \leq \|\mathrm{P_M x}\|$ *for any* $\mathrm{x \in H}$.

Proof. Let $L \subset M$. Then $P_L x \in M$ for any $x \in H$. Hence $P_M P_L x = P_L x$, and (i) follows. If (i) holds, then

$$P_L = P_L^* = (P_M P_L)^* = P_L^* P_M^* = P_L P_M,$$

and (ii) follows. Next, if (ii) holds, then, for any $x \in H$,

$$\|P_L x\| = \|P_L P_M x\| \leq \|P_L\| \, \|P_M x\| \leq \|P_M x\|,$$

so that (iii) holds. It remains to show that if (iii) holds, then $L \subset M$. If this is not true, then there is a point $x_0 \in L$, $x_0 \notin M$. Let $x_0 = y_0 + z_0$ where $y_0 \in M$, $z_0 \perp M$, $z_0 \neq 0$. Then

$$\|P_L x_0\|^2 = \|y_0\|^2 + \|z_0\|^2 > \|y_0\|^2 = \|P_M x_0\|^2.$$

This contradicts (iii).

Lemma 6.3.7 *If* P *is a projection and* I *is the identity operator, then* $I - P$ *is a projection.*

Proof. $I - P$ is self-adjoint and $(I - P)^2 = I - P - P + P^2 = I - P$.

We shall now prove a theorem that contains the last lemma.

Theorem 6.3.8 *The difference* $P = P_M - P_L$ *of two projections* P_M, P_L *is a projection if and only if* P_L *is a part of* P_M. *If this is the case, then* $P = P_{M \ominus L}$.

Proof. If P is a projection, then (by Lemma 6.3.7) so is $I - P = (I - P_M) + P_L$. $I - P_M$ is also a projection. Applying Theorem 6.3.4, we get $(I - P_M)P_L = 0$—that is, $P_L = P_M P_L$. Hence, by Theorem 6.3.6, P_L is a part of P_M. Suppose conversely that P_L is a part of P_M. Then $(I - P_M)P_L = 0$. Hence, by Theorem 6.3.6, $(I - P_M) + P_L$ is a projection. Thus $I - P$ is a projection. By Lemma 6.3.7, also $I - (I - P) = P$ is projection.

Finally, if P_L is a part of P_M, then $P_M - P_L$ and P_L are orthogonal. By Theorem 6.3.4, the subspace Y of the projection $P_M - P_L$ satisfies $Y \oplus L = M$. Hence $Y = M \ominus L$. This completes the proof.

PROBLEMS

6.3.3. If P is self-adjoint and P^2 is a projection, is P a projection?

6.3.4. Let (X,μ) be a measure space and denote by χ_E the characteristic function of a measurable set E. Then the operator $Q_E f = \chi_E f$ defined in $L^2(X,\mu)$ is a projection. Under what condition on E, F is $Q_E + Q_F$ a projection?

6.3.5. Consider the operator $(Qf)(t) = a(t)f(t)$ in $L^2(0,1)$, where $a(t)$ is a scalar function. Find necessary and sufficient conditions on $a(t)$ for Q to be a projection.

6.3.6. Let A be an $n \times n$ matrix with real elements and denote by Q the operator $Qx = Ax$ in R^n. Find necessary and sufficient conditions on A for Q to be (i) self-adjoint, (ii) a projection.

6.3.7. If A is a self-adjoint operator in a Hilbert space, then

$$4(Ax,y) = [(A(x + y),x + y) - (A(x - y),x - y)] + i[(A(x + iy),x + iy) - (A(x - iy),x - iy)].$$

6.3.8. If H is a Hilbert space and A is a bounded linear operator in H such that (Az,z) is real for all $z \in H$, then A is self-adjoint. (This is generally false if H is a real Hilbert space.)

6.4 ORTHONORMAL SETS

A set K in a Hilbert space H is called *orthonormal* if each element of K is a *unit element*—that is, it has norm 1, and if any two distinct elements of H are orthogonal. An orthonormal set K is called *complete* if there exist no nonzero elements that are orthogonal to K—that is, if $K^{\perp} = 0$.

Theorem 6.4.1 *Let* $\{x_n\}$ *be an orthonormal sequence in a Hilbert space* H. *Then, for any* $x \in H$,

$$\sum_{n=1}^{\infty} |(x,x_n)|^2 \leq \|x\|^2. \tag{6.4.1}$$

This inequality is called *Bessel's inequality*. The scalars (x,x_n) are called the *Fourier coefficients* of x with respect to the sequence $\{x_n\}$.

Proof. We have

$$\left\| x - \sum_{n=1}^{m} (x,x_n)x_n \right\|^2 = \|x\|^2 - \left(x, \sum_{n=1}^{m} (x,x_n)x_n \right) - \left(\sum_{n=1}^{m} (x,x_n)x_n, x \right) + \sum_{n=1}^{m} (x,x_n)(x_n,x).$$

Therefore,

$$\left\| x - \sum_{n=1}^{m} (x,x_n)x_n \right\|^2 = \|x\|^2 - \sum_{n=1}^{m} |(x,x_n)|^2. \tag{6.4.2}$$

It follows that

$$\sum_{n=1}^{m} |(x,x_n)|^2 \leq \|x\|^2.$$

Now take $m \to \infty$.

Theorem 6.4.2 *Let* $\{x_n\}$ *be an orthonormal sequence in a Hilbert space* H *and let* $\{\lambda_n\}$ *be any sequence of scalars. Then, for any positive integer* m,

$$\left\| x - \sum_{n=1}^{m} \lambda_n x_n \right\| \geq \left\| x - \sum_{n=1}^{m} (x,x_n)x_n \right\|.$$

Proof. Set $c_n = (x,x_n)$. Then

$$\left\| x - \sum_{n=1}^{m} \lambda_n x_n \right\|^2 = \|x\|^2 - \sum_{n=1}^{m} \bar{\lambda}_n c_n - \sum_{n=1}^{m} \lambda_n \bar{c}_n + \sum_{n=1}^{m} |\lambda_n|^2$$
$$= \|x\|^2 - \sum_{n=1}^{m} |c_n|^2 + \sum_{n=1}^{m} |c_n - \lambda_n|^2$$
$$\geq \|x\|^2 - \sum_{n=1}^{m} |c_n|^2.$$

Now use (6.4.2).

Theorem 6.4.3 *Let* $\{x_n\}$ *be an orthonormal sequence in a Hilbert space* H *and let* $\{\alpha_n\}$ *be a sequence of scalars. Then the series* $\sum \alpha_n x_n$ *is convergent if and only if* $\sum |\alpha_n|^2 < \infty$, *and in that case*

$$\left\| \sum_{n=1}^{\infty} \alpha_n x_n \right\| = \left(\sum_{n=1}^{\infty} |\alpha_n|^2 \right)^{1/2} \tag{6.4.3}$$

and the sum $\sum \alpha_n x_n$ *is independent of the order in which its terms are arranged.*

Proof. If $m > n$,

$$\left\| \sum_{j=n}^{m} \alpha_j x_j \right\|^2 = \sum_{j=n}^{m} |\alpha_j|^2. \tag{6.4.4}$$

From this relation and from the completeness of H, we get the first assertion of the theorem. Taking $n = 1$, $m \to \infty$ in (6.4.4), we obtain the assertion (6.4.3). Now let $z = \sum \alpha_{j_n} x_{j_n}$ be a rearrangement of the series $x = \sum \alpha_j x_j$ and assume that $\sum |\alpha_j|^2 < \infty$. Then

$$\|x - z\|^2 = (x,x) + (z,z) - (x,z) - (z,x). \tag{6.4.5}$$

One easily verifies $(x,x) = (z,z) = \sum |\alpha_j|^2$. Writing

$$s_m = \sum_{j=1}^{m} \alpha_j x_j, \qquad t_m = \sum_{n=1}^{m} \alpha_{j_n} x_{j_n},$$

we also have

$$(x,z) = \lim_m (s_m, t_m) = \sum_{j=1}^{\infty} |\alpha_j|^2.$$

Also $(z,x) = \overline{(x,z)} = (x,z)$. We conclude from (6.4.5) that $\|x - z\|^2 = 0$—that is, $z = x$. This completes the proof of the theorem.

Theorem 6.4.4 *Let* K *be an orthonormal set in a Hilbert space* H. *Then,*

(i) *for any* $x \in H$, $(x,y) = 0$ *for all but a countable number of points* y *of* K;

(ii) *the sum*

$$Ex = \sum_{y \in K_x} (x,y)y \qquad (K_x = \{y; y \in K \text{ and } (x,y) \neq 0\})$$

converges independently of the order by which the terms are taken, and
(iii) E *is the projection on the closed linear space spanned by* K.

Proof. From Bessel's inequality it follows that for any $\varepsilon > 0$ there cannot be more than $\|x\|^2/\varepsilon^2$ points y in K for which $|(x,y)| > \varepsilon$. Using this fact with $\varepsilon = 1, \frac{1}{2}, \frac{1}{3}, \ldots$, the assertion (i) follows. Next, the assertion (ii) follows from Bessel's inequality and Theorem 6.4.3. To prove (iii), denote by $\tilde{K}$ the closed linear subspace spanned by K. If $x \perp \tilde{K}$, then, clearly $Ex = 0$. On the other hand, if $x \in \tilde{K}$, then for any $\varepsilon > 0$ there are scalars $\lambda_1, \ldots, \lambda_n$ and points $y_1, \ldots, y_n$ in K such that

$$\left\| x - \sum_{j=1}^{n} \lambda_j y_j \right\| < \varepsilon.$$

Theorem 6.4.2 then gives

$$\left\| x - \sum_{j=1}^{n} (x,y_j)y_j \right\| < \varepsilon. \tag{6.4.6}$$

We may assume that $y_j \in K_x$ for $1 \leq j \leq n$ [otherwise we strike out from the last inequality those y_j for which $(x,y_j) = 0$]. Arrange the set K_x in a sequence $\{y_j\}$, where $1 \leq j \leq \infty$. From (6.4.2) we see that the left-hand side of (6.4.6) does not increase with n. Hence, taking $n \to \infty$, we get $\|x - Ex\| < \varepsilon$. Since ε is arbitrary, $Ex = x$ for any $x \in \tilde{K}$. We have thus proved that E is the projection on $\tilde{K}$.

Definition A set K is called an *orthonormal basis* of H if K is an orthonormal set and if for any $x \in H$

$$x = \sum_{y \in K_x} (x,y)y. \tag{6.4.7}$$

Theorem 6.4.5 *Let* K *be an orthonormal set in a Hilbert space* H. *Then the following conditions are equivalent*:

(i) K *is complete*;
(ii) *The closed linear subspace spanned by* K *is* H;
(iii) K *is an orthonormal basis*;
(iv) *For any* x ∈ H,

$$\|x\|^2 = \sum_{y \in K_x} |(x,y)|^2. \tag{6.4.8}$$

The relation (6.4.8) is called *Parseval's formula*.

Proof. By Corollary 6.2.3, (i) implies (ii). Suppose (ii) holds. Then, by Theorem 6.4.4, $Ex = x$ for any $x \in H$—that is, K is an orthonormal basis. Suppose next that (iii) holds and arrange the elements of K_x in a sequence $\{x_n\}$. Taking in (6.4.2) $n \to \infty$, we then obtain (6.4.8). Finally, if (iv) holds and if $x \perp K$, then $\|x\|^2 = \sum |(x,y)|^2 = 0$—that is, $x = 0$. Thus (iv) implies (i).

Theorem 6.4.6 *In every Hilbert space there exists an orthonormal basis.*

Proof. Consider the classes of orthonormal sets in H and define a partial order by inclusion. By Zorn's lemma it follows that there exists a maximal orthonormal set K. Since K is maximal, it is complete, and therefore it is also an orthonormal basis.

Lemma 6.4.7 *Any orthonormal basis in a separable Hilbert space is countable.*

Proof. Suppose there exists an orthonormal basis $\{x_\alpha\}$ that is not countable. Denote by B_α the ball with center x_α and radius $\frac{1}{2}$. Since

$$\|x_\alpha - x_\beta\|^2 = \|x_\alpha\|^2 + \|x_\beta\|^2 = 2 \qquad \text{if } \alpha \neq \beta,$$

the balls B_α are mutually disjoint. If there exists a dense sequence $\{y_n\}$ in the space, then there is a ball B_{α_0} that does not contain any of the points y_n. Hence x_{α_0} is not in the closure of $\{y_n\}$—a contradiction.

Theorem 6.4.8 *Any two infinite-dimensional separable Hilbert spaces are isometrically isomorphic.*

Proof. Denote two such spaces by H_1 and H_2. By Lemma 6.4.7, there exist sequences $\{x_n\}$ and $\{y_n\}$ that form orthonormal bases for H_1 and H_2, respectively. For any point $x \in H_1$ and $y \in H_2$ we have

$$x = \sum_{n=1}^{\infty} c_n x_n, \qquad y = \sum_{n=1}^{\infty} d_n y_n, \tag{6.4.9}$$

where $c_n = (x,x_n)$, $d_n = (y,y_n)$. When $\{c_n\}$ and $\{d_n\}$ vary in l^2, the points x and y vary in H_1 and H_2, respectively. We define a map T from H_1 into H_2 as follows: $y = Tx$ if, in (6.4.9), $c_n = d_n$ for all $n \geq 1$. T is clearly linear, and it maps H_1 onto H_2. Since

$$\|Tx\|^2 = \sum_{n=1}^{\infty} |d_n|^2 = \sum_{n=1}^{\infty} |c_n|^2 = \|x\|^2,$$

T is also an isometry. This completes the proof.

Corollary 6.4.9 *Any infinite-dimensional separable Hilbert space is isometrically isomorphic to* $L^2(0,1)$.

PROBLEMS

6.4.1. If T is an isometric isomorphism of a Hilbert space H_1 onto a Hilbert space H_2, then also $(Tx,Ty) = (x,y)$ for all x,y in H_1.

6.4.2. Let H_0 be a linear subspace of a Hilbert space H, spanned by a sequence $\{x_m\}$ of linearly independent elements. Show that there exist scalars λ_{mn} such that the sequence $\{y_m\}$ given by

$$y_m = \sum_{n=1}^{m} \lambda_{mn} x_n$$

is orthonormal and it spans H_0. This process of passing from $\{x_m\}$ to $\{y_m\}$ is called the *Gram–Schmidt process*.

6.4.3. Let $H_1,\dots,H_n$ be Hilbert spaces. Denote by $(\, , \,)_i$ the scalar product in H_i. Then the Cartesian product $H = H_1 \times \cdots \times H_n$ is a Hilbert space with the scalar product

$$(x,y) = \sum_{i=1}^{n} (x_i,y_i)_i,$$

where $x = (x_1,\dots,x_n)$, $y = (y_1,\dots,y_n)$. We call H the *direct sum of the Hilbert spaces* $H_1,\dots,H_n$ and write $H = H_1 \oplus \cdots \oplus H_n$. Show that the map $x \to (0,\dots,0,x_i,0,\dots,0)$ is a projection in H.

6.4.4. Let $f(z),g(z)$ be complex-valued functions in a closed disc D: $|z| \le r$, and assume that $f \in C^0(D)$ and g is analytic in D. Prove that

$$\iint_D f(z)\, \overline{g'(z)}\, dx\, dy = \frac{1}{2i} \int_{\partial D} f(z)\, \overline{g(z)}\, dz.$$

6.4.5. Let D be the disc $|z| < 1$ in the complex plane. Denote by $H^2(D)$ the linear subspace of $L^2(D)$ consisting of all functions holomorphic in D. Prove that $H^2(D)$ is a closed set, so that it is a Hilbert space. Next prove that the sequence $\{\sqrt{n/\pi} z^{n-1}\}$ is orthonormal in $H^2(D)$ and that it is also complete. [*Hint*: the nth Fourier coefficient is equal to

$$a_n = \lim_{r \to 1} \frac{1}{\sqrt{\pi n}} \frac{r^{2n}}{2i} \int_{|z|=r} \frac{f(z)}{z^n}\, dz.$$

On the other hand, if $f(z) = \sum b_n z^n$,

$$b_{n-1} = \frac{1}{2\pi i} \int_{|z|=r} \frac{f(z)}{z^n}\, dz; \qquad \text{thus } a_n = \sqrt{\frac{\pi}{n}}\, b_{n-1}.$$

Now verify Parseval's formula.]

6.4.6. Show that an orthonormal sequence $\{\varphi_n\}$ is complete in $L^2(a,b)$ if $\sum_{n=1}^{\infty} \left(\int_a^x \varphi_n(t)\, dt \right)^2 = x - a$ for all $x \in (a,b)$.

6.4.7. If $\{x_n\}$ is a complete orthonormal sequence in a Hilbert space H and if $\{y_n\}$ is an orthonormal sequence in H satisfying

$$\sum_{n=1}^{\infty} \|x_n - y_n\|^2 < 1,$$

then $\{y_n\}$ is also complete.

6.5 SELF-ADJOINT OPERATORS

Throughout this section we designate by A a self-adjoint operator in a Hilbert space H.

Theorem 6.5.1 $\|A\| = \sup_{\|x\|=1} |(Ax,x)|$.

Proof. Let $\alpha = \sup_{\|x\|=1} |(Ax,x)|$. If $\|x\| = 1$, then

$$|(Ax,x)| \le \|Ax\|\,\|x\| = \|Ax\| \le \|A\|\,\|x\| = \|A\|.$$

Hence $\alpha \le \|A\|$. To prove the converse inequality, let $z \in H$, $Az \ne 0$ and introduce $u = (Az)/\lambda$ where $\lambda = (\|Az\|/\|z\|)^{1/2}$. Then (compare Problem 6.3.7)

$$\begin{aligned}\|Az\|^2 &= (A(\lambda z),u) = \tfrac{1}{4}\{(A(\lambda z + u),\lambda z + u) - (A(\lambda z - u),\lambda z - u)\}\\ &\le \tfrac{1}{4}\alpha(\|\lambda z + u\|^2 + \|\lambda z - u\|^2) = \tfrac{1}{2}\alpha(\|\lambda z\|^2 + \|u\|^2)\\ &= \tfrac{1}{2}\alpha\left(\lambda^2\|z\|^2 + \frac{1}{\lambda^2}\|Az\|^2\right) = \alpha\|z\|\,\|Az\|.\end{aligned}$$

Hence $\|Az\| \le \alpha\|z\|$. It follows that $\|A\| \le \alpha$.

Theorem 6.5.2 *The eigenvalues of a self-adjoint operator are real numbers. Eigenvectors corresponding to distinct eigenvalues are orthogonal.*

Proof. Let $Ax = \lambda x$, $x \ne 0$. Then $(Ax,x) = \lambda\|x\|^2$. Since (Ax,x) is real and $\|x\| \ne 0$, we conclude that λ is real. Next let $Ay = \mu y$, where $y \ne 0$ and $\mu \ne \lambda$. Then μ is a real number, and we have

$$(\lambda - \mu)(x,y) = (\lambda x,y) - (x,\mu y) = (Ax,y) - (x,Ay) = 0.$$

Thus $x \perp y$.

Denote by R_λ the range of the operator $A - \lambda I$ and denote by E_λ the null space of $A - \lambda I$. Then R_λ is a linear subspace of H and E_λ is a closed linear subspace of H.

Lemma 6.5.3 $H = E_\lambda \oplus \overline{R}_\lambda$.

Proof. If λ is not a real number, then, by Theorem 6.5.2, $E_\lambda = \{0\} = E_{\bar{\lambda}}$. Hence the assertion of the lemma is equivalent to the assertion

$$H = E_{\bar{\lambda}} \oplus \bar{R}_\lambda.$$

If $x \in E_{\bar{\lambda}}$, $y \in R_\lambda$, then $(A - \bar{\lambda}I)x = 0$ and $y = (A - \lambda I)z$ for some $z \in H$. It follows that

$$(x,y) = (x,(A - \lambda I)z) = ((A - \bar{\lambda}I)x,z) = 0.$$

If $y \in \bar{R}_\lambda$, then $y = \lim y_n$ for some sequence $\{y_n\}$ in R_λ. Since $(x,y_n) = 0$ for any $x \in E_{\bar{\lambda}}$ and for any $n \geq 1$, we get $(x,y) = \lim (x,y_n) = 0$. We have thus proved that $E_{\bar{\lambda}} \perp \bar{R}_\lambda$. It remains to show that if $x \perp (E_{\bar{\lambda}} \oplus \bar{R}_\lambda)$, then $x = 0$. For any z, $(x,(A - \lambda I)z) = 0$. Hence $((A - \bar{\lambda}I)x,z) = 0$. Since z is arbitrary, $(A - \bar{\lambda}I)x = 0$. Thus $x \in E_{\bar{\lambda}}$. But since $x \perp E_{\bar{\lambda}}$, it follows that $x = 0$.

Corollary 6.5.4 *If* $E_\lambda = 0$, *then* $\bar{R}_\lambda = H$. *Thus, in particular,* A *has no residual spectrum.*

Lemma 6.5.5 *If, for some* λ,

$$\|Ax - \lambda x\| \geq c\|x\| \qquad (c > 0) \tag{6.5.1}$$

for all $x \in H$, *then* λ *belongs to the resolvent set* $\rho(A)$ *of* A.

Proof. The assumption (6.5.1) implies that $E_\lambda = 0$. Hence, by Corollary 6.5.4, $\bar{R}_\lambda = H$. If we prove that R_λ is closed, then, by Theorem 4.6.2, it follows that $(\lambda I - A)^{-1}$ is a bounded operator—that is, $\lambda \in \rho(A)$. To prove that R_λ is closed, let $\{y_n\} \subset R_\lambda$, $y_n \to y$. Then $y_n = (A - \lambda I)x_n$ for some $x_n \in H$, and

$$\|y_n - y_m\| = \|(A - \lambda I)(x_n - x_m)\| \geq c\|x_n - x_m\|.$$

It follows that $\{x_n\}$ is a Cauchy sequence. Let $x = \lim x_n$. Then $y = \lim (A - \lambda I)x_n = (A - \lambda I)x$—that is, $y \in R_\lambda$.

Corollary 6.5.6 *A point* λ *belongs to the spectrum* $\sigma(A)$ *of* A *if and only if there exists a sequence* $\{x_n\}$ *in* H *such that*

$$\|x_n\| = 1, \qquad \lim_n \|Ax_n - \lambda x_n\| = 0.$$

Theorem 6.5.7 *The spectrum of a self-adjoint operator is contained in the real line.*

Proof. Let $\lambda = \mu + i\nu$, $\nu \neq 0$. If $y = (A - \lambda I)x$, then

$$(y,x) = (Ax,x) - \lambda(x,x),$$

$$(x,y) = \overline{(y,x)} = (Ax,x) - \bar{\lambda}(x,x).$$

Hence

$$(x,y) - (y,x) = (\lambda - \bar{\lambda})(x,x) = 2vi\|x\|^2.$$

Therefore

$$2|v|\,\|x\|^2 \le |(x,y)| + |(y,x)| = 2|(x,y)| \le 2\|x\|\,\|y\|,$$

—that is, $|v|\,\|x\| \le \|y\| = \|(A - \lambda I)x\|$. Now apply Lemma 6.5.5.

Definition The *upper bound* of a self-adjoint operator A is the number $M = \sup_{\|x\|=1} (Ax,x)$. The *lower bound* of a self-adjoint operator A is the number $m = \inf_{\|x\|=1} (Ax,x)$.

Note, by Theorem 6.5.1, that $\|A\| = \max(|m|,|M|)$.

Theorem 6.5.8 *The spectrum of a self-adjoint operator* A *lies in the closed interval* [m,M], *where* m *and* M *are the lower and upper bounds of* A, *respectively.*

Proof. Suppose $\lambda > M$. Then

$$-((A - \lambda I)x,x) = -(Ax,x) + \lambda(x,x) \ge (\lambda - M)(x,x).$$

It follows that

$$(\lambda - M)\|x\|^2 \le |((A - \lambda I)x,x)| \le \|(A - \lambda I)x\|\,\|x\|.$$

Hence $\|(A - \lambda I)x\| \ge (\lambda - M)\|x\|$. Lemma 6.5.5 then implies that $\lambda \in \rho(A)$. Similarly one proves that if $\lambda < m$ then $\lambda \in \rho(A)$.

Theorem 6.5.9 *The lower and upper bounds,* m *and* M, *of a self-adjoint operator* A *belong to the spectrum of* A.

Proof. By definition of M, there exists a sequence $\{x_n\}$ in H such that $\|x_n\| = 1$, $(Ax_n,x_n) = M - \delta_n$, $\delta_n \ge 0$, $\delta_n \to 0$. Let $\gamma > \max(|M|,|m|)$. The upper bound of $A + \gamma I$ is $M + \gamma$ and the lower bound of $A + \gamma I$ is $m + \gamma$. Since $M + \gamma \ge m + \gamma > 0$, it follows, by Theorem 6.5.1, that the norm of $A + \gamma I$ is equal to $M + \gamma$. Hence $\|(A + \gamma I)x_n\| \le (M + \gamma)\|x_n\| = M + \gamma$. We now have

$$\begin{aligned}\|Ax_n - Mx_n\|^2 &= \|(A + \gamma I)x_n - (M + \gamma)x_n\|^2 \\ &= \|(A + \gamma I)x_n\|^2 + (M + \gamma)^2\|x_n\|^2 - 2(M + \gamma) \\ &\quad \times ((A + \gamma I)x_n,x_n) \\ &\le (M + \gamma)^2 + (M + \gamma)^2 - 2(M + \gamma)(M - \delta_n + \gamma) \to 0 \\ &\quad \text{if } n \to \infty.\end{aligned}$$

Corollary 6.5.6 then implies that $M \in \sigma(A)$. The proof that $m \in \sigma(A)$ is similar. In fact, one takes a sequence $\{y_n\} \subset H$ with $\|y_n\| = 1$, $(Ay_n, y_n) = m + \varepsilon_n$, $\varepsilon_n \geq 0$, $\varepsilon_n \to 0$, and proves that

$$\|Ay_n - my_n\|^2 = \|(A - \gamma I)y_n - (m - \gamma)y_n\|^2 \to 0 \qquad \text{if } n \to \infty.$$

PROBLEMS

6.5.1. Let $B(x,y)$ be a complex-valued functional defined on $H \times H$. We call it a *bilinear form* if it is linear in x and *antilinear* in y [that is, $\overline{B(x,y)}$ is linear in y]. If $|B(x,y)| \leq C\|x\|\,\|y\|$ for all x,y in H, then we say that $B(x,y)$ is a *bounded form*. If $B(x,y) = \overline{B(y,x)}$, then we say that $B(x,y)$ is a *Hermitian form*. Prove that if $B(x,y)$ is a Hermitian form, then there exists a self-adjoint operator A such that $B(x,y) = (Ax,y)$.

6.5.2. If $B(x,y)$ is a Hermitian bilinear form, then $B(x) = B(x,x)$ is called a *quadratic form*. Prove that

$$4B(x,y) = B(x + y) + B(x - y) + iB(x + iy) - iB(x - iy).$$

6.5.3. Consider the operator $(Ax)(t) = tx(t)$ in $L^2(0,1)$. Prove that $\sigma(A)$ consists of the closed interval $[0,1]$. [*Hint:* If $0 < \lambda < 1$, take $x_n(t) = \sqrt{n}$ for $\lambda < t < \lambda + (1/n)$, $x_n(t) = 0$ if $t \notin [\lambda, \lambda + (1/n)]$ and use Corollary 6.5.6.]

6.5.4. If A is an operator in H and L is a closed linear subspace of H such that $A(L) \subset L$, then L is called an *invariant subspace* of A. Prove that if L is an invariant subspace of a self-adjoint operator A, then also $L^\perp$ is an invariant subspace of A.

6.5.5. Let L be a closed linear subspace, invariant with respect to a self-adjoint operator A. Denote by $\sigma_1(A)$ and $\sigma_2(A)$ the spectra of the restriction of A to L and $L^\perp$, respectively. Prove that $\sigma(A) = \sigma_1(A) \cup \sigma_2(A)$. [*Hint:* Prove $\sigma(A) \subset \sigma_1(A) \cup \sigma_2(A)$ by using Lemma 6.5.5, and prove $\sigma(A) \supset \sigma_1(A) \cup \sigma_2(A)$ by using Corollary 6.5.6.]

6.6 POSITIVE OPERATORS

A self-adjoint operator A is called *positive* if $(Ax,x) \geq 0$ for all $x \in H$. If A is positive, then we write $A \geq 0$. If A and B are self-adjoint operators and if $A - B \geq 0$, then we say that A is *larger* than B and that B is *smaller* than A, and write $A \geq B$ and $B \leq A$. Note that for any bounded linear operator T, TT^* and T^*T are positive operators.

Lemma 6.6.1 *Let* A *be a positive operator. Then there exists a sequence of operators* A_n, *which are polynomials in* A *with real coefficients, such that* *the series* $\sum A_n^2$ *is strongly convergent to* A*—that is, for any* $x \in H$,

$$Ax = \sum_{n=1}^{\infty} A_n^2 x. \tag{6.6.1}$$

Proof. We may assume that $A \neq 0$. Let

$$B_1 = \frac{A}{\|A\|}, \qquad B_{n+1} = B_n - B_n^2 \qquad (n = 2,3,\ldots).$$

The B_n are clearly self-adjoint and $B_n B_k = B_k B_n$ for all $k,\ n \geq 1$. We claim that

$$0 \leq B_n \leq I. \tag{6.6.2}$$

For $n = 1$ this is obvious. Proceeding by induction, we shall assume that (6.6.2) holds for $n = m$ and prove it for $n = m + 1$. Since

$$[B_m^2(I - B_m)x,x] = [(I - B_m)B_m x, B_m x] \geq 0,$$

$$[B_m(I - B_m)^2 x,x] = [B_m(I - B_m)x,(I - B_m)x] \geq 0,$$

it follows that $B_m^2(I - B_m) \geq 0$, $B_m(I - B_m)^2 \geq 0$. Hence

$$B_{m+1} = B_m^2(I - B_m) + B_m(I - B_m)^2 \geq 0.$$

Since also $I - B_{m+1} = (I - B_m) + B_m^2 \geq 0$, the proof of (6.6.2) for $n = m + 1$ is complete.

We have the relations:

$$\sum_{m=1}^{n} B_m^2 = B_1 - B_{n+1} \leq B_1. \tag{6.6.3}$$

Therefore,

$$\sum_{m=1}^{n} (B_m x, B_m x) \leq (B_1 x, x).$$

It follows that $\|B_m x\| \to 0$ as $m \to \infty$. Consequently, (6.6.3) gives

$$\sum_{m=1}^{n} B_m^2 x \to B_1 x \qquad \text{as } n \to \infty.$$

This yields the assertion of the lemma with $A_n = \sqrt{\|A\|}\, B_n$.

Corollary 6.6.2 *If* A *and* B *are positive operators and if* AB = BA, *then* AB *is positive.*

Proof. B commutes with each A_n. Hence

$$(ABx,x) = \sum_{n=1}^{\infty} (BA_n^2 x,x) = \sum_{n=1}^{\infty} (BA_n x, A_n x) \geq 0.$$

Corollary 6.6.3 *If* $\{A_n\}$ *is a monotone-increasing sequence of self-adjoint operators that commute with each other (that is* $A_nA_m = A_mA_n$ *for all* m,n), *and if* B *is a self-adjoint operator that commutes with each* A_n *and that satisfies* $A_n \le B$ *for all* n, *then there exists a self-adjoint operator* A *such that* $\{A_n\}$ *converges strongly to* A.

Proof. The operators $T_n = B - A_n$ are positive and monotone decreasing, and they commute with each other. By Corollary 6.6.2, if $m < n$,

$$(T_m - T_n)T_m \ge 0, \qquad (T_m - T_n)T_n \ge 0.$$

It follows that, for any $x \in H$,

$$(T_m^2 x,x) \ge (T_m T_n x,x) \ge (T_n^2 x,x).$$

Thus $\{(T_m^2 x,x)\}$ is a monotone-decreasing sequence of nonnegative numbers. It therefore has a limit, and

$$\lim_{m,n\to\infty} (T_m T_n x,x) = \lim_{m\to\infty} (T_m^2 x,x).$$

Hence

$$\|T_m x - T_n x\|^2 = ((T_m - T_n)^2 x,x) = (T_m^2 x,x) + (T_n^2 x,x) - 2(T_m T_n x,x) \to 0.$$

If $m,n \to \infty$. Since $\{T_m x\}$ is a Cauchy sequence, it has a limit. But then also $Ax = \lim A_m x$ exists. It is easily seen that A is self-adjoint.

Definition A *square root* of a positive operator A is a self-adjoint operator B satisfying $B^2 = A$.

Theorem 6.6.4 *Every positive operator* A *has a unique positive square root* B. B *commutes with any bounded linear operator that commutes with* A.

Proof. Suppose first that $A \le I$. Set $B_0 = 0$ and

$$B_{n+1} = B_n + \tfrac{1}{2}(A - B_n^2) \qquad (n = 1,2,\ldots). \tag{6.6.4}$$

The B_n are self-adjoint operators and (since they are polynomials in A) they commute with any operator that commutes with A. We have

$$I - B_{n+1} = \tfrac{1}{2}(I - B_n)^2 + \tfrac{1}{2}(I - A). \tag{6.6.5}$$

From this relation we obtain

$$B_{n+1} - B_n = \tfrac{1}{2}[(I - B_{n-1}) + (I - B_n)](B_n - B_{n-1}). \tag{6.6.6}$$

From (6.6.5) it follows that $B_m \le I$ for all m. From (6.6.6) it follows by induction that $B_n \le B_{n+1}$ for all n. By Corollary 6.6.3 there exists a self-adjoint operator B such that $\lim B_n x = Bx$ for any $x \in H$. Applying both sides of (6.6.4) to $x \in H$, and letting $n \to \infty$, we obtain

$$Bx = Bx + \tfrac{1}{2}(A - B^2)x$$

—that is, $B^2 = A$. Since each B_n is positive, also B is positive. Since each B_n commutes with every operator that commutes with A, the same is true of B.

We have assumed, so far, that $A \leq I$. In the general case we consider the operator $\varepsilon^2 A$, where $\varepsilon > 0$, $\varepsilon^2 \|A\| < 1$. Then $\varepsilon^2 A \leq I$, and it therefore has a positive square root B'. Take $B = B'/\varepsilon$.

To prove uniqueness, suppose D is another positive square root of A. Since $A = D^2, D$ commutes with A. Hence B commutes with D. For any $x \in H$, let $y = (B - D)x$. Then

$$\begin{aligned}(By,y) + (Dy,y) &= ((B + D)y,y) = ((B + D)(B - D)x,y)\\ &= ((B^2 - D^2)x,y) = 0.\end{aligned}$$

Since $(By, y) \geq 0$, $(Dy, y) \geq 0$, it follows that $(By, y) = (Dy, y) = 0$. Let C be a positive square root of B. Then

$$\|Cy\|^2 = (C^2 y,y) = (By,y) = 0.$$

Hence $Cy = 0$. We conclude that $By = C(Cy) = 0$. Similarly, $Dy = 0$. Therefore

$$\|Bx - Dx\|^2 = ((B - D)^2 x,x) = ((B - D)y,x) = 0$$

—that is, $Bx = Dx$. Since x is arbitrary, $B = D$.

Lemma 6.6.5 *Let* A *and* B *be self-adjoint operators, and let* AB = BA, $\mathrm{A}^2 = \mathrm{B}^2$. *Denote by* P *the projection operator on the null space* L *of* A – B. *Then*

(i) *Any bounded linear operator that commutes with* A – B *commutes with* P;

(ii) Ax = 0 *implies* Px = x;

(iii) A = (2P – I)B.

Proof. Let D be any bounded linear operator that commutes with $A - B$. If $y \in L$, then $Dy \in L$, since $(A - B)Dy = D(A - B)y = 0$. Therefore $DPx \in L$ for any $x \in X$. Hence $PDPx = DPx$—that is, $PDP = DP$. Since D^* also commutes with $A - B$, $D^*P = PD^*P$. We then get

$$PD = (D^*P)^* = (PD^*P)^* = PDP = DP.$$

This proves (i). To prove (ii), suppose $Ax = 0$. Then

$$\|Bx\|^2 = (Bx,Bx) = (B^2x,x) = (A^2x,x) = \|Ax\|^2 = 0.$$

Thus $Bx = 0$. Therefore also $(A - B)x = 0$. It follows that $Px = x$.

To prove (iii), note that, for any $x \in H$,

$$(A - B)(A + B)x = (A^2 - B^2)x = 0.$$

Hence $(A + B)x \in L$. It follows that $P(A + B)x = (A + B)x$—that is, $P(A + B) = A + B$. Since (compare Lemma 6.5.3) the null space of $A - B$ is orthogonal to the range of $A - B$, $P(A - B) = 0$. Hence

$$P(A + B) - P(A - B) = A + B$$

—that is, $A = (2P - I)B$.

We shall use Lemma 6.6.5 to prove the following result, which will be of fundamental importance for the development of the spectral theory given in the next section.

Lemma 6.6.6 *For any self-adjoint operator* A *there exists a projection* E_+ *having the following properties*:

(i) *any bounded linear operator that commutes with* A *commutes also with* E_+;

(ii) $AE_+ \geq 0$ *and* $A(I - E_+) \leq 0$;

(iii) *If* $Ax = 0$, *then* $E_+x = x$.

Proof. Let B be a positive square root of A^2 and let E_+ be the projection operator on the null space of $A - B$. Then (i) and (iii) follow from Lemma 6.6.5. This lemma also gives $A = (2E_+ - I)B$. Hence

$$AE_+ = BE_+, \qquad A(I - E_+) = -(I - E_+)B.$$

Now use Corollary 6.6.2 to conclude that $AE_+ \geq 0$, $A(I - E_+) \leq 0$.

Definition We write $A_+ = AE_+$ and call it the *positive part* of A. Similarly we write $A_- = A(I - E_+)$ and call it the *negative part* of A. We have $A = A_+ + A_-$.

Notice that $A_+ = \frac{1}{2}(A + B)$, $A_- = \frac{1}{2}(A - B)$.

PROBLEMS

6.6.1. Find the positive square root of the operator $(Ax)(t) = a(t)x(t)$ in $L^2(0,1)$, where $a(t)$ is a nonnegative function.

6.6.2. Find A_+, A_- for the operator A defined in Problem 6.6.1.

6.6.3. If P_L and P_M are projections, then $P_L \leq P_M$ if and only if $L \subset M$.

6.6.4. If A is a self-adjoint operator and if $-\eta I \leq A \leq \eta I$ for some $\eta > 0$, then $\|A\| \leq \eta$.

6.7 SPECTRAL FAMILIES OF SELF-ADJOINT OPERATORS

Let $\{E_\lambda\}$ be a family of projections for $-\infty < \lambda < \infty$ satisfying the following conditions:

(a) $E_\lambda \le E_\mu$ if $\lambda < \mu$;
(b) E_λ is strongly continuous from the left—that is,

$$\lim_{\lambda \nearrow \mu} E_\lambda x = E_\mu x \qquad \text{for any } x \in H;$$

(c) $E_\lambda = 0$ if $\lambda < m$, $E_\lambda = I$ if $\lambda > M$.

We call such a family of projections a *spectral family* or a *resolution of the identity*.

Let $f(\lambda)$ be a continuous complex-valued function for $m \le \lambda \le M$. We extend $f(\lambda)$ to $M < \lambda < M + 1$ so that the extended function, which we again denote by $f(\lambda)$, is continuous for $m \le \lambda < M + 1$. We shall define an integral

$$\int_m^{M+\varepsilon} f(\lambda)\, dE_\lambda \qquad (0 < \varepsilon < 1).$$

Take Π to be any partition of $[m, M + \varepsilon]$: $m = \lambda_0 < \lambda_1 < \cdots < \lambda_{n-1} < \lambda_n = M + \varepsilon$. Let $\Delta_k = [\lambda_{k-1}, \lambda_k]$, $E(\Delta_k) = E(\lambda_k) - E(\lambda_{k-1})$ and take points μ_k in Δ_k. Form the sum

$$S_\Pi = \sum_{k=1}^{n} f(\mu_k) E(\Delta_k).$$

Denote by $|\Pi|$ the mesh of Π—that is $|\Pi| = \max_k (\lambda_k - \lambda_{k-1})$.

Lemma 6.7.1 *If* $|\Pi| \to 0$, S_Π *converges, in the uniform topology, to a bounded linear operator* S *that is independent of* (i) *the continuous extension of* f; (ii) *the choice of* ε; (iii) *the choice of the points* μ_k.

Note that the limit of S_Π, as $|\Pi| \to 0$, is taken in the following sense: for any $\eta > 0$ there exists a $\delta > 0$ such that $\|S_\Pi - S\| < \eta$ whenever $|\Pi| < \delta$.

We write

$$S = \int_m^{M+\varepsilon} f(\lambda)\, dE_\lambda$$

and call it the *integral* of *f with respect to the spectral family* $\{E_\lambda\}$.

Proof. It suffices to consider the case where $f(\lambda)$ is real-valued. For any $\eta > 0$ there exists a $\delta > 0$ such that $|f(\lambda) - f(\lambda')| < \eta$ if $|\lambda - \lambda'| < \delta$. Let Π_1, Π_2 be two partitions of $[m, M + \varepsilon]$ with $|\Pi_1| < \delta$, $|\Pi_2| < \delta$. Let Π

be the partition whose points are obtained by combining the points that occur in both partitions Π_1 and Π_2. We shall estimate $\|S_\Pi - S_{\Pi_1}\|$.

Let Π_1 be given by $\lambda_0 < \lambda_1 < \cdots < \lambda_n$ and let Π be given by

$$\lambda_0 < \lambda_{0,1} < \cdots < \lambda_{0,k_0} = \lambda_1 < \lambda_{1,1} < \cdots < \lambda_{1,k_1} = \lambda_2 < \cdots < \lambda_{n-1,k_{n-1}} = \lambda_n.$$

In S_{Π_1} there occur the points $\mu_1 \leq \mu_2 \leq \cdots \leq \mu_n$ with $\mu_k \in [\lambda_{k-1},\lambda_k]$. Analogously in S_Π there occur the points

$$\mu_{0,1} \leq \mu_{0,2} \leq \cdots \leq \mu_{0,k_0} \leq \mu_{1,1} \leq \mu_{1,2} \leq \cdots \leq \mu_{1,k_1} \leq \mu_{2,1} \leq \cdots \leq \mu_{n-1,k_{n-1}}.$$

We have

$$S_\Pi = \sum_{i=0}^{n-1} \sum_{j=1}^{k_i} f(\mu_{i,j})[E(\lambda_{i,j}) - E(\lambda_{i,j-1})],$$

where $\lambda_{0,0} = \lambda_0, \lambda_{1,0} = \lambda_1, \ldots, \lambda_{n-1,0} = \lambda_{n-1}$. We can write S_{Π_1} in the form

$$S_{\Pi_1} = \sum_{i=0}^{n-1} \sum_{j=1}^{k_i} f(\mu_{i+1})[E(\lambda_{i,j}) - E(\lambda_{i,j-1})].$$

Since $|\mu_{i+1} - \mu_{i,j}| < \delta$, we have

$$|f(\mu_{i+1}) - f(\mu_{i,j})| < \eta.$$

Hence

$$S_{\Pi_1} \leq S_\Pi + \eta \sum_{i=0}^{n-1} \sum_{j=1}^{k_i} [E(\lambda_{i,j}) - E(\lambda_{i,j-1})] = S_\Pi + \eta I.$$

Similarly $S_\Pi \leq S_{\Pi_1} + \eta I$. Hence $-\eta I \leq S_\Pi - S_{\Pi_1} \leq \eta I$. It follows (compare Problem 6.6.4) that $\|S_\Pi - S_{\Pi_1}\| \leq \eta$. Similarly $\|S_\Pi - S_{\Pi_2}\| \leq \eta$. We conclude that $\|S_{\Pi_1} - S_{\Pi_2}\| \leq 2\eta$. Thus $S = \lim_{|\Pi| \to 0} S_\Pi$ exists, in the uniform topology. Furthermore, S is independent of the choice of the points μ_k. Since $E_\lambda - E_\mu = 0$ if $M < \mu < \lambda$, it is readily seen that S is independent of the choice of ε and of the extension of f. This completes the proof.

Note that $I = \int_m^{M+\varepsilon} dE_\mu$.

We now state the main result of the present section.

Theorem 6.7.2 *To every self-adjoint operator* A *there corresponds a unique family* $\{E_\lambda\}$ *of projections,* $-\infty < \lambda < \infty$, *satisfying the following conditions*:

(i) *Any bounded linear operator that commutes with* A *commutes also with the* E_λ;

(ii) $E_\lambda \leq E_\mu$ *if* $\lambda < \mu$;

(iii) E_λ *is strongly continuous from the left—that is, for any* $x \in H$, $\lim_{\lambda \nearrow \mu} E_\lambda x = E_\mu x$. *Furthermore,* $E_{\mu+0}x = \lim_{\lambda \searrow \mu} E_\lambda x$ *exists for any* $x \in H$.

(iv) $E_\lambda = 0$ *if* $\lambda < m$ *and* $E_\lambda = I$ *if* $\lambda > M$, *where* m *and* M *are the lower and upper bounds of* A;

(v) A *is given by*

$$A = \int_m^{M+\varepsilon} \lambda \, dE_\lambda \qquad (0 < \varepsilon < 1). \tag{6.7.1}$$

The family $\{E_\lambda\}$ is called the *spectral family of A*, or the *resolution of the identity corresponding to A*.

Proof. Denote by $E_+(\lambda)$ the projection operator constructed by Lemma 6.6.6 for $A - \lambda I$. The operators $E_\lambda = I - E_+(\lambda)$ are projections. We shall prove that they satisfy the conditions (i)–(v). Since $E_+(\lambda)$ commutes with any operator that commutes with A, the same is true of E_λ. Hence (i) is satisfied. In particular it follows from (i) that $E_\lambda E_\mu = E_\mu E_\lambda$ for all λ,μ.

To prove (ii), let $\lambda < \mu$ and set $P = E_\lambda(I - E_\mu)$. We have

$$E_\lambda P = P, \qquad (I - E_\mu)P = P. \tag{6.7.2}$$

From the definition of E_λ we also have

$$(A - \lambda I)E_\lambda \leq 0, \qquad (A - \mu I)(I - E_\mu) \geq 0. \tag{6.7.3}$$

Take any $x \in H$ and let $y = Px$. Then (6.7.2) implies that

$$E_\lambda y = E_\lambda Px = Px = y$$

and, similarly, $(I - E_\mu)y = y$. Using (6.7.3), we get

$$((A - \lambda I)y,y) = ((A - \lambda I)E_\lambda y,y) \leq 0,$$
$$((A - \mu I)y,y) = ((A - \mu I)(I - E_\mu)y,y) \geq 0.$$

It follows that $(\lambda - \mu)(y,y) \geq 0$. Since $\lambda < \mu$, $(y,y) \leq 0$—that is, $Px = y = 0$. We have thus proved that $P = 0$—that is, $E_\lambda = E_\lambda E_\mu$. But this implies (compare Problem 6.6.3) that $E_\lambda \leq E_\mu$. Thus (ii) is satisfied.

Let $\lambda < \mu$, $\Delta = [\lambda,\mu]$, $E(\Delta) = E_\mu - E_\lambda$. Then

$$E_\mu E(\Delta) = E(\Delta), \qquad (I - E_\lambda)E(\Delta) = E(\Delta).$$

Therefore, by (6.7.3) (with λ and μ interchanged) and the fact that $E(\Delta) \geq 0$,

$$(A - \mu I)E(\Delta) = (A - \mu I)E_\mu E(\Delta) \leq 0,$$
$$(A - \lambda I)E(\Delta) = (A - \lambda I)(I - E_\lambda)E(\Delta) \geq 0.$$

Consequently,

$$\lambda E(\Delta) \leq AE(\Delta) \leq \mu E(\Delta). \tag{6.7.4}$$

We shall now prove (iii). Let $x \in H$. Then, by (ii), $(E_\lambda x,x)$ is a nondecreasing function of λ. Therefore $\lim_{\lambda \nearrow \mu} (E_\lambda x,x)$ exists. It follows that

$$\|E_\nu x - E_\lambda x\|^2 = ((E_\nu - E_\lambda)(E_\nu - E_\lambda)x,x) = ((E_\nu - E_\lambda)x,x) \to 0$$

if $\lambda < \nu < \mu$, $\lambda \to \mu$. Thus

$$\lim_{\lambda \nearrow \mu} E_\lambda x = E_{\mu-0} x \qquad \text{exists.}$$

Similarly,

$$\lim_{\lambda \searrow \mu} E_\lambda x = E_{\mu+0} x \qquad \text{exists.}$$

To complete the proof of (iii) it remains to show that the operator $E(\Delta_0) = E_\mu - E_{\mu-0}$ is zero.

For any $x \in H$,

$$E(\Delta)x = (E_\mu - E_\lambda)x \to E(\Delta_0)x \qquad \text{as } \lambda \nearrow \mu.$$

From (6.7.4) we then conclude that the operator $\mu E(\Delta_0) - AE(\Delta_0)$ is both positive and negative. Hence (cf. Problem 6.6.4) $\mu E(\Delta_0) - AE(\Delta_0) = 0$. Hence, for any $x \in H$, and $y = E(\Delta_0)x$, $(A - \mu I)y = 0$. By the assertion (iii) of Lemma 6.6.6 we conclude that $E_\mu y = 0$—that is, $E_\mu E(\Delta_0)x = 0$. Hence

$$E(\Delta_0)x = \lim_{\lambda \nearrow \mu} E(\Delta)x = \lim_{\lambda \nearrow \mu} E_\mu E(\Delta)x = E_\mu E(\Delta_0)x = 0.$$

We have thus proved that $E(\Delta_0) = 0$. This completes the proof of (iii).

To prove (iv), let $\lambda < m$ and suppose that $E_\lambda \neq 0$. Then there exists a point $x \in H$ such that $E_\lambda x \neq 0$. Let $y = E_\lambda x$. We may assume that $\|y\| = 1$. Then

$$(Ay,y) - \lambda = ((A - \lambda I)y,y) = ((A - \lambda I)E_\lambda y,y) \leq 0$$

by (6.7.3). Thus $m \leq (Ay,y) \leq \lambda$—a contradiction. Thus $E_\lambda = 0$ if $\lambda < m$. Suppose next that $\lambda > M$. If $E_\lambda \neq I$, then there exists a point $x \in H$ such that $z = (I - E_\lambda)x \neq 0$. We may assume that $\|z\| = 1$. Then

$$(Az,z) - \lambda = ((A - \lambda I)z,z) = ((A - \lambda I)(I - E_\lambda)z,z) \geq 0$$

by (6.7.3). Hence $M \geq (Az,z) \geq \lambda$—a contradiction.

We shall now prove (v). Take a partition

$$m = \lambda_0 < \lambda_1 < \cdots < \lambda_{n-1} < \lambda_n = M + \varepsilon \qquad (0 < \varepsilon < 1),$$

and set $\Delta_k = [\lambda_{k-1},\lambda_k]$, $E(\Delta_k) = E(\lambda_k) - E(\lambda_{k-1})$. From (6.7.4) and the relation $I = \sum_{k=1}^{n} E(\Delta_k)$, we get

$$\sum_{k=1}^{n} \lambda_{k-1} E(\Delta_k) \leq A \leq \sum_{k=1}^{n} \lambda_k E(\Delta_k).$$

Now take a sequence of partitions with mesh going to zero and use Lemma 6.7.1.

It remains to prove uniqueness. Before proving it, we shall establish the following lemma.

Lemma 6.7.3 *Let* A *be a self-adjoint operator and let* $\{E_\lambda\}$ *satisfy* (i)–(v). *Then, for any polynomial* $p(\lambda)$,

$$p(A) = \int_m^{M+\varepsilon} p(\lambda)\, dE_\lambda. \tag{6.7.5}$$

Proof. For any $\eta > 0$ there is a $\delta > 0$ such that any finite sum $S = \sum_{k=1}^{n} \lambda_k E(\Delta_k)$ with $\max_k (\lambda_k - \lambda_{k-1}) \leq \delta$ satisfies

$$\|A - \sum_{k=1}^{n} \lambda_k E(\Delta_k)\| \leq \eta,$$

or,

$$\| [A - \sum_{k=1}^{n} \lambda_k E(\Delta_k)]^2 \| \leq \eta^2.$$

Thus

$$\|A^2 + \left[\sum_{k=1}^{n} \lambda_k E(\Delta_k)\right]^2 - 2AS\| \leq \eta^2.$$

Using the relations

$$E(\Delta_k)E(\Delta_j) = (E_{\lambda_k} - E_{\lambda_{k-1}})(E_{\lambda_j} - E_{\lambda_{j-1}}) = 0$$

if $k \neq j$ and $\|AS - A^2\| \leq \|A\|\eta$, we find that

$$\| A^2 - \sum_{k=1}^{n} \lambda_k^2 E(\Delta_k) \| \leq c\eta.$$

This gives (6.7.5) with $p(\lambda) = \lambda^2$. Similarly one can prove by induction that (6.7.5) is valid for $p(\lambda) = \lambda^n$, n any positive integer. From this (6.7.5) follows for any polynomial.

Corollary 6.7.4 *For any* $x \in H$ *and for any polynomial* $p(\lambda)$,

$$(p(A)x,x) = \int_m^{M+\varepsilon} p(\lambda)\, d(E_\lambda x,x). \tag{6.7.6}$$

The integral on the right is defined analogously to the definition of $\int_m^{M+\varepsilon} p(\lambda)\, dE_\lambda$. Since $(E_\lambda x,x)$ is monotone nondecreasing, this integral coincides with the Riemann-Stieltjes integral (defined in Problem 2.11.8).

To prove (6.7.6) we take sums $S_\Pi = \sum p(\mu_k)E(\Delta_k)$ that converge to $p(A)$ as $|\Pi| \to 0$, and notice that

$$(S_\Pi x,x) \to (p(A)x,x),$$

$$(S_\Pi x,x) = \sum p(\mu_k)(E(\Delta_k)x,x) \to \int_m^{M+\varepsilon} p(\lambda)\, d(E_\lambda x,x),$$

as $|\Pi| \to 0$.

We can now prove the uniqueness assertion of Theorem 6.7.2. Suppose $\{F_\lambda\}$ is another family satisfying (i)–(v). By Corollary 6.7.4, for any polynomial $p(\lambda)$

$$\int_m^{M+\varepsilon} p(\lambda)\, d\sigma(\lambda) = 0, \tag{6.7.7}$$

where $\sigma(\lambda) = (E_\lambda x,x) - (F_\lambda x,x)$ is a function of bounded variation. Since any continuous function can be approximated uniformly by polynomials, we obtain (using Problem 2.11.10) the relation (6.7.7) for any continuous function $p(\lambda)$. Since $\sigma(m) = 0$ and since $\sigma(\lambda)$ is continuous from the left, it follows (by Problem 4.14.3) that $\sigma(\lambda) \equiv 0$. Thus $(E_\lambda x,x) = (F_\lambda x,x)$. The operator $E_\lambda - F_\lambda$ is self-adjoint, and we have just proved that its lower and upper bounds are equal to zero. Hence, by Theorem 6.5.1, $E_\lambda - F_\lambda = 0$. This completes the proof.

Notation Let $f(\lambda)$ be any continuous function for $m \le \lambda \le M$. We define

$$f(A) = \int_m^{M+\varepsilon} f(\lambda)\, dE_\lambda. \tag{6.7.8}$$

From Lemma 6.7.3 it follows that if $f(\lambda)$ is a polynomial $\sum a_n \lambda^n$, then $f(A) = \sum a_n A^n$.

One easily gets, by the argument used in proving Corollary 6.7.4,

$$f(A)x = \int_m^{M+\varepsilon} f(\lambda)\, dE_\lambda x, \tag{6.7.9}$$

$$(f(A)x,x) = \int_m^{M+\varepsilon} f(\lambda)\, d(E_\lambda x,x). \tag{6.7.10}$$

Theorem 6.7.5 *Let* $f(\lambda)$, $f_1(\lambda)$, $f_2(\lambda)$ *be continuous functions and let* a_1, a_2 *be scalars.*

(i) *If* $f(\lambda) = a_1 f_1(\lambda) + a_2 f_2(\lambda)$, *then* $f(A) = a_1 f_1(A) + a_2 f_2(A)$.
(ii) *If* $f(\lambda) = f_1(\lambda) f_2(\lambda)$, *then* $f(A) = f_1(A) f_2(A)$.
(iii) $[f(A)]^* = \bar{f}(A)$, *where* $\bar{f}(\lambda) = \overline{f(\lambda)}$.
(iv) $\|f(A)\| \le \max_{m \le \lambda \le M} |f(\lambda)|$ *if f is real-valued.*

(v) *If* B *is any bounded linear operator that commutes with* A, *then* B *also commutes with* f(A).

Proof. Consider partitions Π: $m = \lambda_0 < \lambda_1 < \cdots < \lambda_{n-1} < \lambda_n = M + \varepsilon$ with $|\Pi| \to 0$, and let $E(\Delta_k) = E_{\lambda_k} - E_{\lambda_{k-1}}$. Then

$$a_1 f_1(A) + a_2 f(A) = \lim_{|\Pi|\to 0} \sum_{k=1}^{n} a_1 f_1(\lambda_k)E(\Delta_k) + \lim_{|\Pi|\to 0} \sum_{k=1}^{n} a_2 f_2(\lambda_k)E(\Delta_k)$$

$$= \lim_{|\Pi|\to 0} \sum_{k=1}^{n} [a_1 f_1(\lambda_k) + a_2 f_2(\lambda_k)]E(\Delta_k)$$

$$= \lim_{|\Pi|\to 0} \sum_{k=1}^{n} f(\lambda_k)E(\Delta_k) = f(A).$$

This proves (i). To prove (ii), note that $E(\Delta_k)E(\Delta_j) = 0$ if $k \neq j$. Hence

$$f_1(A)f_2(A) = \left\{\lim_{|\Pi|\to 0} \sum_{k=1}^{n} f_1(\lambda_k)E(\Delta_k)\right\}\left\{\lim_{|\Pi|\to 0} \sum_{k=1}^{n} f_2(\lambda_k)E(\Delta_k)\right\}$$

$$= \lim_{|\Pi|\to 0} \sum_{k=1}^{n} \sum_{j=1}^{n} f_1(\lambda_k)f_2(\lambda_j)E(\Delta_k)E(\Delta_j)$$

$$= \lim_{|\Pi|\to 0} \sum_{k=1}^{n} f(\lambda_k)E(\Delta_k) = f(A).$$

The assertion (iii) follows from

$$[\sum f(\lambda_k)E(\Delta_k)]^* = \sum \overline{f(\lambda_k)}E(\Delta_k).$$

To prove (iv) note that $f(A)$ is self-adjoint. Also, by (6.7.10),

$$|(f(A)x,x)| = \left|\int_m^{M+\varepsilon} f(\lambda)\, d(E_\lambda x,x)\right| \leq \left\{\max_{m\leq\lambda\leq M+\varepsilon} |f(\lambda)|\right\} \int_m^{M+\varepsilon} d(E_\lambda x,x).$$

Since the last integral is equal to $\|x\|$, we obtain, upon taking $\varepsilon \to 0$,

$$\sup_{\|x\|=1} |(f(A)x,x)| \leq \max_{m\leq\lambda\leq M} |f(\lambda)|.$$

Now use Theorem 6.5.1.

The assertion (v) follows from the fact that B commutes with each operator E_λ.

PROBLEMS

6.7.1. Let A be a self-adjoint operator and let $\{E_\lambda\}$ be its spectral family. Let $E(\Delta) = E(\beta) - E(\alpha)$, where $\alpha < \beta$. Prove that for any continuous function $f(\lambda)$,

$$E(\Delta)\left[\int_m^{M+\varepsilon} f(\lambda)\, dE_\lambda\right] = \int_\alpha^\beta f(\lambda)\, dE_\lambda.$$

6.7.2. Let A be a self-adjoint operator and let $\{E_\gamma\}$ be its spectral family. Then, for any continuous functions f,g,

$$\left(\int_m^{M+\varepsilon} f(\lambda)\, dE_\lambda x, \int_m^{M+\varepsilon} g(\lambda)\, dE_\lambda x\right) = \int_m^{M+\varepsilon} f(\lambda)\overline{g(\lambda)}\, d(E_\lambda x,x).$$

6.7.3. A compact self-adjoint operator in a Hilbert space H has at least one eigenvalue.

6.7.4. Denote by N_λ the space of all solutions of $(A - \lambda I)x = 0$, A self-adjoint. Denote by N the closed linear subspace spanned by the class of all subspaces N_λ. If all the points of the spectrum of A are eigenvalues, and if $N = H$, then we say that A has a *pure point spectrum*. Prove that a compact self-adjoint operator A has a pure point spectrum. [*Hint:* $A(N) \subset N$, hence $A(N^\perp) \subset N^\perp$. If $N^\perp \neq 0$, then A has in $N^\perp$ an eigenvector x_0.]

6.7.5. Let A be a self-adjoint operator in a separable Hilbert space, having pure point spectrum. Then there is an orthonormal basis $\{x_n\}$ of eigenvectors. Denote by λ_n the eigenvalue corresponding to x_n. Denote by P_n the projection $P_n x = (x,x_n)x_n$. Then $P_n P_m = 0$ if $n \neq m$. Prove, without using Theorem 6.7.2:

$$x = \sum_{n=1}^{\infty} (x,x_n)x_n = \sum_{n=1}^{\infty} P_n x;$$

$$Ax = \sum_{n=1}^{\infty} \lambda_n(x,x_n)x_n = \sum_{n=1}^{\infty} \lambda_n P_n x;$$

$$(Ax,x) = \sum_{n=1}^{\infty} \lambda_n |(x,x_n)|^2 = \sum_{n=1}^{\infty} \lambda_n (P_n x,x).$$

If $d = \inf_n |\lambda - \lambda_n| > 0$, then

$$R(\lambda;A)x = \sum_{n=1}^{\infty} \frac{(x,x_n)}{\lambda_n - \lambda} x_n = \sum_{n=1}^{\infty} \frac{1}{\lambda_n - \lambda} P_n x;$$

$$\|R(\lambda;A)\| \leq \frac{1}{d}.$$

6.7.6. Find the spectral family of the operator $(Ax)(t) = tx(t)$ in $L^2(0,1)$.

6.8 THE RESOLVENT OF SELF-ADJOINT OPERATORS

Throughout this section A denotes a self-adjoint operator in a Hilbert space H. The lower and upper bounds of A will be denoted by m and M, respectively. We denote by $\sigma(A)$ the spectrum of A and by $\rho(A)$ the resolvent of A. Recall that $\lambda \in \rho(A)$ if and only if the resolvent $R(\lambda;A) = (\lambda I - A)^{-1}$ exists (as a bounded linear operator in H).

By Theorem 6.5.8 the resolvent exists for all complex λ not in the real interval $[m,M]$. By Theorem 6.5.9, the resolvent does not exist at the points $\lambda = m$ and $\lambda = M$—that is, m and M belong to $\sigma(A)$. We shall characterize, in this section, the points λ of $[m,M]$ for which $R(\lambda;A)$ exists—that is, the points λ which belong to $\rho(A)$.

Theorem 6.8.1 *Let* $\lambda_0 \in [m,M]$. *If there exists an interval* (α,β) *with* $\alpha < \lambda_0 < \beta$, *such that* $E_\lambda = \text{const.}$ *for* $\alpha < \lambda < \beta$, *then* $\lambda_0 \in \rho(A)$ *and*

$$R(\lambda_0;A) = \int_m^{M+\varepsilon} \psi(\lambda)\, dE_\lambda, \tag{6.8.1}$$

where $\psi(\lambda)$ *is any continuous function satisfying* $\psi(\lambda) = 1/(\lambda - \lambda_0)$ *if* $\lambda \le \alpha$ *and if* $\lambda \ge \beta$.

Proof. It is easily seen that if $f(\lambda),\hat{f}(\lambda)$ are continuous functions, and and if $f(\lambda) = \hat{f}(\lambda)$ for $\lambda \le \alpha$ and for $\lambda \ge \beta$, then

$$\int_m^{M+\varepsilon} f(\lambda)\, dE_\lambda = \int_m^{M+\varepsilon} \hat{f}(\lambda)\, dE_\lambda. \tag{6.8.2}$$

Consider now the integral

$$R = \int_m^{M+\varepsilon} \psi(\lambda)\, dE_\lambda.$$

By Theorem 6.7.5(ii),

$$R(A - \lambda_0 I) = (A - \lambda_0 I)R = \int_m^{M+\varepsilon} \psi(\lambda)(\lambda - \lambda_0) dE_\lambda.$$

By the remark of (6.8.2), the last integral is equal to I. This proves the theorem.

The integral

$$\int_m^{M+\varepsilon} \frac{dE_\lambda}{\lambda - \lambda_0}$$

for $\lambda_0 \in [m,M]$ has not been defined, since $1/(\lambda - \lambda_0)$ is not continuous. However, under the assumptions of Theorem 6.8.1 it can be defined as a limit of sums corresponding to partitions Π, $|\Pi| \to 0$, provided we agree to take $(\mu_k - \lambda_0)^{-1}E(\Delta_k) = 0$ whenever $E(\Delta_k) = 0$ (even when $\mu_k = \lambda_0$). With this agreement one then easily verifies that

$$R(\lambda_0;A) = \int_m^{M+\varepsilon} \frac{dE_\lambda}{\lambda - \lambda_0}. \tag{6.8.3}$$

We shall now prove the converse of Theorem 6.8.1.

Theorem 6.8.2 *If* $\lambda_0 \in [m,M]$ *and* $R(\lambda_0; A)$ *exists, then there exists an interval* (α,β) *such that* $\alpha < \lambda_0 < \beta$ *and* $E_\lambda = \text{const.}$ *for all* $\alpha < \lambda < \beta$.

Proof. Let $\Delta = [\alpha,\beta]$ be an interval with λ_0 as its midpoint, and let $E(\Delta) = E(\beta) - E(\alpha)$. Applying $R(\lambda_0;A)E(\Delta)$ to both sides of the identity

$$(A - \lambda_0 I)x = \int_m^{M+\varepsilon} (\lambda - \lambda_0)\, dE_\lambda x \qquad (x \in H),$$

we obtain (compare Problem 6.7.1)

$$E(\Delta)x = R(\lambda_0;A) \int_\alpha^\beta (\lambda - \lambda_0)\, dE_\lambda x.$$

Hence

$$\|E(\Delta)x\| \leq \|R(\lambda_0;A)\| \left\| \int_\alpha^\beta (\lambda - \lambda_0)\, dE_\lambda x \right\|.$$

By Problem 6.7.2,

$$\left\| \int_\alpha^\beta (\lambda - \lambda_0)\, dE_\lambda x \right\|^2 = \int_\alpha^\beta (\lambda - \lambda_0)^2\, d(E_\lambda x,x) \leq \frac{(\beta - \alpha)^2}{4} \int_\alpha^\beta d(E_\lambda x,x)$$

$$= \frac{(\beta - \alpha)^2}{4} \|E(\Delta)x\|^2.$$

Hence

$$\|E(\Delta)x\| \leq \frac{\beta - \alpha}{2} \|R(\lambda_0;A)\|\, \|E(\Delta)x\|.$$

If $(\beta - \alpha) < 2\|R(\lambda_0;A)\|^{-1}$, then $\|E(\Delta)x\| = 0$—that is, $E_\lambda = \text{const.}$ for $\alpha < \lambda < \beta$. This completes the proof.

So far we have characterized the points of the spectrum as follows: $\lambda_0 \in \sigma(A)$ if and only if $\lambda_0 \in [m,M]$ and there is no open interval containing λ_0 on which E_λ is constant. We shall now characterize the eigenvalues of A.

Theorem 6.8.3 *A point* λ_0 *in* $[m,M]$ *is an eigenvalue of* A *if and only if* E_λ *is discontinuous at* $\lambda = \lambda_0$*—that is,* $E_{\lambda_0 + 0} \neq E_{\lambda_0}$.

Proof. Suppose λ_0 is an eigenvalue and let $Ax_0 - \lambda_0 x_0 = 0$, $x_0 \neq 0$. Then $((A - \lambda_0 I)^2 x_0,x_0) = 0$. It follows that

$$\int_m^{M+\varepsilon} (\lambda - \lambda_0)^2\, d(E_\lambda x_0,x_0) = 0.$$

Since the integrand is nonnegative and $(E_\lambda x_0, x_0)$ is nondecreasing,

$$\int_{\lambda_0+\eta}^{M+\varepsilon} (\lambda - \lambda_0)^2 \, d(E_\lambda x_0, x_0) = 0 \qquad \text{for any } \eta > 0.$$

Since $\lambda - \lambda_0 \geq \eta$ in the last integral, we get

$$\int_{\lambda_0+\eta}^{M+\varepsilon} d(E_\lambda x_0, x_0) = 0,$$

—that is, $(x_0, x_0) - (E_{\lambda_0+\eta} x_0, x_0) = 0$. Thus

$$\|(I - E_{\lambda_0+\eta})x_0\|^2 = ((I - E_{\lambda_0+\eta})x_0, x_0) = 0.$$

It follows that $(I - E_{\lambda_0+\eta})x_0 = 0$—that is, $E_{\lambda_0+\eta} x_0 = x_0$. Similarly we obtain $E_{\lambda_0-\eta} x_0 = 0$. Taking $\eta \to 0$, we find that

$$(E_{\lambda_0+0} - E_{\lambda_0})x_0 = x_0. \tag{6.8.4}$$

Hence, $E_{\lambda_0+0} \neq E_{\lambda_0}$.

Suppose, conversely, that $E_{\lambda_0+0} \neq E_{\lambda_0}$. Since $E_{\lambda_0} \leq E_{\lambda_0+0}$, there exists a point $x_0 \neq 0$ that belongs to the subspace of the projection E_{λ_0+0} but that is orthogonal to the subspace of the projection E_{λ_0}. Thus

$$E_{\lambda_0+0} x_0 = x_0, \qquad E_{\lambda_0} x_0 = 0.$$

We then also have $E_\lambda x_0 = x_0$ if $\lambda > \lambda_0$. Hence

$$E(\Delta)x_0 = x_0 \qquad \text{if } \Delta = [\lambda_0, \lambda_0 + \eta]$$

where $\eta > 0$. We deduce that

$$Ax_0 = AE(\Delta)x_0 = \int_{\lambda_0}^{\lambda_0+\eta} \lambda \, dE_\lambda x_0.$$

Also,

$$\lambda_0 x_0 = \lambda_0 E(\Delta)x_0 = \int_{\lambda_0}^{\lambda_0+\eta} \lambda_0 \, dE_\lambda x_0.$$

Therefore,

$$\|Ax_0 - \lambda_0 x_0\|^2 = \left\| \int_{\lambda_0}^{\lambda_0+\eta} (\lambda - \lambda_0) \, dE_\lambda x_0 \right\|^2 = \int_{\lambda_0}^{\lambda_0+\eta} (\lambda - \lambda_0)^2 \, d(E_\lambda x_0, x_0)$$
$$\leq \eta^2 \|E(\Delta)x\|^2.$$

Since η is arbitrary, $Ax_0 - \lambda_0 x_0 = 0$. This completes the proof.

Corollary 6.8.4 *The space of the eigenvectors corresponding to* λ_0 *coincides with the subspace of the projection operator* $E_{\lambda_0+0} - E_{\lambda_0}$.

PROBLEMS

6.8.1. Let $f(\lambda)$ be a continuous function of λ, with values in a Banach space X, where λ varies in an open set G of the complex plane. Let Γ be a continuously differentiable curve in G. Show that the contour integral $\int_\Gamma f(\lambda)\,d\lambda$ can be defined in the same way as for complex-valued functions $f(\lambda)$. If Γ is oriented counterclockwise, then we write $\int_\Gamma f(\lambda)\,d\lambda$ also in the form $\oint_\Gamma f(\lambda)\,d\lambda$. Prove: If A is a bounded operator in X, then $A[\int_\Gamma f(\lambda)\,d\lambda] = \int_\Gamma [Af(\lambda)]\,d\lambda$.

6.8.2. Let T be a bounded linear operator in a Banach space X and let Γ be a continuously differentiable simple closed curve containing the spectrum $\sigma(T)$ of T in its interior. Let $f(\lambda)$ be an entire complex analytic function. Define

$$f(T) = \frac{1}{2\pi i}\oint_\Gamma f(\lambda)R(\lambda;T)\,d\lambda. \tag{6.8.5}$$

Use Cauchy's theorem to show that the definition is independent of Γ. Prove that if $f(\lambda) = \lambda^n$, then $f(T) = T^n$. [*Hint:* $R(\lambda;T) = \sum T^m/\lambda^{m+1}$.]

6.8.3. Let f,g be entire complex analytic functions and let α,β be scalars. Prove:

(i) $(\alpha f + \beta g)(T) = \alpha f(T) + \beta g(T)$.

(ii) $(fg)(T) = f(T)g(T)$. [*Hint:* Modify the contour Γ for $g(T)$, replacing it by a contour Γ_2 that contains the contour $\Gamma = \Gamma_1$ for $f(T)$, and use (5.3.1).]

(iii) $[f(T)]^* = \bar{f}(T^*)$, where $\bar{f}(\lambda) = \overline{f(\bar{\lambda})}$.

6.8.4. If T is self-adjoint, then $f(T)$ as defined in (6.8.5) coincides with $f(T)$ defined by $\int_m^{M+\varepsilon} f(\lambda)\,dE_\lambda$, where $\{E_\lambda\}$ is the spectral family of T.

6.8.5. If A is self-adjoint with upper bound M, and if $\operatorname{Re}\lambda > M$, then

$$R(\lambda;A) = \int_0^\infty e^{-\lambda t}e^{tA}\,dt.$$

6.9 EIGENVALUE PROBLEMS FOR DIFFERENTIAL EQUATIONS

Let A be a compact self-adjoint operator in a separable Hilbert space H. From Problems 6.7.4, 6.7.5 we know that there exists an orthonormal basis $\{x_m\}$, where x_m are eigenvectors.

We shall consider now the case where A is an *integral operator* in $L^2(\Omega)$:

$$(Af)(x) = \int_\Omega K(x,y)f(y)\,dy. \tag{6.9.1}$$

Here Ω is a bounded open set in R^n. The function $K(x,y)$ is called a *kernel.* If $K(x,y)$ is measurable in $\Omega \times \Omega$ and if

$$\int_\Omega \int_\Omega |K(x,y)|^2 \, dx \, dy < \infty,$$

then A is a compact operator in $L^2(\Omega)$ (see Problem 5.1.4). Further, it is easily verified that A is self-adjoint if and only if the kernel $K(x,y)$ is *symmetric* —that is, $K(x,y) = \overline{K(y,x)}$.

We shall assume from now on that $K(x,y)$ is symmetric, that Ω is a bounded open set, and that

$$\sup_\Omega \int_\Omega |K(x,y)|^2 \, dy < \infty. \tag{6.9.2}$$

The eigenvectors of A are functions in $L^2(\Omega)$. We therefore call them also *eigenfunctions.*

Since A is compact and self-adjoint, it has an orthonormal basis of eigenfunctions $\{x_m\}$ with eigenvalues $\{\mu_m\}$. Some of the eigenvalues μ_n may be equal to zero. We shall denote by $\{\varphi_m\}$ the subsequence of $\{x_m\}$ consisting of those eigenfunctions with non-zero eigenvalues. We arrange the φ_m in such a way that the corresponding eigenvalues λ_m satisfy: $|\lambda_m| \geq |\lambda_{m+1}|$. The sequence $\{\varphi_m\}$ may be finite or infinite.

The following theorem is called the *Hilbert-Schmidt theorem.*

Theorem 6.9.1 *Let* $f \in L^2(\Omega)$ *have the form* $f = Ag$*—that is,*

$$f(x) = \int_\Omega K(x,y)g(y) \, dy \qquad [g \in L^2(\Omega)]. \tag{6.9.3}$$

Then the series $\sum c_m \varphi_m$ $[c_m = (f,\varphi_m)]$ *converges uniformly and absolutely to* $f(x)$.

Proof. Note that f is orthogonal to the null space of A. Hence the series $\sum c_m \varphi_m$ converges to f in $L^2(\Omega)$. If we prove that

$$\sum |c_m \varphi_m(x)| \text{ is uniformly convergent,} \tag{6.9.4}$$

then the assertions of the theorem follow. Of course we shall assume here that the series consists of an infinite number of terms. We have

$$\begin{aligned} c_m &= \int f(x)\overline{\varphi_m(x)} \, dx = \iint K(x,y)g(y)\overline{\varphi_m(x)} \, dy \, dx \\ &= \int g(y) \overline{\int K(y,x)\varphi_m(x) \, dx} \, dy \\ &= \lambda_m \int g(y)\overline{\varphi_m(y)} \, dy = \lambda_m g_m, \end{aligned}$$

where $g_m = (g,\varphi_m)$. By Bessel's inequality

$$\sum_{m=1}^{\infty} |g_m|^2 < \infty. \tag{6.9.5}$$

The function $K(x,y)$ as a function of x, has the Fourier coefficients

$$\int K(x,y)\overline{\varphi_m(x)}\,dx = \overline{\int K(y,x)\varphi_m(x)\,dx} = \lambda_m \overline{\varphi_m(y)}.$$

Hence, by Bessel's inequality,

$$\sum_{m=1}^{\infty} |\lambda_m|^2\,|\varphi_m(y)|^2 \le \int |K(x,y)|^2\,dx \le C < \infty, \tag{6.9.6}$$

where C is a constant. Using (6.9.5), (6.9.6) we get, for $j > i$,

$$\begin{aligned}\sum_{m=i}^{j} |c_m\,\varphi_m(x)| &= \sum_{m=i}^{j} |\lambda_m\,g_m\,\varphi_m(x)| \\ &\le \sum_{m=i}^{j} |g_m|^2 \sum_{m=1}^{\infty} |\lambda_m|^2\,|\varphi_m(x)|^2 \\ &\le C \sum_{m=i}^{j} |g_m|^2 \to 0\end{aligned}$$

if $i \to \infty$. This proves (6.9.4).

We shall give an important application in case Ω is a bounded open set in R^1. Consider the ordinary differential operator

$$Lx = p_0(t)\,\frac{d^n x}{dt^n} + p_1(t)\,\frac{d^{n-1}x}{dt^{n-1}} + \cdots + p_n(t)x$$

with complex-valued coefficients $p_j(t)$ in $C^{n-j}[a,b]$, and with $p_0(t) \neq 0$ for all $t \in [a,b]$. Here $-\infty < a < b < \infty$.

Let $M_{jk}, N_{jk} (1 \le j \le n,\ 1 \le k \le n)$, be complex numbers and consider the *boundary operators*

$$U_j x = \sum_{k=1}^{n} [M_{jk}\,x^{(k-1)}(a) + N_{jk}\,x^{(k-1)}(b)] \qquad (1 \le j \le n).$$

We denote by Ux the vector $(U_1 x, \ldots, U_n x)$. Consider the problem of solving

$$Lx = \lambda x \qquad (a < t < b), \tag{6.9.7}$$

$$Ux = 0, \tag{6.9.8}$$

where λ is a complex number. We call this problem an *eigenvalue problem.*

L is called *self-adjoint* if

$$(Lu,v) = (u,Lv) \tag{6.9.9}$$

for all u,v in $C_0^\infty(a,b)$. The pair (L,U) is called *self-adjoint* if (6.9.9) holds for all u,v in $C^n[a,b]$ for which $Uu = 0$, $Uv = 0$. We then say that (6.9.7), (6.9.8) form a *self-adjoint eigenvalue problem*. If the eigenvalue problem (6.9.7), (6.9.8) has a solution $x(t) \not\equiv 0$, for some λ, then we call λ an *eigenvalue* and $x(t)$ an *eigenfunction*.

PROBLEM

6.9.1. If (L,U) is self-adjoint, then all the eigenvalues are real numbers. Furthermore, eigenfunctions corresponding to distinct eigenvalues are orthogonal in $L^2(a,b)$.

We shall assume from now on that (L,U) *is self-adjoint and* $\lambda = 0$ *is not an eigenvalue*. It can then be shown that there exists a function $G(t,\tau)$ such that, for any continuous function $f(t)$,

$$x(t) = \int_a^b G(t,\tau)f(\tau)\,d\tau \tag{6.9.10}$$

is the unique solution of $Lx = f$ in (a,b), $Ux = 0$. G has the following properties:

The derivatives $\partial^j G(t,\tau)/\partial t^j$ are continuous in (t,τ) for $0 \le j \le n-2$, and are uniformly continuous in (t,τ) for $a \le t < \tau$ and $\tau < t \le b$ if $j = n-1$, $j = n$. But $\partial^{n-1}G(t,\tau)/\partial t^{n-1}$ has a jump $1/p_0(\tau)$ at $t = \tau$. As a function of t, $G(t,\tau)$ satisfies $LG(t,\tau) = 0$ for $t \ne \tau$, and $UG = 0$ for each fixed $\tau \in (a,b)$.

The function $G(t,\tau)$ is uniquely determined by all the properties stated in the last paragraph. It is called *Green's function* for the pair (L,U).

In terms of Green's function we can reduce the system (6.9.7), (6.9.8) to the single equation

$$x(t) = \lambda \int_a^b G(t,\tau)x(\tau)\,d\tau. \tag{6.9.11}$$

It can be shown that $G(t,\tau) = \overline{G(\tau,t)}$. Hence we can apply the Hilbert-Schmidt theorem.

We arrange the sequence $\{\lambda_m\}$ of all the eigenvalues of (L,U) such that $|\lambda_m| \le |\lambda_{m+1}|$, and denote by $\{\varphi_m\}$ an orthonormal sequence of eigenfunctions, with φ_m corresponding to λ_m. (An eigenvalue λ_{m_0} will occur k times in the sequence $\{\lambda_m\}$ if there are k linearly independent solutions of (6.9.7), (6.9.8) for $\lambda = \lambda_{m_0}$.) The notation here is slightly different from that used in the Hilbert-Schmidt theorem, since the numbers λ_m that occur here are the reciprocals of the eigenvalues of the integral operator with kernel $G(t,\tau)$.

If u is in $C^n[a,b]$ and if $Uu = 0$, then set $f = Lu$. It follows that

$$u = \int_a^b G(t,\tau)f(\tau)\,d\tau.$$

Applying the Hilbert-Schmidt theorem we conclude:

Theorem 6.9.2 *Every function* u *in* $C^n[a,b]$ *that satisfies* $Uu = 0$ *can be written in the form*

$$u = \sum c_m \varphi_m \qquad [c_m = (u,\varphi_m)], \tag{6.9.12}$$

where the series is convergent uniformly and absolutely.

Since the functions in $C^n[a,b]$ that vanish in some neighborhoods of $t = a$ and $t = b$ form a dense subset in $L^2[a,b]$, it follows that the closed linear subspace spanned by $\{\varphi_m\}$ coincides with $L^2(a,b)$. Thus

Corollary 6.9.3 *The sequence* $\{\varphi_m\}$ *of the orthonormal eigenfunctions of the self-adjoint system* (6.9.7), (6.9.8) *is complete in* $L^2[a,b]$.

Recall that it was assumed in this corollary that $\lambda = 0$ is not an eigenvalue.

Since $L^2[a,b]$ is an infinite-dimensional space, we conclude:

Corollary 6.9.4 *The self-adjoint problem* (6.9.7), (6.9.8) *has an infinite number of eigenvalues.*

PROBLEMS

6.9.2. Consider the eigenvalue problem

$$Lx = -\frac{d^2x}{dt^2} = \lambda x,$$

$$x(0) = 0, \qquad x(1) = 0.$$

Verify that

$$G(t,\tau) = \begin{cases} t(1-\tau), & \text{if } 0 \le t < \tau, \\ (1-t)\tau, & \text{if } \tau < t \le 1, \end{cases}$$

is Green's function. Prove also that for any continuous f there is a unique solution of $Lu = f$ in (0,1), $u(0) = u(1) = 0$, and it is given by

$$u(t) = \int_0^1 G(t,\tau)f(\tau)\,d\tau.$$

6.9.3. In the eigenvalue problem of Problem 6.9.2, show that the eigenvalues are given by $\lambda_m = \pi^2 m^2$ $(m = 1,2,\ldots)$ and the orthonormalized eigenfunctions are given by $\varphi_m(t) = \sqrt{2}\sin \pi m t$. Applying Corollary 6.9.3, it follows that $\{\sqrt{2}\sin \pi m t\}$ is an orthonormal basis in $L^2(0,1)$.

6.9.4. Consider the eigenvalue problem

$$Lx = i\frac{dx}{dt} - \alpha x = \lambda x \qquad (\alpha \text{ not an integer}),$$

$$x(0) - x(1) = 0.$$

This is a self-adjoint problem. Prove that the function

$$G(t,\tau) = \begin{cases} \dfrac{ie^{i\alpha(\tau - t)}}{1 - e^{i\alpha}}, & \text{if } 0 \le t < \tau, \\ \dfrac{ie^{i\alpha(1+\tau-t)}}{1 - e^{i\alpha}}, & \text{if } \tau < t \le 1, \end{cases}$$

is Green's function and verify the assertion of (6.9.10). Prove also that the eigenvalues are $2\pi m - \alpha$ ($m = 0, \pm 1, \pm 2, \ldots$) and the corresponding eigenfunctions are $e^{-i(2\pi m - \alpha)t}$. Corollary 6.9.3 implies that $\{e^{2\pi i m}\}$ ($m = 0, \pm 1, \pm 2, \ldots$) is complete in $L^2(0,1)$.

6.9.5. Let $Lx = -[p(t)x']' + q(t)x'$, $p \in C^1[a,b]$, $q \in C^0[a,b]$ and $p \neq 0$. Let the boundary conditions be

$$\alpha x(a) + \beta x'(a) = 0, \qquad \gamma x(b) + \delta x'(b) = 0.$$

If p,q are real-valued functions and if $\alpha,\beta,\gamma,\delta$ are real numbers, then the system is self-adjoint.

In the following problems, A and K are as in the Hilbert-Schmidt theorem.

6.9.6. A kernel $K(x,y)$ is called *degenerate* if there exists a finite number of functions $a_i(x), b_i(y)$ such that $K(x,y) = \sum a_i(x)b_i(y)$. Prove that $K(x,y)$ is a degenerate kernel if and only if A has a finite number of eigenvalues different from zero. [*Hint:* If A has a finite number of eigenvalues $\lambda_1, \ldots, \lambda_m$ different from zero, and if $\{\varphi_1, \ldots, \varphi_m\}$ is an orthonormal set and $A\varphi_j = \lambda_j \varphi_j$, then, by the Hilbert-Schmidt theorem,

$$\int \left[K(x,y) - \sum_{j=1}^{m} \lambda_j \overline{\varphi_j(x)} \varphi_j(y) \right] h(x)\, dx = 0$$

for any function h in $L^2(\Omega)$. Hence

$$K(x,y) = \sum_{j=1}^{m} \lambda_j \overline{\varphi_j(x)} \varphi_j(y).]$$

6.9.7. Let $\{\varphi_m\}$, $\{\lambda_m\}$ be as in the Hilbert-Schmidt theorem. If $K_2(x,y) = \int K(x,z)K(z,y)\, dz$, then

$$K_2(x,y) = \sum \lambda_m \overline{\varphi_m(x)} \varphi_m(y),$$

where the convergence is uniform in x for each fixed y. [*Hint:* Use the Hilbert-Schmidt theorem.]

6.9.8. Let $\{\varphi_m\}$, $\{\lambda_m\}$ be as in Problem 6.9.7. Then, for any fixed y,

$$\int_\Omega \left| K(x,y) - \sum_{j=1}^{m} \lambda_j \overline{\varphi_j(x)} \varphi_j(y) \right|^2 dx \to 0 \qquad \text{as } m \to \infty.$$

6.9.9. Prove *Dini's theorem*: if a monotone-increasing sequence of continuous functions on a compact metric space S converges to a continuous function on S, then the convergence is uniform on S.

6.9.10. Assume that $K(x,y)$ is continuous in (x,y) for $x \in \overline{\Omega}$, $y \in \overline{\Omega}$, $x \neq y$. Prove that the $\varphi_m(x)$ are continuous on $\overline{\Omega}$ and that $K_2(x,y)$ is continuous in (x,y) for $x \in \overline{\Omega}$, $y \in \overline{\Omega}$. By Problems 6.9.7, 6.9.8 it follows that

$$\sum_{m=1}^{k} \lambda_m^2 |\varphi_m(y)|^2 \to K_2(y,y) \qquad \text{as } k \to \infty,$$

uniformly with respect to $y \in \overline{\Omega}$ (if the sequence $\{\varphi_m\}$ is infinite). Hence,

$$\sum \lambda_m^2 = \int K_2(y,y)\, dy = \iint |K(x,y)|^2\, dx\, dy.$$

Bibliography

We have made no attempt to provide a systematic and detailed bibliography. References [1]–[31] include most of the relatively recent books that deal with the subject matter of the book in a substantial way. The reader will find a proof of Zorn's lemma in [33], [35]. More details on the application given in Section 5.4 can be found in [36]. A detailed theory on eigenvalue problems for ordinary differential equations is available in [32]; for further applications to ordinary differential equations, see [34], [38]. The construction of Green's function in Section 4.9 is taken from [37].

MEASURE AND INTEGRATION

[1] Asplund, E., and Bungart, L., *A First Course in Integration.* New York: Holt, Rinehart and Winston, Inc., 1961.

[2] Carathéodory, C., *Vorlesungen über Reele Funktionen*, 2nd ed. New York: Chelsea Publishing Company, 1948.

[3] Halmos, P. R., *Measure Theory.* Princeton, N.J.: D. Van Nostrand Company, 1959.

[4] Hildebrant, T. H., *Introduction to the Theory of Integration.* New York: Academic Press, Inc., 1963.

[5] Graves, L. M., *The Theory of Functions of Real Variables.* New York: McGraw-Hill, Inc., 1946.

[6] McShane, E.J., *Integration.* Princeton, N.J.: Princeton University Press, 1944.

[7] Munroe, M. E., *Introduction to Measure Theory and Integration.* Reading, Mass., Addison-Wesley Publishing Company, Inc., 1959.
[8] Natanson, I. P., *Theory of Functions of Real Variables.* New York: Fredrick Ungar Publishing Company, vol. 1, 1955; vol. 2, 1961.
[9] Rudin, W., *Principles of Mathematical Analysis,* 2nd ed. New York: McGraw-Hill, Inc., 1964.
[10] Saks, S., *Theory of the Integral.* New York: Hafner Publishing Company, 1937.
[11] Titchmarsh, E. C., *The Theory of Functions,* 2nd ed. London: Oxford University Press, 1950.
[12] Williamson, J. H., *Lebesgue Integration.* New York: Holt, Rinehart and Winston, Inc., 1962.

FUNCTIONAL ANALYSIS

[13] Banach, S., *Théorie des opérations lineáres,* 2nd ed. New York: Chelsea Publishing Company, 1963.
[14] Davis, M., *First Course in Functional Analysis.* New York: Gordon and Breach, 1966.
[15] Day, M. M., *Normed Linear Spaces.—Ergebnisse der Mathematik,* vol. 21. Berlin: Springer-Verlag, 1958.
[16] Dunford, N., and Schwartz, J. T., *Linear Operators.* New York: Interscience Publishers, vol. 1, 1958; vol. 2, 1963.
[17] Edwards, R. E., *Functional Analysis, Theory and Applications.* New York: Holt, Rinehart and Winston, Inc., 1965.
[18] Goffman, C., and Pedrick, G., *First Course in Functional Analysis.* Englewood Cliffs, N.J.: Prentice-Hall, Inc., 1965.
[19] Halmos, P. R., *Introduction to Hilbert Space.* New York: Chelsea Publishing Company, 1957.
[20] Kantrovich, L. V., and Akilov, G. P., *Functional Analysis in Normed Spaces.* New York: Pergamon Press, Inc., 1964.
[21] Kolmogorov, A. N., and Fomin, S. V., *Elements of the Theory of Functions and Functional Analysis.* Rochester, N.Y.: Graglock Press, vol. 1, 1957.
[22] Liusternik, L. A., and Sobolev, V. J., *Elements of Functional Analysis.* New York: Fredrick Ungar Publishing Company, 1961.
[23] Sz.-Nagy, B., *Spektraldarstellung Linearer Transformationen des Hilbertschen Raumes. Ergebnisse der Mathematik,* vol. 39. Berlin: Springer-Verlag, 1967.
[24] Naimark, M. A., *Normed Rings.* Gronigen, Netherlands: P. Noordhoff, N.V., 1959.
[25] Riesz, F., and Sz.-Nagy, B., *Functional Analysis.* New York: Fredrick Ungar Publishing Company, 1955.

[26] Royden, H. L., *Real Analysis*, 2nd ed. New York: The Macmillan Company, 1968.

[27] Taylor, A. E., *Introduction to Functional Analysis*. New York: John Wiley & Sons, Inc., 1958.

[28] Vulikh, B. Z., *Introduction to Functional Analysis*. New York: Pergamon Press, Inc., 1963.

[29] Wilansky, A., *Functional Analysis*. Waltham, Mass.: Blaisdell Publishing Company, 1964.

[30] Yosida, K., *Functional Analysis*. Berlin: Springer-Verlag, 1965.

[31] Zaanen, A. S., *Linear Analysis*. Amsterdam: North-Holland Publishing Company, 1953.

MISCELLANEOUS

[32] Coddington, E. A. and Levinson, N., *Theory of Ordinary Differential Equations*. New York: McGraw-Hill, Inc., 1955.

[33] Hausdorff, F., *Mengenlehre*. New York: Dover Publications, Inc., 1944.

[34] Hellwig, G., *Differential Operators of Mathematical Physics*. Berlin: Springer-Verlag, 1964. (Addison-Wesley Publishing Company, Inc.)

[35] Kelley, J. L., *General Topology*. Princeton, N.J.: D. Van Nostrand Company, 1955.

[36] Kellogg, O. D., *Foundation of Potential Theory*. New York: Dover Publications, Inc., 1953.

[37] Lax, P. D., "On the existence of Green's function," *Proc. Amer. Math. Soc.*, vol. 3 (1952), pp. 526–531.

[38] Naimark, M. A., *Linear Differential Operators*, part II. New York: Fredrick Ungar Publishing Company, 1968.

INDEX

A CATALOGUE OF SELECTED DOVER BOOKS IN ALL FIELDS OF INTEREST

A CATALOGUE OF SELECTED DOVER BOOKS IN ALL FIELDS OF INTEREST

THE EARLY WORK OF AUBREY BEARDSLEY, Aubrey Beardsley. 157 plates, 2 in color: *Manon Lescaut, Madame Bovary, Morte Darthur, Salome,* other. Introduction by H. Marillier. 182pp. 8⅛ x 11. 21816-3 Pa. $4.50

THE LATER WORK OF AUBREY BEARDSLEY, Aubrey Beardsley. Exotic masterpieces of full maturity: *Venus and Tannhauser, Lysistrata, Rape of the Lock, Volpone,* Savoy material, etc. 174 plates, 2 in color. 186pp. 8⅛ x 11. 21817-1 Pa. $5.95

THOMAS NAST'S CHRISTMAS DRAWINGS, Thomas Nast. Almost all Christmas drawings by creator of image of Santa Claus as we know it, and one of America's foremost illustrators and political cartoonists. 66 illustrations. 3 illustrations in color on covers. 96pp. 8⅜ x 11¼. 23660-9 Pa. $3.50

THE DORÉ ILLUSTRATIONS FOR DANTE'S DIVINE COMEDY, Gustave Doré. All 135 plates from Inferno, Purgatory, Paradise; fantastic tortures, infernal landscapes, celestial wonders. Each plate with appropriate (translated) verses. 141pp. 9 x 12. 23231-X Pa. $4.50

DORÉ'S ILLUSTRATIONS FOR RABELAIS, Gustave Doré. 252 striking illustrations of *Gargantua and Pantagruel* books by foremost 19th-century illustrator. Including 60 plates, 192 delightful smaller illustrations. 153pp. 9 x 12. 23656-0 Pa. $5.00

LONDON: A PILGRIMAGE, Gustave Doré, Blanchard Jerrold. Squalor, riches, misery, beauty of mid-Victorian metropolis; 55 wonderful plates, 125 other illustrations, full social, cultural text by Jerrold. 191pp. of text. 9⅜ x 12¼. 22306-X Pa. $7.00

THE RIME OF THE ANCIENT MARINER, Gustave Doré, S. T. Coleridge. Dore's finest work, 34 plates capture moods, subtleties of poem. Full text. Introduction by Millicent Rose. 77pp. 9¼ x 12. 22305-1 Pa. $3.50

THE DORE BIBLE ILLUSTRATIONS, Gustave Doré. All wonderful, detailed plates: Adam and Eve, Flood, Babylon, Life of Jesus, etc. Brief King James text with each plate. Introduction by Millicent Rose. 241 plates. 241pp. 9 x 12. 23004-X Pa. $6.00

THE COMPLETE ENGRAVINGS, ETCHINGS AND DRYPOINTS OF ALBRECHT DURER. "Knight, Death and Devil"; "Melencolia," and more—all Dürer's known works in all three media, including 6 works formerly attributed to him. 120 plates. 235pp. 8⅜ x 11¼. 22851-7 Pa. $6.50

MECHANICK EXERCISES ON THE WHOLE ART OF PRINTING, Joseph Moxon. First complete book (1683-4) ever written about typography, a compendium of everything known about printing at the latter part of 17th century. Reprint of 2nd (1962) Oxford Univ. Press edition. 74 illustrations. Total of 550pp. 6⅛ x 9¼. 23617-X Pa. $7.95

CATALOGUE OF DOVER BOOKS

THE COMPLETE WOODCUTS OF ALBRECHT DURER, edited by Dr. W. Kurth. 346 in all: "Old Testament," "St. Jerome," "Passion," "Life of Virgin," Apocalypse," many others. Introduction by Campbell Dodgson. 285pp. 8½ x 12¼. 21097-9 Pa. $7.50

DRAWINGS OF ALBRECHT DURER, edited by Heinrich Wolfflin. 81 plates show development from youth to full style. Many favorites; many new. Introduction by Alfred Werner. 96pp. 8⅛ x 11. 22352-3 Pa. $5.00

THE HUMAN FIGURE, Albrecht Dürer. Experiments in various techniques—stereometric, progressive proportional, and others. Also life studies that rank among finest ever done. Complete reprinting of *Dresden Sketchbook.* 170 plates. 355pp. 8⅜ x 11¼. 21042-1 Pa. $7.95

OF THE JUST SHAPING OF LETTERS, Albrecht Dürer. Renaissance artist explains design of Roman majuscules by geometry, also Gothic lower and capitals. Grolier Club edition. 43pp. 7⅞ x 10¾ 21306-4 Pa. $3.00

TEN BOOKS ON ARCHITECTURE, Vitruvius. The most important book ever written on architecture. Early Roman aesthetics, technology, classical orders, site selection, all other aspects. Stands behind everything since. Morgan translation. 331pp. 5⅜ x 8½. 20645-9 Pa. $4.50

THE FOUR BOOKS OF ARCHITECTURE, Andrea Palladio. 16th-century classic responsible for Palladian movement and style. Covers classical architectural remains, Renaissance revivals, classical orders, etc. 1738 Ware English edition. Introduction by A. Placzek. 216 plates. 110pp. of text. 9½ x 12¾. 21308-0 Pa. $10.00

HORIZONS, Norman Bel Geddes. Great industrialist stage designer, "father of streamlining," on application of aesthetics to transportation, amusement, architecture, etc. 1932 prophetic account; function, theory, specific projects. 222 illustrations. 312pp. 7⅞ x 10¾. 23514-9 Pa. $6.95

FRANK LLOYD WRIGHT'S FALLINGWATER, Donald Hoffmann. Full, illustrated story of conception and building of Wright's masterwork at Bear Run, Pa. 100 photographs of site, construction, and details of completed structure. 112pp. 9¼ x 10. 23671-4 Pa. $5.50

THE ELEMENTS OF DRAWING, John Ruskin. Timeless classic by great Viltorian; starts with basic ideas, works through more difficult. Many practical exercises. 48 illustrations. Introduction by Lawrence Campbell. 228pp. 5⅜ x 8½. 22730-8 Pa. $3.75

GIST OF ART, John Sloan. Greatest modern American teacher, Art Students League, offers innumerable hints, instructions, guided comments to help you in painting. Not a formal course. 46 illustrations. Introduction by Helen Sloan. 200pp. 5⅜ x 8½. 23435-5 Pa. $4.00

THE ANATOMY OF THE HORSE, George Stubbs. Often considered the great masterpiece of animal anatomy. Full reproduction of 1766 edition, plus prospectus; original text and modernized text. 36 plates. Introduction by Eleanor Garvey. 121pp. 11 x 14¾. 23402-9 Pa. $6.00

BRIDGMAN'S LIFE DRAWING, George B. Bridgman. More than 500 illustrative drawings and text teach you to abstract the body into its major masses, use light and shade, proportion; as well as specific areas of anatomy, of which Bridgman is master. 192pp. 6½ x 9¼. (Available in U.S. only) 22710-3 Pa. $3.50

ART NOUVEAU DESIGNS IN COLOR, Alphonse Mucha, Maurice Verneuil, Georges Auriol. Full-color reproduction of *Combinaisons ornementales* (c. 1900) by Art Nouveau masters. Floral, animal, geometric, interlacings, swashes—borders, frames, spots—all incredibly beautiful. 60 plates, hundreds of designs. 9⅜ x 8-1/16. 22885-1 Pa. $4.00

FULL-COLOR FLORAL DESIGNS IN THE ART NOUVEAU STYLE, E. A. Seguy. 166 motifs, on 40 plates, from *Les fleurs et leurs applications decoratives* (1902): borders, circular designs, repeats, allovers, "spots." All in authentic Art Nouveau colors. 48pp. 9⅜ x 12¼. 23439-8 Pa. $5.00

A DIDEROT PICTORIAL ENCYCLOPEDIA OF TRADES AND INDUSTRY, edited by Charles C. Gillispie. 485 most interesting plates from the great French Encyclopedia of the 18th century show hundreds of working figures, artifacts, process, land and cityscapes; glassmaking, papermaking, metal extraction, construction, weaving, making furniture, clothing, wigs, dozens of other activities. Plates fully explained. 920pp. 9 x 12. 22284-5, 22285-3 Clothbd., Two-vol. set $40.00

HANDBOOK OF EARLY ADVERTISING ART, Clarence P. Hornung. Largest collection of copyright-free early and antique advertising art ever compiled. Over 6,000 illustrations, from Franklin's time to the 1890's for special effects, novelty. Valuable source, almost inexhaustible.

Pictorial Volume. Agriculture, the zodiac, animals, autos, birds, Christmas, fire engines, flowers, trees, musical instruments, ships, games and sports, much more. Arranged by subject matter and use. 237 plates. 288pp. 9 x 12. 20122-8 Clothbd. $14.50

Typographical Volume. Roman and Gothic faces ranging from 10 point to 300 point, "Barnum," German and Old English faces, script, logotypes, scrolls and flourishes, 1115 ornamental initials, 67 complete alphabets, more. 310 plates. 320pp. 9 x 12. 20123-6 Clothbd. $15.00

CALLIGRAPHY (CALLIGRAPHIA LATINA), J. G. Schwandner. High point of 18th-century ornamental calligraphy. Very ornate initials, scrolls, borders, cherubs, birds, lettered examples. 172pp. 9 x 13. 20475-8 Pa. $7.00

ART FORMS IN NATURE, Ernst Haeckel. Multitude of strangely beautiful natural forms: Radiolaria, Foraminifera, jellyfishes, fungi, turtles, bats, etc. All 100 plates of the 19th-century evolutionist's *Kunstformen der Natur* (1904). 100pp. 9⅜ x 12¼. 22987-4 Pa. $5.00

CHILDREN: A PICTORIAL ARCHIVE FROM NINETEENTH-CENTURY SOURCES, edited by Carol Belanger Grafton. 242 rare, copyright-free wood engravings for artists and designers. Widest such selection available. All illustrations in line. 119pp. 8⅜ x 11¼. 23694-3 Pa. $4.00

WOMEN: A PICTORIAL ARCHIVE FROM NINETEENTH-CENTURY SOURCES, edited by Jim Harter. 391 copyright-free wood engravings for artists and designers selected from rare periodicals. Most extensive such collection available. All illustrations in line. 128pp. 9 x 12. 23703-6 Pa. $4.50

ARABIC ART IN COLOR, Prisse d'Avennes. From the greatest ornamentalists of all time—50 plates in color, rarely seen outside the Near East, rich in suggestion and stimulus. Includes 4 plates on covers. 46pp. 9⅜ x 12¼. 23658-7 Pa. $6.00

AUTHENTIC ALGERIAN CARPET DESIGNS AND MOTIFS, edited by June Beveridge. Algerian carpets are world famous. Dozens of geometrical motifs are charted on grids, color-coded, for weavers, needleworkers, craftsmen, designers. 53 illustrations plus 4 in color. 48pp. 8¼ x 11. (Available in U.S. only) 23650-1 Pa. $1.75

DICTIONARY OF AMERICAN PORTRAITS, edited by Hayward and Blanche Cirker. 4000 important Americans, earliest times to 1905, mostly in clear line. Politicians, writers, soldiers, scientists, inventors, industrialists, Indians, Blacks, women, outlaws, etc. Identificatory information. 756pp. 9¼ x 12¾. 21823-6 Clothbd. $40.00

HOW THE OTHER HALF LIVES, Jacob A. Riis. Journalistic record of filth, degradation, upward drive in New York immigrant slums, shops, around 1900. New edition includes 100 original Riis photos, monuments of early photography. 233pp. 10 x 7⅞. 22012-5 Pa. $7.00

NEW YORK IN THE THIRTIES, Berenice Abbott. Noted photographer's fascinating study of city shows new buildings that have become famous and old sights that have disappeared forever. Insightful commentary. 97 photographs. 97pp. 11⅜ x 10. 22967-X Pa. $5.00

MEN AT WORK, Lewis W. Hine. Famous photographic studies of construction workers, railroad men, factory workers and coal miners. New supplement of 18 photos on Empire State building construction. New introduction by Jonathan L. Doherty. Total of 69 photos. 63pp. 8 x 10¾. 23475-4 Pa. $3.00

HOLLYWOOD GLAMOUR PORTRAITS, edited by John Kobal. 145 photos capture the stars from 1926-49, the high point in portrait photography. Gable, Harlow, Bogart, Bacall, Hedy Lamarr, Marlene Dietrich, Robert Montgomery, Marlon Brando, Veronica Lake; 94 stars in all. Full background on photographers, technical aspects, much more. Total of 160pp. 8⅜ x 11¼. 23352-9 Pa. $6.00

THE NEW YORK STAGE: FAMOUS PRODUCTIONS IN PHOTOGRAPHS, edited by Stanley Appelbaum. 148 photographs from Museum of City of New York show 142 plays, 1883-1939. *Peter Pan, The Front Page, Dead End, Our Town,* O'Neill, hundreds of actors and actresses, etc. Full indexes. 154pp. 9½ x 10. 23241-7 Pa. $6.00

DIALOGUES CONCERNING TWO NEW SCIENCES, Galileo Galilei. Encompassing 30 years of experiment and thought, these dialogues deal with geometric demonstrations of fracture of solid bodies, cohesion, leverage, speed of light and sound, pendulums, falling bodies, accelerated motion, etc. 300pp. 5⅜ x 8½. 60099-8 Pa. $4.00

THE GREAT OPERA STARS IN HISTORIC PHOTOGRAPHS, edited by James Camner. 343 portraits from the 1850s to the 1940s: Tamburini, Mario, Caliapin, Jeritza, Melchior, Melba, Patti, Pinza, Schipa, Caruso, Farrar, Steber, Gobbi, and many more—270 performers in all. Index. 199pp. 8⅜ x 11¼. 23575-0 Pa. $7.50

J. S. BACH, Albert Schweitzer. Great full-length study of Bach, life, background to music, music, by foremost modern scholar. Ernest Newman translation. 650 musical examples. Total of 928pp. 5⅜ x 8½. (Available in U.S. only) 21631-4, 21632-2 Pa., Two-vol. set $11.00

COMPLETE PIANO SONATAS, Ludwig van Beethoven. All sonatas in the fine Schenker edition, with fingering, analytical material. One of best modern editions. Total of 615pp. 9 x 12. (Available in U.S. only) 23134-8, 23135-6 Pa., Two-vol. set $15.50

KEYBOARD MUSIC, J. S. Bach. Bach-Gesellschaft edition. For harpsichord, piano, other keyboard instruments. English Suites, French Suites, Six Partitas, Goldberg Variations, Two-Part Inventions, Three-Part Sinfonias. 312pp. 8⅛ x 11. (Available in U.S. only) 22360-4 Pa. $6.95

FOUR SYMPHONIES IN FULL SCORE, Franz Schubert. Schubert's four most popular symphonies: No. 4 in C Minor ("Tragic"); No. 5 in B-flat Major; No. 8 in B Minor ("Unfinished"); No. 9 in C Major ("Great"). Breitkopf & Hartel edition. Study score. 261pp. 9⅜ x 12¼. 23681-1 Pa. $6.50

THE AUTHENTIC GILBERT & SULLIVAN SONGBOOK, W. S. Gilbert, A. S. Sullivan. Largest selection available; 92 songs, uncut, original keys, in piano rendering approved by Sullivan. Favorites and lesser-known fine numbers. Edited with plot synopses by James Spero. 3 illustrations. 399pp. 9 x 12. 23482-7 Pa. $9.95

PRINCIPLES OF ORCHESTRATION, Nikolay Rimsky-Korsakov. Great classical orchestrator provides fundamentals of tonal resonance, progression of parts, voice and orchestra, tutti effects, much else in major document. 330pp. of musical excerpts. 489pp. 6½ x 9¼. 21266-1 Pa. **$7.50**

TRISTAN UND ISOLDE, Richard Wagner. Full orchestral score with complete instrumentation. Do not confuse with piano reduction. Commentary by Felix Mottl, great Wagnerian conductor and scholar. Study score. 655pp. 8⅛ x 11. 22915-7 Pa. $13.95

REQUIEM IN FULL SCORE, Giuseppe Verdi. Immensely popular with choral groups and music lovers. Republication of edition published by C. F. Peters, Leipzig, n. d. German frontmaker in English translation. Glossary. Text in Latin. Study score. 204pp. 9⅜ x 12¼.
23682-X Pa. $6.00

COMPLETE CHAMBER MUSIC FOR STRINGS, Felix Mendelssohn. All of Mendelssohn's chamber music: Octet, 2 Quintets, 6 Quartets, and Four Pieces for String Quartet. (Nothing with piano is included). Complete works edition (1874-7). Study score. 283 pp. 9⅜ x 12¼.
23679-X Pa. **$7.50**

POPULAR SONGS OF NINETEENTH-CENTURY AMERICA, edited by Richard Jackson. 64 most important songs: "Old Oaken Bucket," "Arkansas Traveler," "Yellow Rose of Texas," etc. Authentic original sheet music, full introduction and commentaries. 290pp. 9 x 12. 23270-0 Pa. **$7.95**

COLLECTED PIANO WORKS, Scott Joplin. Edited by Vera Brodsky Lawrence. Practically all of Joplin's piano works—rags, two-steps, marches, waltzes, etc., 51 works in all. Extensive introduction by Rudi Blesh. Total of 345pp. 9 x 12. 23106-2 Pa. $14.95

BASIC PRINCIPLES OF CLASSICAL BALLET, Agrippina Vaganova. Great Russian theoretician, teacher explains methods for teaching classical ballet; incorporates best from French, Italian, Russian schools. 118 illustrations. 175pp. 5⅜ x 8½. 22036-2 Pa. $2.50

CHINESE CHARACTERS, L. Wieger. Rich analysis of 2300 characters according to traditional systems into primitives. Historical-semantic analysis to phonetics (Classical Mandarin) and radicals. 820pp. 6⅛ x 9¼.
21321-8 Pa. $10.00

EGYPTIAN LANGUAGE: EASY LESSONS IN EGYPTIAN HIEROGLYPHICS, E. A. Wallis Budge. Foremost Egyptologist offers Egyptian grammar, explanation of hieroglyphics, many reading texts, dictionary of symbols. 246pp. 5 x 7½. (Available in U.S. only)
21394-3 Clothbd. $7.50

AN ETYMOLOGICAL DICTIONARY OF MODERN ENGLISH, Ernest Weekley. Richest, fullest work, by foremost British lexicographer. Detailed word histories. Inexhaustible. Do not confuse this with *Concise Etymological Dictionary*, which is abridged. Total of 856pp. 6½ x 9¼.
21873-2, 21874-0 Pa., Two-vol. set $12.00

A MAYA GRAMMAR, Alfred M. Tozzer. Practical, useful English-language grammar by the Harvard anthropologist who was one of the three greatest American scholars in the area of Maya culture. Phonetics, grammatical processes, syntax, more. 301pp. 5⅜ x 8½. 23465-7 Pa. $4.00

THE JOURNAL OF HENRY D. THOREAU, edited by Bradford Torrey, F. H. Allen. Complete reprinting of 14 volumes, 1837-61, over two million words; the sourcebooks for *Walden*, etc. Definitive. All original sketches, plus 75 photographs. Introduction by Walter Harding. Total of 1804pp. 8½ x 12¼. 20312-3, 20313-1 Clothbd., Two-vol. set $70.00

CLASSIC GHOST STORIES, Charles Dickens and others. 18 wonderful stories you've wanted to reread: "The Monkey's Paw," "The House and the Brain," "The Upper Berth," "The Signalman," "Dracula's Guest," "The Tapestried Chamber," etc. Dickens, Scott, Mary Shelley, Stoker, etc. 330pp. 5⅜ x 8½. 20735-8 Pa. $4.50

SEVEN SCIENCE FICTION NOVELS, H. G. Wells. Full novels. *First Men in the Moon, Island of Dr. Moreau, War of the Worlds, Food of the Gods, Invisible Man, Time Machine, In the Days of the Comet*. A basic science-fiction library. 1015pp. 5⅜ x 8½. (Available in U.S. only) 20264-X Clothbd. $8.95

ARMADALE, Wilkie Collins. Third great mystery novel by the author of *The Woman in White* and *The Moonstone*. Ingeniously plotted narrative shows an exceptional command of character, incident and mood. Original magazine version with 40 illustrations. 597pp. 5⅜ x 8½. 23429-0 Pa. $6.00

MASTERS OF MYSTERY, H. Douglas Thomson. The first book in English (1931) devoted to history and aesthetics of detective story. Poe, Doyle, LeFanu, Dickens, many others, up to 1930. New introduction and notes by E. F. Bleiler. 288pp. 5⅜ x 8½. (Available in U.S. only) 23606-4 Pa. $4.00

FLATLAND, E. A. Abbott. Science-fiction classic explores life of 2-D being in 3-D world. Read also as introduction to thought about hyperspace. Introduction by Banesh Hoffmann. 16 illustrations. 103pp. 5⅜ x 8½. 20001-9 Pa. $2.00

THREE SUPERNATURAL NOVELS OF THE VICTORIAN PERIOD, edited, with an introduction, by E. F. Bleiler. Reprinted complete and unabridged, three great classics of the supernatural: *The Haunted Hotel* by Wilkie Collins, *The Haunted House at Latchford* by Mrs. J. H. Riddell, and *The Lost Stradivarious* by J. Meade Falkner. 325pp. 5⅜ x 8½. 22571-2 Pa. $4.00

AYESHA: THE RETURN OF "SHE," H. Rider Haggard. Virtuoso sequel featuring the great mythic creation, Ayesha, in an adventure that is fully as good as the first book, *She*. Original magazine version, with 47 original illustrations by Maurice Greiffenhagen. 189pp. 6½ x 9¼. 23649-8 Pa. $3.50

UNCLE SILAS, J. Sheridan LeFanu. Victorian Gothic mystery novel, considered by many best of period, even better than Collins or Dickens. Wonderful psychological terror. Introduction by Frederick Shroyer. 436pp. 5⅜ x 8½. 21715-9 Pa. $6.00

JURGEN, James Branch Cabell. The great erotic fantasy of the 1920's that delighted thousands, shocked thousands more. Full final text, Lane edition with 13 plates by Frank Pape. 346pp. 5⅜ x 8½. 23507-6 Pa. $4.50

THE CLAVERINGS, Anthony Trollope. Major novel, chronicling aspects of British Victorian society, personalities. Reprint of Cornhill serialization, 16 plates by M. Edwards; first reprint of full text. Introduction by Norman Donaldson. 412pp. 5⅜ x 8½. 23464-9 Pa. $5.00

KEPT IN THE DARK, Anthony Trollope. Unusual short novel about Victorian morality and abnormal psychology by the great English author. Probably the first American publication. Frontispiece by Sir John Millais. 92pp. 6½ x 9¼. 23609-9 Pa. $2.50

RALPH THE HEIR, Anthony Trollope. Forgotten tale of illegitimacy, inheritance. Master novel of Trollope's later years. Victorian country estates, clubs, Parliament, fox hunting, world of fully realized characters. Reprint of 1871 edition. 12 illustrations by F. A. Faser. 434pp. of text. 5⅜ x 8½. 23642-0 Pa. $5.00

YEKL and THE IMPORTED BRIDEGROOM AND OTHER STORIES OF THE NEW YORK GHETTO, Abraham Cahan. Film *Hester Street* based on *Yekl* (1896). Novel, other stories among first about Jewish immigrants of N.Y.'s East Side. Highly praised by W. D. Howells—Cahan "a new star of realism." New introduction by Bernard G. Richards. 240pp. 5⅜ x 8½. 22427-9 Pa. $3.50

THE HIGH PLACE, James Branch Cabell. Great fantasy writer's enchanting comedy of disenchantment set in 18th-century France. Considered by some critics to be even better than his famous *Jurgen*. 10 illustrations and numerous vignettes by noted fantasy artist Frank C. Pape. 320pp. 5⅜ x 8½. 23670-6 Pa. $4.00

ALICE'S ADVENTURES UNDER GROUND, Lewis Carroll. Facsimile of ms. Carroll gave Alice Liddell in 1864. Different in many ways from final Alice. Handlettered, illustrated by Carroll. Introduction by Martin Gardner. 128pp. 5⅜ x 8½. 21482-6 Pa. $2.50

FAVORITE ANDREW LANG FAIRY TALE BOOKS IN MANY COLORS, Andrew Lang. The four Lang favorites in a boxed set—the complete *Red, Green, Yellow* and *Blue* Fairy Books. 164 stories; 439 illustrations by Lancelot Speed, Henry Ford and G. P. Jacomb Hood. Total of about 1500pp. 5⅜ x 8½. 23407-X Boxed set, Pa. $15.95

HOUSEHOLD STORIES BY THE BROTHERS GRIMM. All the great Grimm stories: "Rumpelstiltskin," "Snow White," "Hansel and Gretel," etc., with 114 illustrations by Walter Crane. 269pp. 5⅜ x 8½.
21080-4 Pa. $3.50

SLEEPING BEAUTY, illustrated by Arthur Rackham. Perhaps the fullest, most delightful version ever, told by C. S. Evans. Rackham's best work. 49 illustrations. 110pp. 7⅞ x 10¾. 22756-1 Pa. $2.50

AMERICAN FAIRY TALES, L. Frank Baum. Young cowboy lassoes Father Time; dummy in Mr. Floman's department store window comes to life; and 10 other fairy tales. 41 illustrations by N. P. Hall, Harry Kennedy, Ike Morgan, and Ralph Gardner. 209pp. 5⅜ x 8½. 23643-9 Pa. $3.00

THE WONDERFUL WIZARD OF OZ, L. Frank Baum. Facsimile in full color of America's finest children's classic. Introduction by Martin Gardner. 143 illustrations by W. W. Denslow. 267pp. 5⅜ x 8½.
20691-2 Pa. $3.50

THE TALE OF PETER RABBIT, Beatrix Potter. The inimitable Peter's terrifying adventure in Mr. McGregor's garden, with all 27 wonderful, full-color Potter illustrations. 55pp. 4¼ x 5½. (Available in U.S. only)
22827-4 Pa. $1.25

THE STORY OF KING ARTHUR AND HIS KNIGHTS, Howard Pyle. Finest children's version of life of King Arthur. 48 illustrations by Pyle. 131pp. 6⅛ x 9¼. 21445-1 Pa. $4.95

CARUSO'S CARICATURES, Enrico Caruso. Great tenor's remarkable caricatures of self, fellow musicians, composers, others. Toscanini, Puccini, Farrar, etc. Impish, cutting, insightful. 473 illustrations. Preface by M. Sisca. 217pp. 8⅜ x 11¼. 23528-9 Pa. $6.95

PERSONAL NARRATIVE OF A PILGRIMAGE TO ALMADINAH AND MECCAH, Richard Burton. Great travel classic by remarkably colorful personality. Burton, disguised as a Moroccan, visited sacred shrines of Islam, narrowly escaping death. Wonderful observations of Islamic life, customs, personalities. 47 illustrations. Total of 959pp. 5⅜ x 8½.
21217-3, 21218-1 Pa., Two-vol. set $12.00

INCIDENTS OF TRAVEL IN YUCATAN, John L. Stephens. Classic (1843) exploration of jungles of Yucatan, looking for evidences of Maya civilization. Travel adventures, Mexican and Indian culture, etc. Total of 669pp. 5⅜ x 8½. 20926-1, 20927-X Pa., Two-vol. set $7.90

AMERICAN LITERARY AUTOGRAPHS FROM WASHINGTON IRVING TO HENRY JAMES, Herbert Cahoon, et al. Letters, poems, manuscripts of Hawthorne, Thoreau, Twain, Alcott, Whitman, 67 other prominent American authors. Reproductions, full transcripts and commentary. Plus checklist of all American Literary Autographs in The Pierpont Morgan Library. Printed on exceptionally high-quality paper. 136 illustrations. 212pp. 9⅛ x 12¼. 23548-3 Pa. $12.50